Embedded Systems

SO-ARM-598

Series Editors

Nikil D. Dutt, Department of Computer Science, Donald Bren School
  of Information and Computer Sciences, University of California, Irvine,
  Zot Code 3435, Irvine, CA 92697-3435, USA
Peter Marwedel, Informatik 12, TU Dortmund, Otto-Hahn-Str. 16,
  44227 Dortmund, Germany
Grant Martin, Tensilica Inc., 3255-6 Scott Blvd., Santa Clara, CA 95054, USA

For other titles published in this series, go to
www.springer.com/series/8563

Peter Marwedel

Embedded System Design

Embedded Systems Foundations of Cyber-Physical Systems

2nd Edition

 Springer

Dr. Peter Marwedel
TU Dortmund
Informatik 12
Otto-Hahn-Str. 16
44221 Dortmund
Germany
peter.marwedel@tu-dortmund.de

ISBN 978-94-007-0256-1   e-ISBN 978-94-007-0257-8
DOI 10.1007/978-94-007-0257-8
Springer Dordrecht Heidelberg London New York

Cover design: VTEX, Vilnius

Printed on acid-free paper

Springer is part of Springer Science+Business Media (www.springer.com)

Contents

Preface

Definitions and scope

Until the late 1980s, information processing was associated with large main-frame computers and huge tape drives. During the 1990s, this shifted towards information processing being associated with personal computers, PCs. The trend towards miniaturization continues and the majority of information processing devices will be small portable computers, many of which will be integrated into larger products. Their presence in these larger products, such as telecommunication equipment, will be less obvious than for the PC. Usually, technical products must be technologically advanced to attract customers' interest. Cars, cameras, TV sets, mobile phones, etc. can hardly be sold any more in technologically advanced countries unless they come with built-in computers. Hence and according to several forecasts (see, for example [National Research Council, 2001]), the future of information and communication technologies (ICT) is characterized by terms such as

1 ubiquitous computing [Weiser, 2003],

2 pervasive computing [Hansmann, 2001], [Burkhardt, 2001],

3 ambient intelligence [Koninklijke Philips Electronics N.V., 2003], [Marzano and Aarts, 2003],

4 the disappearing computer [Weiser, 2003],

5 and the post-PC era.

The first term reflects the fact that computing (and communication) will be everywhere. The expectation is that *information* will be available *anytime, anywhere*. The predicted penetration of our day-to-day life with computing

devices led to the term "pervasive computing". For ambient intelligence, there
is some emphasis on communication technology in future homes and smart
buildings. These three terms focus on only slightly different aspects of future
information technology. Ubiquitous computing focuses more on the long term
goal of providing information anytime, anywhere, whereas pervasive computing focuses more on practical aspects and the exploitation of already available
technology. The fourth term refers to the expectation that processors and software will be used in much smaller systems and will in many cases even be
invisible. The term **post-PC era** denotes the fact that in the future, standard-
PCs will be less dominant hardware platforms.

Two basic technologies are needed for next-generation ICT systems:

- **embedded systems**,

- and **communication technologies**.

Fig. 0.1 shows a graphical representation of how ubiquitous computing is influenced by embedded systems and by communication technology.

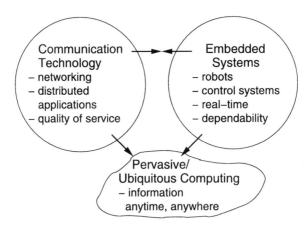

Figure 0.1.   Influence of embedded systems on ubiquitous computing (©European Commission)

For example, ubiquitous computing devices -like embedded systems- must
meet real-time and dependability requirements of embedded systems while using fundamental techniques of communication technology, such as networking.

A comprehensive coverage of communication technologies would require a
separate book. Therefore, this book does not cover communication technologies, except as a minor topic in few subsections. What are "embedded systems"
anyway? They can be defined as follows [Marwedel, 2003]:

Definition: Embedded systems are information processing systems embedded into enclosing products.

Examples include embedded systems in cars, trains, planes, and telecommunication or fabrication equipment. Such systems come with a large number of common characteristics, including real-time constraints, and dependability as well as efficiency requirements. For such systems, the link to physics and physical systems is rather important. This link is emphasized in the following citation [Lee, 2006]:

"Embedded software is software integrated with physical processes. The technical problem is managing time and concurrency in computational systems".

This citation could be used as a definition of the term "embedded software" and could be extended into a definition of "embedded systems" by just replacing "software" by "system". However, the strong link to physics has recently been stressed even more by the introduction of the term "cyber-physical systems" (CPS or "cy-phy" systems for short). Cy-phy systems can be defined as follows:

Definition: *"Cyber-Physical Systems (CPS) are integrations of computation and physical processes"* [Lee, 2007].

The new term emphasizes the link to physical quantities such as time, energy and space. Emphasizing this link makes a lot of sense, since it is frequently ignored in a world of applications running on PCs. For cy-phy systems, we may be expecting models to include models of the physical environment as well. **In this sense, we may think of cy-phy systems to comprise embedded systems (the information processing part) and the physical environment.** We will refer to the new term whenever we want to emphasize the link to physics and the environment. In the future, links to chemistry and biology are likely to be important as well.

This book provides an overview of key concepts for embedded systems as they are needed for cyber-physical systems. The scope includes specification techniques, hardware components, system software, application mapping, evaluation and validation, as well as exemplary optimizations and test methods.

Importance of embedded and cyber-physical systems

Following the success of ICT for office and work flow applications, embedded and cyber-physical systems are considered to be **the** most important application area of ICT during the coming years. The number of processors in embedded systems already exceeds the number of processors in PCs, and this trend is expected to continue. According to forecasts, the size of embedded software will also increase at a large rate. Another kind of Moore's law was predicted:

For many products in the area of consumer electronics the amount of code is doubling every two years [Vaandrager, 1998]. The increasing importance of embedded systems is also reflected in a report of the National Research Council in the United States [National Research Council, 2001]. According to the introduction of this report, "*Information technology (IT) is on the verge of another revolution. ... networked systems of embedded computers ... have the potential to change radically the way people interact with their environment by linking together a range of devices and sensors that will allow information to be collected, shared, and processed in unprecedented ways. ... The use ... throughout society could well dwarf previous milestones in the information revolution.*"

Statistics regarding the size of the embedded systems market can be found on relevant web sites. Sites such as "IT facts" [IT Facts, 2010] demonstrate the importance of the embedded system market. The size of the embedded system market can be analyzed from a variety of perspectives. Many of the embedded processors are 8-bit processors, but despite this, even the majority of all 32-bit processors are integrated into embedded systems [Stiller, 2000]. Already in 1996, it was estimated that the average American came into contact with 60 microprocessors per day [Camposano and Wolf, 1996]. Some high-end cars contain more than 100 processors[1]. These numbers are much larger than what is typically expected, since most people do not realize that they are using processors. The importance of embedded systems was also stated by journalist Margaret Ryan [Ryan, 1995]:

"*... embedded chips form the backbone of the electronics driven world in which we live. ... they are part of almost everything that runs on electricity*".

According to quite a number of forecasts, the embedded system market will be much larger than the market for PC-like systems.

In the United States, the National Science Foundation is supporting research on cyber-physical systems [National Science Foundation, 2010]. In Europe, the Sixth and the Seventh Framework Programme [European Commission Cordis, 2010] support research and development of embedded systems. Also, the ARTEMIS joint undertaking [ARTEMIS Joint Undertaking, 2010] was created as a public/private partnership between government institutions and companies in order to move research and development in embedded computing ahead. This initiative demonstrates the huge interest of the European commercial sector in this technology. Similar initiatives exist on other continents as well.

---

[1] According to personal communication.

This importance of embedded/cyber-physical systems is so far not well reflected in many of the current curricula. This book is intended as an aid for changing this situation. It provides the material for a first course on such systems. Therefore, it has been designed as a textbook. However, it provides more references than typical textbooks and also helps to structure the area. Hence, this book should also be useful for faculty members and engineers. For students, the inclusion of a rich set of references facilitates access to relevant sources of information.

Audience for this book

This book is intended for the following audience:

- Computer science (CS), computer engineering (CE), and electrical engineering (EE) students as well as students in other ICT-related areas who would like to specialize in embedded/cyber-physical systems. The book should be appropriate for third year students who do have a basic knowledge of computer hardware and software. This means that the book primarily targets senior undergraduate students. However, it can also be used at the graduate level if embedded system design is not part of the undergraduate program. This book is intended to pave the way for **more advanced topics** that should be **covered in follow-up courses**. The book assumes a basic knowledge of computer science. EE students may have to read some additional material in order to fully understand the topics of this book. This should be compensated by the fact that some material covered in this book may already be known to EE students.

- Engineers who have so far worked on systems hardware and who have to move more towards software of embedded systems. This book should provide enough background to understand the relevant technical publications.

- PhD students who would like to get a quick, broad overview of key concepts in embedded system technology before focusing on a specific research area.

- Professors designing a new curriculum for embedded systems.

Curriculum integration of embedded systems

Unfortunately, embedded systems are hardly covered in the latest edition of the Computer Science Curriculum, as published by ACM and the IEEE Computer Society [ACM/IEEE, 2008]. However, the growing number of applications results in the need for more education in this area. This education should help to overcome the limitations of currently available design technologies for embedded systems. For example, there is still a need for better specification languages, models, tools generating implementations from specifications, timing verifiers, system software, real-time operating systems, low-power design techniques, and design techniques for dependable systems. This book should help teaching the essential issues and should be a stepping stone for starting more research in the area.

Areas covered in this book

This book covers hardware as well as software aspects of embedded systems. This is in-line with the ARTIST guidelines for curricula: *"The development of embedded systems cannot ignore the underlying hardware characteristics. Timing, memory usage, power consumption, and physical failures are important."* [Caspi et al., 2005].

The book focuses on the fundamental bases of software and hardware. Specific products and tools are mentioned only if they have outstanding characteristics. Again, this is in-line with the ARTIST guidelines: *"It seems that fundamental bases are really difficult to acquire during continuous training if they haven't been initially learned, and we must focus on them."* [Caspi et al., 2005]. As a consequence, this book goes beyond teaching embedded system design by programming micro-controllers. With this approach, we would like to make sure that the material taught will not be outdated too soon. The concepts covered in this book should be relevant for a number of years to come.

The proposed positioning of the current textbook in computer science and computer engineering curricula is explained in a paper [Marwedel, 2005]. A key goal of this book is to provide an overview of embedded system design and to relate the most important topics in embedded system design to each other. This way, we avoid a problem mentioned in the ARTIST guidelines: *"The lack of maturity of the domain results in a large variety of industrial practices, often due to cultural habits. ... curricula ... concentrate on one technique and do not present a sufficiently wide perspective. .. As a result, industry has difficulty finding adequately trained engineers, fully aware of design choices"* [Caspi et al., 2005].

The book should also help to bridge the gap between practical experiences with programming micro-controllers and more theoretical issues. Furthermore, it should help to motivate students and teachers to look at more details. While the book covers a number of topics in detail, others are covered only briefly. These brief sections have been included in order to put a number of related issues into perspective. Furthermore, this approach allows lecturers to have appropriate links in the book for adding complementary material of their choice. The book includes more references than textbooks would normally do. This way, the book can also be used as a comprehensive tutorial, providing pointers for additional reading. Such references can also stimulate taking benefit of the book during labs, projects, and independent studies as well as a starting point for research.

Additional information related to the book can be obtained from the following web page:

http://ls12-www.cs.tu-dortmund.de/~marwedel/es-book.

This page includes links to slides, simulation tools, error corrections, and other related material. Readers who discover errors or who would like to make comments on how to improve the book should send an e-mail to:

peter.marwedel at tu-dortmund.de

Assignments could also use the information in complementary books (e.g. [Wolf, 2001], [Buttazzo, 2002], and [Gajski et al., 2009]).

Prerequisites

The book assumes a basic understanding in several areas:

- electrical networks at the high-school level (e.g. Kirchhoff's laws),

- operational amplifiers (optional),

- computer organization, for example at the level of the introductory book by J.L. Hennessy and D.A. Patterson [Hennessy and Patterson, 2008],

- fundamental digital circuits such as gates and registers,

- computer programming (including foundations of software engineering),

- fundamentals of operating systems,

- fundamentals of computer networks,

- finite state machines,

- some first experience with programming micro-controllers,

- fundamental mathematical concepts (such as tuples, integrals, and linear equations), and welcome knowledge in statistics and Fourier series,

- algorithms (graph algorithms and optimization algorithms such as branch and bound),

- the concept of NP-completeness.

These prerequisites can be grouped into courses as shown in the top row in fig. 0.2.

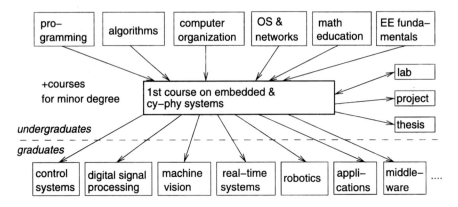

Figure 0.2.   Positioning of the topics of this book

Recommended additional teaching

A course using this textbook should be complemented by an exciting lab, using, for example, small robots, such as Lego Mindstorms^{TM} or similar robots. Another option is to let students gain some practical experience with finite state machine tools.

The book should be complemented by follow-up courses providing a more specialized knowledge in some of the following areas (see the bottom row in fig. 0.2)[2]:

- control systems,

- digital signal processing,

- machine vision,

- real-time systems, real-time operating systems, and scheduling,

- middleware,

- application areas such as telecommunications, automotive, medical equipment, and smart homes,

---

[2]The partitioning between undergraduate courses and graduate courses may differ between universities.

- robotics,

- sensors and actuators,

- specification languages for embedded systems,

- computer-aided design tools for application-specific hardware,

- formal verification of hardware systems,

- testing of hardware and software systems,

- performance evaluation of computer systems,

- low-power design techniques,

- security and dependability of computer systems,

- ubiquitous computing,

- impact of embedded systems.

History of the book

The first edition of this book was published in 2003. The field of embedded systems is moving fast and many new results have become available since then. Also, there are areas for which the emphasis has shifted. In some cases, a more detailed treatment of the topic became desirable. New developments have been taken up when the first German edition of the book was published in 2007. Therefore it became necessary to publish a major new English release, the current second edition.

Names used in this book without any reference to copyrights or trademarks may still be legally protected.

Please enjoy reading the book!

Dortmund (Germany), August 2010

Peter Marwedel

This book is dedicated
to my family members
Veronika, Malte,
Gesine, and Ronja.

Acknowledgments

My PhD students, in particular Lars Wehmeyer, did an excellent job in proof-reading a preliminary version of this book. Also, the students attending my courses provided valuable help. Corrections were contributed by David Hec, Thomas Wiederkehr, Thorsten Wilmer and Henning Garus. In addition, the following colleagues and students gave comments or hints which were incorporated into this book: R. Dömer, N. Dutt (UC Irvine), A. B. Kahng (UC San Diego), W. Kluge, R. von Hanxleden (U. Kiel), P. Buchholz, M. Engel, H. Krumm, O. Spinczyk (TU Dortmund), W. Müller, F. Rammig (U. Paderborn), W. Rosenstiel (U. Tübingen), L. Thiele (ETH Zürich), and R. Wilhelm (Saarland University). Material from the following persons was used to prepare this book: G. C. Buttazzo, D. Gajski, R. Gupta, J. P. Hayes, H. Kopetz, R. Leupers, R. Niemann, W. Rosenstiel, H. Takada, L. Thiele, and R. Wilhelm. PhD students of my group contributed to the assignments included in this book. Of course, the author is responsible for all errors and mistakes.

I do acknowledge the support of the European Commission through projects MORE, Artist2, ArtistDesign, Hipeac(2), PREDATOR, MNEMEE and MADNESS, which provided an excellent context for writing the second edition of this book.

The book has been produced using the LATEXtype setting system from the TeXnicCenter user interface. I would like to thank the authors of this software for their contribution to this work.

Acknowledgments also go to all those who have patiently accepted the author's additional workload during the writing of this book and his resulting reduced availability for professional as well as personal partners.

Finally, it should be mentioned that the Springer company has supported the publication of the book. Their support has been stimulating during the work on this book.

Chapter 1

INTRODUCTION

1.1 Application areas and examples

Embedded and cy-phy systems are present in quite diverse areas. The following list comprises key areas in which such systems are used:

- **Automotive electronics:** Modern cars can be sold in technologically advanced countries only if they contain a significant amount of electronics. These include air bag control systems, engine control systems, anti-braking systems (ABS), electronic stability programs (ESP) and other safety features, air-conditioning, GPS-systems, anti-theft protection, and many more. Embedded systems can help to reduce the impact on the environment.

- **Avionics:** A significant amount of the total value of airplanes is due to the information processing equipment, including flight control systems, anti-collision systems, pilot information systems, and others. Embedded systems can decrease emissions (such as carbon-dioxide) from airplanes. Dependability is of utmost importance.

- **Railways:** For railways, the situation is similar to the one discussed for cars and airplanes. Again, safety features contribute significantly to the total value of trains, and dependability is extremely important.

- **Telecommunication:** Mobile phones have been one of the fastest growing markets in the recent years. For mobile phones, radio frequency (RF) design, digital signal processing and low power design are key aspects. Other forms of telecommunication are also important.

- **Health sector:** The importance of healthcare products is increasing, in particular in aging societies. There is a huge potential for improving the med-

P. Marwedel, *Embedded System Design*, Embedded Systems,
DOI 10.1007/978-94-007-0257-8_1, © Springer Science+Business Media B.V. 2011

ical service by taking advantage of information processing within medical equipment. There are very diverse techniques that can be applied in this area.

- **Security:** The interest in various kinds of security is also increasing. Embedded systems can be used to improve security in many ways. This includes secure identification/authentication of people, for example with finger print sensors or face recognition systems.

The SMARTpen® [IMEC, 1997] is another example, providing authentication of payments (see fig. 1.1).

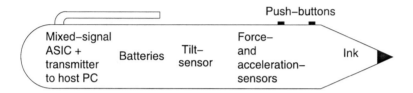

Figure 1.1.   SMARTpen (Original version)

The SMARTpen is a pen-like instrument analyzing physical parameters while its user is signing. Physical parameters include the tilt, force and acceleration. These values are transmitted to a host PC and compared with information available about the user. As a result, it can be checked if both the image of the signature as well as the way it has been produced coincide with the stored information. More recently, smart pens locally recording written patterns became commercially available and these devices are not necessarily used for authentications.

- **Consumer electronics:** Video and audio equipment is a very important sector of the electronics industry. The information processing integrated into such equipment is steadily growing. New services and better quality are implemented using advanced digital signal processing techniques. Many TV sets (in particular high-definition TV sets), multimedia phones, and game consoles comprise powerful high-performance processors and memory systems. They represent special cases of embedded systems.

- **Fabrication equipment:** Fabrication equipment is a very traditional area in which embedded/cyber-physical systems have been employed for decades. Safety is very important for such systems, the energy consumption is less important. As an example, fig. 1.2 (taken from Kopetz [Kopetz, 1997]) shows a container with an attached drain pipe. The pipe includes a valve and a sensor. Using the readout from the sensor, a computer may have to control the amount of liquid leaving the pipe.

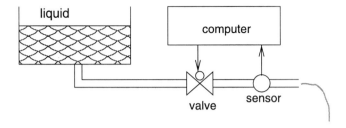

Figure 1.2. Controlling a valve

The valve is an example of an actuator (see definition on page 8).

- **Smart buildings:** Information processing can be used to increase the comfort level in buildings, can reduce the energy consumption within buildings, and can improve safety and security. Subsystems which traditionally were unrelated must be connected for this purpose. There is a trend towards integrating air-conditioning, lighting, access control, accounting and distribution of information into a single system. Tolerance levels of air conditioning subsystems can be increased for empty rooms, and the lighting can be automatically reduced. Air condition noise can be reduced to a level required for the actual operating conditions. Intelligent usage of blinds can also optimize lighting and air-conditioning. Available rooms can be displayed at appropriate places, simplifying ad-hoc meetings and cleaning. Lists of non-empty rooms can be displayed at the entrance of the building in emergency situations (provided the required power is still available). This way, energy can be saved on cooling, heating and lighting. Also safety can be improved. Initially, such systems might mostly be present in high-tech office buildings, but the trend toward energy-efficient buildings also affects the design of private homes. One of the goals is to design so-called **zero-energy-buildings** (buildings which produce as much energy as they consume) [Northeast Sustainable Energy Association, 2010]. Such a design would be one contribution towards a reduction of the global carbon-dioxide footprint and global warming.

- **Logistics:** There are several ways in which embedded/cyber-physical system technology can be applied to logistics. Radio frequency identification (RFID) technology provides easy identification of each and every object, worldwide. Mobile communication allows unprecedented interaction. The need of meeting real-time constraints and scheduling are linking embedded systems and logistics. The same is true of energy minimization issues.

- **Robotics:** Robotics is also a traditional area in which embedded/cyber-physical systems have been used. Mechanical aspects are very impor-

tant for robots. Most of the characteristics described above also apply to
robotics. Recently, some new kinds of robots, modeled after animals or
human beings, have been designed. Fig. 1.3 shows such a robot.

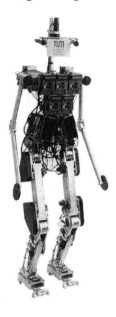

Figure 1.3. Robot "Johnnie" (courtesy H. Ulbrich, F. Pfeiffer, Lehrstuhl für Angewandte
Mechanik, TU München), ©TU München

- **Military applications:** Information processing has been used in military
  equipment for many years. In fact, some of the very first computers analyzed military radar signals.

This set of examples demonstrates the huge variety of embedded and cyber-
physical systems. Why does it make sense to consider all these types of embedded systems in one book? It makes sense because information processing
in these systems has many common characteristics, despite being physically so
different.

1.2   Common characteristics

Common characteristics of these systems are the following:

- Cyber-physical systems must be **dependable**.

  Many cyber-physical systems are safety-critical and therefore must be dependable. Nuclear power plants are examples of extremely safety-critical

systems that are at least partially controlled by software. Dependability is, however, also important in other systems, such as cars, trains, airplanes etc. A key reason for being safety-critical is that these systems are directly connected to the physical environment and have an immediate impact on the environment.

Dependability encompasses the following aspects of a system:

1 **Reliability:** Reliability is the probability that a system will not fail[1].

2 **Maintainability:** Maintainability is the probability that a failing system can be repaired within a certain time-frame.

3 **Availability:** Availability is the probability that the system is available. Both the reliability and the maintainability must be high in order to achieve a high availability.

4 **Safety:** This term describes the property that a system will not cause any harm.

5 **Security:** This term describes the property that confidential data remains confidential and that authentic communication is guaranteed.

Designers may be tempted to focus just on the functionality of systems initially, assuming that dependability can be added once the design is working. Typically, this approach does not work, since certain design decisions will not allow achieving the required dependability in the aftermath. For example, if the physical partitioning is done in the wrong way, redundancy may be impossible. Therefore, *"making the system dependable must not be an after-thought"*, it must be considered from the very beginning [Kopetz, 1997].

Even perfectly designed systems can fail if the assumptions about the workload and possible errors turn out to be wrong [Kopetz, 1997]. For example, a system might fail if it is operated outside the initially assumed temperature range.

■ Embedded systems must be **efficient**. The following metrics can be used for evaluating the efficiency of embedded systems:

1 **Energy:** Computational energy efficiency is a key characteristic of execution platform technologies. A comparison between technologies and changes over time (corresponding to a certain fabrication technology) can be seen from fig. 1.4 (approximating information provided by H. De Man [Man, 2007] and based on information provided by Philips).

---

[1] A formal definition of this term is provided in Chapter 5 of this book.

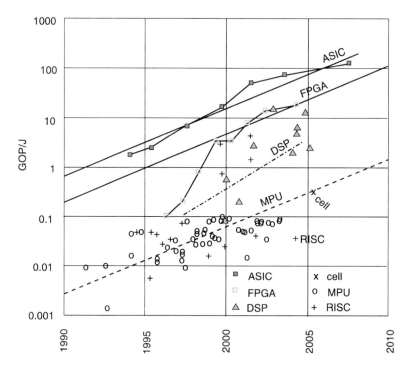

Figure 1.4. Energy efficiency as a function of time and technology (©Philips, Hugo de Man, 2007)

Obviously, the number of operations[2] per Joule is increasing as technology advances to smaller and smaller feature sizes of integrated circuits. However, for any given technology, the number of operations per Joule is largest for hardwired application specific integrated circuits (ASICs). For reconfigurable logic usually coming in the form of field programmable gate arrays (FPGAs; see page 152), this value is about one order of magnitude less. For programmable processors, it is even lower. However, processors offer the largest amount of flexibility, resulting from the flexibility of software. There is also some flexibility for reconfigurable logic, but it is limited to the size of applications that can be mapped to such logic. For hardwired designs, there is no flexibility. The trade-off between flexibility and efficiency also applies to processors: For processors optimized for the application domain, such as processors optimized for digital signal processing

---

[2] In this context, operations could be 32 bit additions.

(DSP), power-efficiency values approach those of reconfigurable logic. For general standard microprocessors, the values for this figure of merit are the worst. This can be seen from fig. 1.4, comprising values for microprocessors such as x86-like processors (see "MPU" entries), RISC processors and the cell processor designed by IBM and Sony.

As a rule of thumb, we can consider devices such as smart phones to be limited to a power consumption[3] of about two Watts and that about half of this power is required for radio frequency (RF) transmissions, displays and audio amplifiers, leaving about 1 Watt for computations. This limitation is caused both by the available battery technology and by the need to keep devices at comfortable temperatures. Improving battery technology would allow us to consume power over longer periods, but the thermal limitation prevents us from going significantly beyond the two Watts in the near future. Of course, a larger power consumption is feasible for larger devices. Nevertheless, environmental concerns also result in the need to keep the power consumption low.

Computational requirements are increasing at a rapid rate, especially for multimedia applications. De Man and Philips estimated that advanced multimedia applications need about 10 to 100 billion operations per second. Fig. 1.4 demonstrates that advanced hardware technologies provide us with this number of operations per Joule (= Ws). This means that the most power efficient platform technologies hardly provide the efficiency which is needed. It also means that we really must use all sources of efficiency improvements. Standard processors (entries for MPU and RISC) are hopelessly inefficient.

This situation leads to forecasts (see, for example, the ITRS Roadmap for Semiconductors [ITRS Organization, 2009]) predicting that availability of energy will be a key limitation for new mobile applications. According to this road map, "... *these trends imply that computation performance, in some suitable metric, must be increased by one-to-two orders of magnitude by 2020. This raises the question of the maximum attainable performance per joule and suggests a rapprochement between information theory and thermodynamics*".

2 **Run-time efficiency:** Embedded systems should exploit the available hardware architecture as much as possible. Inefficiencies, resulting from a poor mapping of applications to platforms, should be avoided. For example, compilers should not introduce overhead, since this would lead to wasted energy and possibly higher than necessary clock rates.

---

[3]Strictly speaking, we are not really "consuming" power or the closely related energy. Rather, we are converting electrical energy into thermal energy. However, *electrical* energy is really disappearing.

3 **Code size:** Dynamically loading additional code to be executed on embedded system devices is still an exception and limited to cases such as smart phones and set-top boxes. It is likely to remain an exception, due to limited connectivity and safety concerns. Therefore, the code to be run on an embedded system typically has to be stored with the system. Typically, there are no hard disks for this. Therefore, code-size should be as small as possible for the intended application. This is especially true for **systems on a chip** (SoCs), systems for which all the information processing circuits are included on a single chip. If the instruction memory is to be integrated onto this chip, it should be used very efficiently. However, the importance of this design goal might change, when larger memory densities (measured in bits per volume unit) become available. Flash-based memories will potentially have a large impact.

4 **Weight:** All portable systems must be lightweight. A low weight is frequently an important argument for buying a particular system.

5 **Cost:** For high-volume embedded systems in mass markets, especially in consumer electronics, competitiveness on the market is an extremely crucial issue, and efficient use of hardware components and the software development budget are required. A minimum amount of resources should be used for implementing the required functionality. We should be able to meet requirements using the least amount of hardware resources and energy. In order to reduce the energy consumption, clock frequencies and supply voltages should be as small as possible. Also, only the necessary hardware components should be present. Components which do not improve the worst case execution time (such as many caches or memory management units) can frequently be omitted.

- Frequently, embedded systems are connected to the physical environment through **sensors** collecting information about that environment and **actuators** controlling that environment.

  Definition: Actuators are devices converting numerical values into physical effects.

  This link to the physical environment also motivated the introduction of the term "cyber-physical system". Embedded system education focusing on the programming of micro-controllers is frequently neglecting this link. In this respect, the new term helps **liberating embedded system design from the programming of micro-controllers**.

- Many cyber-physical systems must meet **real-time constraints**. Not completing computations within a given time-frame can result in a serious loss of the quality provided by the system (for example, if the audio or video quality is affected) or may cause harm to the user (for example, if cars, trains or planes do not operate in the predicted way). Some time constraints are called hard time constraints:

Definition: *"A time-constraint is called hard if not meeting that constraint could result in a catastrophe"* [Kopetz, 1997].

All other time constraints are called **soft time constraints**.

Many of today's information processing systems are using techniques for speeding-up information processing *on the average*. For example, caches improve the average performance of a system. In other cases, reliable communication is achieved by repeating certain transmissions. These cases include Ethernet protocols: they typically rely on resending messages whenever the original messages have been lost. On the average, such repetitions result in a (hopefully only) small loss of performance, even though for a certain message the communication delay can be orders of magnitude larger than the normal delay. In the context of real-time systems, arguments about the average performance or delay cannot be accepted. *"A guaranteed system response has to be explained without statistical arguments"* [Kopetz, 1997].

- Typically, embedded systems are **reactive systems**. They can be defined as follows:

  Definition: *"A reactive system is one that is in continual interaction with its environment and executes at a pace determined by that environment"* [Bergé et al., 1995].

  Reactive systems can be thought of as being in a certain state, waiting for an input. For each input, they perform some computation and generate an output and a new state. Therefore, automata are very good models of such systems. Mathematical functions, which describe the problems solved by most algorithms, would be an inappropriate model.

- Many embedded systems are **hybrid systems** in the sense that they include analog and digital parts. Analog parts use continuous signal values in continuous time, whereas digital parts use discrete signal values in discrete time.

- Most embedded systems do not use keyboards, mice and large computer monitors for their user-interface. Instead, there is a **dedicated user-interface** consisting of push-buttons, steering wheels, pedals etc. Because of this, the user hardly recognizes that information processing is involved. Due to this, the new era of computing has also been characterized by the term **disappearing computer**.

- These systems are frequently **dedicated towards a certain application**.

For example, processors running control software in a car or a train will always run that software, and there will be no attempt to run a computer game or spreadsheet program on the same processor. There are mainly two reasons for this:

1 Running additional programs would make those systems less dependable.

2 Running additional programs is only feasible if resources such as memory are unused. No unused resources should be present in an efficient system.

However, the situation is slowly changing for systems such as smart phones. Smart phones are becoming more PC-like and can hardly be called cyber-physical systems. Also, the situation is becoming a bit more dynamic in the automotive industry as well, as demonstrated by the AUTOSAR initiative [AUTOSAR, 2010].

■ Embedded systems are **under-represented in teaching and in public discussions**. One of the problems in teaching embedded system design is the comprehensive equipment which is needed to make the topic interesting and practical. Also, real embedded systems are very complex and hence difficult to teach.

Due to this set of common characteristics (except for the last one), it does make sense to analyze common approaches for designing embedded systems, instead of looking at the different application areas only in isolation.

Actually, not every embedded system will have all the above characteristics. We can define the term "embedded system" also in the following way: **Information processing systems meeting most of the characteristics listed above are called *embedded systems*.** This definition includes some fuzziness. However, it seems to be neither necessary nor possible to remove this fuzziness.

1.3   Challenges in Embedded System Design

Embedded systems comprise a large amount of software. Nevertheless, embedded system design is not just a special case of software design. Many additional design goals must be taken into account. For example:

1 Embedded systems **really** must be dependable. The level of dependability goes far beyond the traditional level reached for PC-like systems. Examples of serious cases of undependability include the following:

   ■ In one case, the voice control system at Los Angeles airport was lost for more than 3 hours [Broesma, 2004]. The problem resulted from a server running an operating system. A counter in the operating system kept track of the time since the last reboot. This counter was overflowing after about 48 days. Therefore, the maintenance staff was instructed to reboot the server every month. Once, the staff forgot to reboot and this resulted in a system crash.

- Many other cases of failing computer systems are reported to *the risks digest, forum on the risks to the public in computers and related Systems* (see [Neumann, 2010] for the most recent edition).

2 Due to efficiency targets, software designs cannot be done independently of the underlying hardware. Therefore, software and hardware must be taken into account during the design steps. This, however, is difficult, since such integrated approaches are typically not taught at educational institutes. The cooperation between electrical engineering and computer science has not yet reached the required level. A mapping of specifications to hardware would provide the best energy efficiency. However, hardware implementations are very expensive and require long design times. Therefore, hardware designs do not provide the flexibility to change designs as needed. We need to find a good compromise between efficiency and flexibility.

3 Embedded systems must meet many *non-functional requirements* such as real-time constraints, energy/power efficiency and dependability requirements. Many objectives must be taken into account during the design. Just capturing non-functional requirements is already difficult.

4 The link to physics has additional implications. For example, we must check if we will definitely meet real-time constraints. Managing time is one of the largest challenges [Lee, 2006].

5 Real systems are profoundly concurrent. Managing concurrency is therefore another major challenge.

6 Real embedded systems are complex. Therefore, they comprise various **components** and we are interested in **compositional design**. This means, we would like to study the impact of combining components. For example, we would like to know whether we could add a GPS system to the sources of information in a car without overloading the communication bus.

7 Traditional sequential programming languages are not the best way to describe concurrent, timed systems.

The table in fig. 1.5 highlights some distinguishing features between the design of PC-like systems and embedded systems during the mapping of applications to hardware platforms.

	Embedded	PC-like
Architectures	Frequently heterogeneous very compact	Mostly homogeneous not compact (x86 etc)
x86 compatibility	Less relevant	Very relevant
Architecture fixed?	Sometimes not	Yes
Model of computation (MoCs)	C+multiple models (data flow, discrete events, ...)	Mostly von Neumann (C, C++, Java)
Optim. objectives	Multiple (energy, size, ...)	Average performance dominates
Real-time relevant	Yes, very!	Hardly
Applications	Several concurrent apps.	Mostly single application
Apps. known at design time	Most, if not all	Only some (e.g. WORD)

Figure 1.5. Scope of mapping applications to PC-like and Embedded Systems hardware

Compatibility with traditional instruction sets employed for PCs is less important for embedded systems, since it is typically possible to compile software applications for architectures at hand. Sequential programming languages do not match well with the need to describe concurrent real-time systems, and other ways of modeling applications may be preferred. Several objectives must be considered during the design of embedded/cyber-physical systems. In addition to the average performance, the worst case execution time, energy consumption, weight, reliability, operating temperatures, etc. may have to be optimized. Meeting real-time constraints is very important for cyber-physical systems, but hardly ever for PC-like systems. Meeting time constraints can be verified at design time only, if all the applications are known at design time. Also, it must be known, which applications should run concurrently. For example, designers must ensure that a GPS-application, a phone call, and concurrent data transfers can be executed at the same time without losing voice samples. In contrast, there is no need to guarantee time constraints for multiple, concurrently running software media-players. For PC-like systems, such knowledge is almost never available.

1.4   Design Flows

The design of embedded systems is a rather complex task, which has to be broken down into a number of subtasks to be tractable. These subtasks must be performed one after the other and some of them must be repeated.

The design information flow starts with ideas in people's heads. These ideas should incorporate knowledge about the application area. These ideas must be captured in a design specification. In addition, standard hardware and system software components are typically available and should be reused whenever possible (see fig. 1.6).

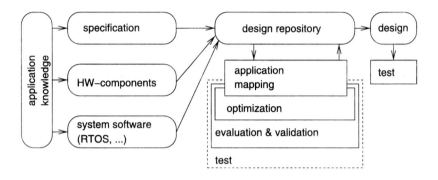

Figure 1.6. Simplified design flow

In this diagram (as well as in other similar diagrams in this book), **we are using boxes with rounded corners for stored information and rectangles for transformations on data**. In particular, information is stored in the **design repository**. The repository allows keeping track of design models. In most cases, the repository should provide version management or "revision control", such as CVS [Cederqvist, 2006] or SVN [Collins-Sussman et al., 2008]. A good design repository should also come with a design management interface which would also keep track of the applicability of design tools and sequences, all integrated into a comfortable graphical user interface (GUI). The design repository and the GUI can be extended into an **integrated development environment** (IDE), also called **design framework** (see, for example [Liebisch and Jain, 1992]). An integrated development environment keeps track of dependencies between tools and design information.

Using the repository, design decisions can be taken in an iterative fashion. At each step, design model information must be retrieved. This information is then considered.

During design iterations, **applications are mapped** to execution platforms and new (partial) design information is generated. The generation comprises the mapping of operations to concurrent tasks, the mapping of operations to either hardware or software (called hardware/software partitioning), compilation, and scheduling.

Designs should be **evaluated** with respect to various objectives including performance, dependability, energy consumption, manufacturability etc. At the current state of the art, none of the design steps can be guaranteed to be correct. Therefore, it is also necessary to **validate** the design. Validation consists of checking intermediate or final design descriptions against other descriptions. Thus, each new design should be evaluated and validated.

Due to the importance of the efficiency of embedded systems, **optimizations** are important. There is a large number of possible optimizations, including high-level transformations (such as advanced loop transformations) and energy-oriented optimizations.

Design iterations could also include **test** generation and an evaluation of the testability. Testing needs to be included in the design iterations if testability issues are already considered during the design steps. In fig. 1.6, test generation has been included as optional step of design iterations (see dashed box). If test generation is not included in the iterations, it must be performed after the design has been completed.

At the end of each step, the repository should be updated.

Details of the flow between the repository, application mapping, evaluation, validation, optimization, testability considerations and storage of design information may vary. These actions may be interleaved in many different ways, depending on the design methodology used.

This book presents embedded system design from a broad perspective, and it is not tied towards particular design flows or tools. Therefore, we have not indicated a particular list of design steps. For any particular design environment, we can "unroll" the loop in fig. 1.6 and attach names to particular design steps. For example, this leads to the particular case of the SpecC [Gajski et al., 2000] design flow shown in fig. 1.7.

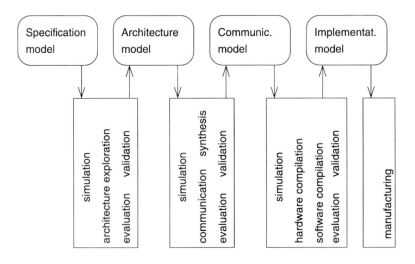

Figure 1.7. Design flow for SpecC tools (simplified)

In this case, a particular set of design steps, such as architecture exploration, communication synthesis and software and hardware compilation are included. The precise meaning of these terms is not relevant in this book. In the case of fig. 1.7, validation and evaluation are explicitly shown for each of the steps, but are wrapped into one larger box.

A second instance of an unfolded fig. 1.6 is shown in fig. 1.8. It is the V-model of design flows [V-Modell XT Authors, 2010], which has to be adhered to for many German IT projects, especially in the government sector.

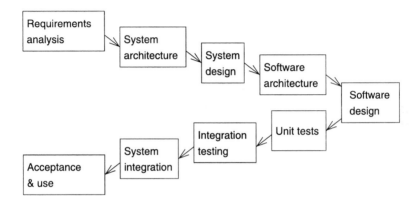

Figure 1.8.   Design flow for the V-model (rotated standard view)

Fig. 1.8 very clearly shows the different steps that must be performed. The steps correspond to certain phases during the software development process (the precise meaning is again not relevant in the context of this book). Note that taking design decisions, evaluating and validating designs is lumped into a single box in this diagram. Application knowledge, system software and system hardware are not explicitly shown. The V-model also includes a model of the integration and testing phase (lower "wing") of the diagram. This corresponds to an inclusion of testing in the loop of fig. 1.6. The shown model corresponds to the V-model version "97". The more recent V-model XT allows a more general set of design steps. This change matches very well our interpretation of design flows in fig. 1.6. Other iterative approaches include the **waterfall model** and the **spiral model**. More information about software engineering for embedded systems can be found in a book by J. Cooling [Cooling, 2003].

Our generic design flow model is also consistent with flow models used in hardware design. For example, Gajski's Y-chart [Gajski and Kuhn, 1983] (see fig. 1.9) is a very popular model.

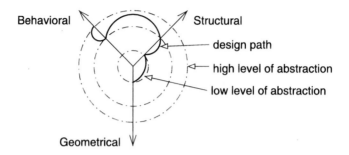

Figure 1.9. Gajski's Y-chart and design path (in bold)

Gajski considers design information in three dimensions: behavior, structure and layout. The first dimension just reflects the behavior. A high-level model would describe the overall behavior, while finer-grained models would describe the behavior of components. Models at the second dimension include structural information, such as information about hardware components. High-level descriptions in this dimension could correspond to processors, low-level descriptions to transistors. The third dimension represents geometrical layout information of chips. Design paths will typically start with a coarse-grained behavioral description and finish with a fine-grained geometrical description. Along this path, each step corresponds to one iteration of our generic design flow model. In the example of fig. 1.9, an initial refinement is done in the behavioral domain. The second design step maps the behavior to structural elements, and so on. Finally, a detailed geometrical description of the chip layout is obtained.

The previous three diagrams demonstrate that a number of design flows are using the iterative flow of fig. 1.6. The nature of the iterations in fig. 1.6 can be a source of discussions. Ideally, we would like to describe the properties of our system and then let some smart tool do the rest. Automatic generation of design details is called **synthesis**.

Definition: **Synthesis** is the process of generating the description of a system in terms of related lower-level components from some high-level description of the expected behavior [Marwedel, 1990].

Synthesis, if successful, avoids many manual design steps. The goal of using this paradigm for designing systems has been considered in the "describe-and-synthesize" paradigm by Gajski [Gajski et al., 1994]. This paradigm is in contrast to the traditional "specify-explore-refine" approach, also known as "design-and-simulate" approach. This second term stresses the fact that manual design typically has to be combined with simulation, for example for catching design errors. Simulation is more important than in automatic synthesis.

1.5   Structure of this book

Consistent with the design information flow shown above, this book is structured as follows: Chapter 2 provides an overview of specification techniques, languages and models. Key hardware components of embedded systems are presented in Chapter 3. Chapter 4 deals with system software components, particularly embedded operating systems. Chapter 5 contains the essentials of embedded system design evaluation and verification. Mapping applications to execution platforms is one of the key steps in the design process of embedded systems. Standard techniques achieving such a mapping are listed in Chapter 6. This Chapter also includes standard scheduling techniques. Due to the need for generating efficient designs, many optimization techniques are needed. From among the abundant set of available optimization techniques, several groups are mentioned in Chapter 7. Chapter 8 contains a brief introduction to testing mixed hardware/software systems. The appendix comprises a description of a standard optimization technique and some prerequisites for understanding one of the circuits in Chapter 3.

It may be necessary to design special purpose hardware or to optimize processor architectures for a given application. However, hardware design is not covered in this book. Coussy and Morawiec [Coussy and Morawiec, 2008] provide an overview of recent high-level hardware synthesis techniques.

The content of this book is different from the content of most other books on embedded systems design. Traditionally, the focus of many books on embedded systems is on explaining the use of micro-controllers, including their memory, I/O and interrupt structure. There are many such books [Ball, 1996], [Heath, 2000], [Ball, 1998], [Barr, 1999], [Ganssle, 2000], [Barrett and Pack, 2005], [Ganssle, 2008], [Ganssle et al., 2008], and [Labrosse, 2000].

We believe that, due to the increasing complexity of embedded systems, this focus has to be extended to include at least different specification paradigms, fundamentals of hardware building blocks, the mapping of applications to execution platforms, as well as evaluation, validation and optimization techniques. In the current book, we will be covering all these areas. The goal is to provide students with an introduction to embedded systems, enabling students to put the different areas into perspective.

For further details, we recommend a number of sources (some of which have also been used in preparing this book):

- There is a large number of sources of information on specification languages. These include earlier books by Young [Young, 1982], Burns and Wellings [Burns and Wellings, 1990], Bergé [Bergé et al., 1995] and de Micheli [De Micheli et al., 2002]. There is a huge amount of information on new languages such as SystemC [Müller et al., 2003], SpecC [Gajski et al., 2000], and Java [Wellings, 2004], [Dibble, 2008], [Bruno and Bollella, 2009], [Java Community Process, 2002], [Anonymous, 2010b].

- Approaches for designing and using real-time operating systems (RTOSes) are presented in a book by Kopetz [Kopetz, 1997].

- Real-time scheduling is covered comprehensively in the books by Buttazzo [Buttazzo, 2002] and by Krishna and Shin [Krishna and Shin, 1997].

- Other sources of information about embedded systems include books by Laplante [Laplante, 1997], Vahid [Vahid, 2002], the ARTIST road map [Bouyssounouse and Sifakis, 2005], the "Embedded Systems Handbook" [Zurawski, 2006], and recent books by Gajski et al. [Gajski et al., 2009], and Popovici et al. [Popovici et al., 2010].

- Approaches for embedded system education are covered in the Workshops on Embedded Systems Education (WESE); see [Jackson et al., 2009] for results from the most recent workshop.

- The website of the European network of excellence on embedded and real-time systems [Artist Consortium, 2010] provides numerous links for the area.

- The web page of a special interest group of ACM [ACM SIGBED, 2010] focuses on embedded systems.

- Symposia dedicated towards embedded/cyber-physical systems include the Embedded Systems Week (see www.esweek.org) and the Cyber-Physical Systems Week (see www.cpsweek.org).

- Robotics is an area that is closely linked to embedded and cyber-physical systems. We recommend the book by Fu, Gonzalez and Lee [Fu et al., 1987] for information on robotics.

1.6   Assignments

1 Please list possible definitions of the term "embedded system"!

2 How would you define the term "cyber-physical system"?

3 Use the sources available to you to demonstrate the importance of embedded systems!

4 Compare the curriculum of your educational program with the description of the curriculum in this introduction. Which prerequisites are missing in your program? Which advanced courses are available?

5 Please enumerate application areas of embedded systems and indicate up to 5 examples of embedded systems!

6 Please enumerate up to six characteristics of embedded systems!

7 How do different hardware technologies differ with respect to their energy efficiency?

8 Suppose that your mobile uses a lithium battery rated at 720 mAh. The nominal voltage of the battery is 3.7 V. Assuming a constant power consumption of 1 W, how long would it take to empty the battery? All secondary effects such as decreasing voltages should be ignored in this calculation.

9 The computational efficiency is sometimes also measured in terms of billions of operations per second per Watt. How is this different from the figure of merit used in fig. 1.4?

10 Which real-time constraints are called "hard real-time constraints"?

11 How could you define the term "reactive system"?

Chapter 2

SPECIFICATIONS AND MODELING

2.1 Requirements

Consistent with the simplified design flow (see fig. 1.6), we will now describe requirements and approaches for specifying and modeling embedded systems.

Specifications for embedded systems provide **models** of the system under design (SUD). Models can be defined as follows [Jantsch, 2004]:

Definition: *"A model is a simplification of another entity, which can be a physical thing or another model. The model contains exactly those characteristics and properties of the modeled entity that are relevant for a given task. A model is minimal with respect to a task if it does not contain any other characteristics than those relevant for the task".*

Models are described in languages. Languages should be capable of representing the following features[1]:

- **Hierarchy:** Human beings are generally not capable of comprehending systems containing many objects (states, components) having complex relations with each other. The description of all real-life systems needs more objects than human beings can understand. Hierarchy (in combination with **abstraction**) is a key mechanism helping to solve this dilemma. Hierarchies can be introduced such that humans need to handle only a small number of objects at any time.

  There are two kinds of hierarchies:

---

[1]Information from the books of Burns et al. [Burns and Wellings, 1990], Bergé et al. [Bergé et al., 1995] and Gajski et al. [Gajski et al., 1994] is used in this list.

P. Marwedel, *Embedded System Design*, Embedded Systems,
DOI 10.1007/978-94-007-0257-8_2, © Springer Science+Business Media B.V. 2011

- **Behavioral hierarchies:** Behavioral hierarchies are hierarchies containing objects necessary to describe the system behavior. States, events and output signals are examples of such objects.

- **Structural hierarchies:** Structural hierarchies describe how systems are composed of physical components.

  For example, embedded systems can be comprised of components such as processors, memories, actuators and sensors. Processors, in turn, include registers, multiplexers and adders. Multiplexers are composed of gates.

- **Component-based design** [Sifakis, 2008]: It must be "easy" to derive the behavior of a system from the behavior of its components. If two components are connected, the resulting new behavior should be predictable. Example: suppose that we add another component (say, some GPS unit) to a car. The impact of the additional processor on the overall behavior of the system (including buses etc.) should be predictable.

- **Concurrency:** Real-life systems are distributed, concurrent systems composed of components. It is therefore necessary to be able to specify concurrency conveniently. Unfortunately, human beings are not very good at understanding concurrent systems and many problems with real systems are actually a result of an incomplete understanding of possible behaviors of concurrent systems.

- **Synchronization and communication:** Components must be able to communicate and to synchronize. Without communication, components could not cooperate and we would use each of them in isolation. It must also be possible to agree on the use of resources. For example, it is necessary to express mutual exclusion.

- **Timing-behavior:** Many embedded systems are real-time systems. Therefore, explicit timing requirements are one of the characteristics of embedded systems. The need for explicit modeling of time is even more obvious from the term "cyber-physical system". Time is one of the key dimensions of physics. Hence, timing requirements **must** be captured in the specification of embedded/cyber-physical systems.

  However, standard theories in computer science model time only in a very abstract way. The O-notation is one of the examples. This notation just reflects growth rates of functions. It is frequently used to model run-times of algorithms, but it fails to describe real execution times. In physics, quantities have units, but the O-notation does not even have units. So, it would not distinguish between femtoseconds and centuries. A similar remark applies to termination properties of algorithms. Standard theories are concerned with proving that a certain algorithm *eventually* terminates. For real-time systems, we need to show that an algorithm terminates in a given amount of time.

The resulting problems are very clearly formulated in a statement made by E. Lee: *"The lack of timing in the core abstraction (of computer science) is a flaw, from the perspective of embedded software"* [Lee, 2005].

According to Burns and Wellings [Burns and Wellings, 1990], modeling time must be possible in the following four contexts:

- Techniques for measuring **elapsed time**:

  For many applications it is necessary to check, how much time has elapsed since some computation was performed. Access to a timer would provide a mechanism for this.

- Means for **delaying of processes** for a specified time:

  Typically, real-time languages provide some delay construct. Unfortunately, typical implementations of embedded systems in software do not guarantee precise delays. Let us assume that task T should be delayed by some amount δ^2. Usually, this delay is implemented by changing task T's state in the operating system from "ready" or "run" to "suspended". At the end of this time interval, T's state is changed from "suspended" to "ready". This does not mean that the task actually executes. If some higher priority task is executing or if preemption is not used, the delayed task will be delayed longer than δ.

- Possibility to specify **timeouts**:

  There are many situations in which we must wait for a certain event to occur. However, this event may actually not occur within a given time interval and we would like to be notified about this. For example, we might be waiting for a response from some network connection. We would like to be notified if this response is not received within some amount of time, say δ. This is the purpose of **timeouts**. Real-time languages usually also provide some timeout construct. Implementations of timeouts frequently come with the same problems which we mentioned for delays.

- Methods for specifying **deadlines** and **schedules**:

  For many applications it is necessary to complete certain computations in a limited amount of time. For example, if the sensors of some car signal an accident, air-bags must be ignited within about ten milliseconds. In this context, we must guarantee that the software will decide

---

[2] In this book, we will not distinguish between threads, processes and tasks.

whether or not to ignite the air-bags in that given amount of time. The air-bags could harm passengers, if they go off too late. Unfortunately, most languages do not allow to specify timing constraints. If they can be specified at all, they must be specified in separate control files, pop-up menus etc. But the situation is still bad even if we are able to specify these constraints: many modern hardware platforms do not have a very predictable timing behavior. Caches, stalled pipelines, speculative execution, task preemption, interrupts, etc. may have an impact on the execution time which is very difficult to predict. Accordingly, **timing analysis** (verifying the timing constraints) is a very hard design task.

- **State-oriented behavior:** It was already mentioned in Chapter 1 that automata provide a good mechanism for modeling reactive systems. Therefore, the state-oriented behavior provided by automata should be easy to describe. However, classical automata models are insufficient, since they cannot model timing and since hierarchy is not supported.

- **Event-handling:** Due to the reactive nature of embedded systems, mechanisms for describing events must exist. Such events may be external events (caused by the environment) or internal events (caused by components of the SUD).

- **Exception-oriented behavior:** In many practical systems exceptions do occur. In order to design dependable systems, it must be possible to describe actions to handle exceptions easily. It is not acceptable that exceptions must be indicated for each and every state (such as in the case of classical state diagrams). Example: In fig. 2.1, input k might correspond to an exception.

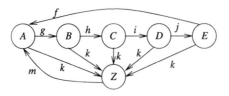

Figure 2.1.   State diagram with exception k

Specifying this exception at each state makes the diagram very complex. The situation would get worse for larger state diagrams with many transitions. Below, we will show how all the transitions can be replaced by a single one.

- **Presence of programming elements:** Popular programming languages have proven to be a convenient means of expressing computations that have to be performed. Hence, programming language elements should be available in the specification technique used. Classical state diagrams do not meet this requirement.

- **Executability:** Specifications are not automatically consistent with the ideas in people's heads. Executing the specification is a means of plausibility checking. Specifications using programming languages have a clear advantage in this context.

- **Support for the design of large systems:** There is a trend towards large and complex embedded software programs. Software technology has found mechanisms for designing such large systems. For example, object-orientation is one such mechanism. It should be available in the specification methodology.

- **Domain-specific support:** It would of course be nice if the same specification technique could be applied to all the different types of embedded systems, since this would minimize the effort for developing specification techniques and tool support. However, due to the wide range of application domains, there is little hope that one language can be used to efficiently represent specifications in all domains. For example, control-dominated, data-dominated, centralized and distributed applications-domains can all benefit from language features dedicated towards those domains.

- **Readability:** Of course, specifications must be readable by human beings. Otherwise, it would not be feasible to validate whether or not the specification meets the real intent of the persons specifying the SUD. All design documents should also be machine-readable into order to process them in a computer. Therefore, specifications should be captured in languages which are readable by humans and by computers.

  Initially, such specifications could use a natural language such as English or Japanese. Even this natural language description should be captured in a design document, so that the final implementation can be checked against the original document. However, natural languages are not sufficient for later design phases, since natural languages lack key requirements for specification techniques: it is necessary to check specifications for completeness, absence of contradictions and it should be possible to derive implementations from the specification in a systematic way. Natural languages do not meet these requirements.

- **Portability and flexibility:** Specifications should be independent of specific hardware platforms so that they can be easily used for a variety of target platforms. Ideally, changing the hardware platform should have no impact on the specification. In practice, small changes may have to be tolerated.

- **Termination:** It should be feasible to identify processes that will terminate from the specification. This means that we would like to use specifications for which the halting problem (the problem of figuring out whether or not a certain algorithm will terminate; see, for example [Sipser, 2006]) is decidable.

- **Support for non-standard I/O-devices:** Many embedded systems use I/O-devices other than those typically found in a PC. It should be possible to describe inputs and outputs for those devices conveniently.

- **Non-functional properties:** Actual SUDs must exhibit a number of non-functional properties, such as fault-tolerance, size, extendibility, expected lifetime, power consumption, weight, disposability, user friendliness, electromagnetic compatibility (EMC) etc. There is no hope that all these properties can be defined in a formal way.

- **Support for the design of dependable systems:** Specification techniques should provide support for designing dependable systems. For example, specification languages should have unambiguous semantics, facilitate formal verification and be capable of describing security and safety requirements.

- **No obstacles to the generation of efficient implementations:** Since embedded systems must be efficient, no obstacles prohibiting the generation of efficient realizations should be present in the specification.

- **Appropriate model of computation (MoC):** The von-Neumann model of sequential execution combined with some communication technique is a commonly used MoC. However, this model has a number of serious problems, in particular for embedded system applications. Problems include:

  – Facilities for describing timing are lacking.
  – Von-Neumann computing is implicitly based on accesses to globally shared memory (such as in Java). It has to guarantee mutually exclusive access to shared resources. Otherwise, multi-threaded applications allowing pre-emptions at any time can lead to very unexpected program behaviors[3]. Using primitives for ensuring mutually exclusive access can, however, very easily lead to deadlocks. Possible deadlocks may be difficult to detect and may remain undetected for many years.

---

[3]Examples are typically provided in courses on operating systems.

Lee [Lee, 2006] provided a very alarming example in this direction. Lee studied implementations of a simple observer pattern in Java. For this pattern, changes of values must be propagated from some producer to a set of subscribed observers. This is a very frequent pattern in embedded systems, but is difficult to implement correctly in a multi-threaded von-Neumann environment with preemptions. Lee's code is a possible implementation of the observer pattern in Java for a multi-threaded environment:

```
public synchronized void addListener(listener) {...}
public synchronized void setValue(newvalue) {
    myvalue=newvalue;
    for (int i=0; i<mylisteners.length; i++) {
        myListeners[i].valueChanged(newvalue)
    }
}
```

Method addListener subscribes new observers, method setValue propagates new values to subscribed observers. In general, in a multithreaded environment, threads can be pre-empted any time, resulting in an arbitrarily interleaved execution of these threads. Adding observers while setValue is already active could result in complications, i.e. we would not know if the new value had reached the new listener. Moreover, the set of observers constitutes a global data structure of this class. Therefore, these methods are synchronized in order to avoid changing the set of observers while values are already partially propagated. This way, only one of the two methods can be active at a given time. This mutual exclusion is necessary to prevent unwanted interleavings of the execution of methods in a multithreaded environment. Why is this code problematic? It is problematic since valueChanged could attempt to get exclusive access to some resource (say, R). If that resource is allocated to some other method (say, A), then this access is delayed until A releases R. If A calls (possibly indirectly) addListener or setValue before releasing R, then these methods will be in a deadlock: setValue waits for R, releasing R requires A to proceed, A cannot proceed before its call of setValue or addListener is serviced. Hence, we will have a deadlock.

This example demonstrates the existence of deadlocks resulting from using multiple threads which can be arbitrarily pre-empted and therefore require mutual exclusion for their access to critical resources. Lee showed [Lee, 2006] that many of the proposed "solutions" of the problem are problematic themselves. So, even this very simple pattern is difficult to implement correctly in a multi-threaded von-Neumann environment. This example shows that concurrency is really difficult to understand for humans and there may be the risk of oversights, even after very rigorous code inspections.

Lee came to the drastic conclusion that *"nontrivial software written with threads, semaphores, and mutexes is incomprehensible to humans"* and that *"threads as a concurrency model are a poor match for embedded systems. ... they work well only ... where best-effort scheduling policies are sufficient"* [Lee, 2005].

The underlying reasons for deadlocks have been studied in detail in the context of operating systems (see, for example, [Stallings, 2009]). From this context, it is well-known that four conditions must hold at run-time to get into a deadlock: mutual exclusion, no pre-emption of resources, holding resources while waiting for more, and a cyclic dependency between threads. Obviously, all four conditions are met in the above example. The theory of operating systems provides no general way out of this problem. Rare deadlocks may be acceptable for a PC, but they are clearly unacceptable for a safety-critical system.

We would like to specify SUDs such that we do not have to care about possible deadlocks. Therefore, it makes sense to study non-von-Neumann MoCs avoiding this problem. We will study such MoCs from the next section onwards. It will be shown that the observer pattern can be easily implemented in other MoCs.

From the list of requirements, it is already obvious that there will not be any single formal language capable of meeting all these requirements. Therefore, in practice, we must live with compromises and possibly also with a mixture of languages (each of which would be appropriate for describing a certain type of problems). The choice of the language used for an actual design will depend on the application domain and the environment in which the design has to be performed. In the following, we will present a survey of languages that can be used for actual designs. These languages will demonstrate the essential features of the corresponding model of computation.

2.2 Models of computation

Models of computation (MoCs) describe the mechanism assumed for performing computations. In the general case, we must consider systems comprising components. It is now common practice to strictly distinguish between the computations performed in the components and communication. Accordingly, MoCs define (see also [Lee, 1999], [Janka, 2002], [Jantsch, 2004], [Jantsch, 2006]):

- **Components** and the organization of computations in such components: Procedures, processes, functions, finite state machines are possible components.

- **Communication protocols:** These protocols describe methods for communication between components. Asynchronous message passing and rendez-vous based communication are examples of communication protocols.

Relations between components can be captured in graphs. In such graphs, we will refer to the computations also as processes or tasks. Accordingly, relations between these will be captured by **task graphs** and **process networks**. Nodes in the graph represent components performing computations. Computations map input data streams to output data streams. Computations are sometimes implemented in high-level programming languages. Typical computations contain (possibly non-terminating) iterations. In each cycle of the iteration, they consume data from their inputs, process the data received, and generate data on their output streams. Edges represent relations between components. We will now introduce these graphs at a more detailed level.

The most obvious relation between computations is their causal dependence: Many computations can only be executed after other computations have terminated. This dependence is typically captured in **dependence graphs**. Fig. 2.2 shows a dependence graph for a set of computations.

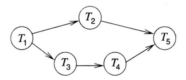

Figure 2.2. Dependence graph

Definition: A dependence graph is a directed graph $G = (V, E)$, where V is the set of **vertices** or **nodes** and E is the set of **edges**. $E \subseteq V \times V$ imposes a relation on V. If $(v_1, v_2) \in E$, then v_1 is called an **immediate predecessor** of v_2 and v_2 is called an **immediate successor** of v_1. Suppose E^* is the transitive closure of E. If $(v_1, v_2) \in E^*$, then v_1 is called a **predecessor** of v_2 and v_2 is called a **successor** of v_1.

Such dependence graphs form a special case of task graphs. Task graphs may contain more information than modeled in fig. 2.2. For example, task graphs may include the following extensions of dependence graphs:

1 **Timing information:** Tasks may have arrival times, deadlines, periods, and execution times. In order to take these into account while scheduling

computations, it may be useful to include this information in the graphs. Adopting the notation used in the book by Liu [Liu, 2000], we include possible execution intervals in fig. 2.3. Computations T_1 to T_3 are assumed to be independent. The first number in brackets is the arrival time, the second the deadline (execution times are not explicitly shown). For example, T_1 is assumed to be available at time 0 and should be completed no later than at time 7.

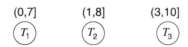

Figure 2.3.   Graphs including timing information

Significantly more complex combinations of timing and dependence relations can exist.

2 **Distinction between different types of relations** between computations: Precedence relations just model constraints for possible execution sequences. At a more detailed level, it may be useful to distinguish between constraints for scheduling and communication between computations. Communication can again be described by edges, but additional information may be available for each of the edges, such as the time of the communication and the amount of information exchanged. Precedence edges may be kept as a separate type of edges, since there could be situations in which computations must execute sequentially even though they do not exchange information.

In fig. 2.2, input and output (I/O) is not explicitly described. Implicitly it is assumed that computations without any predecessor in the graph might be receiving input at some time. Also, they might generate output for the successor and that this output could be available only after the computation has terminated. It is often useful to describe input and output more explicitly. In order to do this, another kind of relation is required. Using the same symbols as Thoen [Thoen and Catthoor, 2000], we use partially filled circles for denoting input and output. In fig. 2.4, partially filled circles identify I/O edges.

3 **Exclusive access to resources:** Computations may be requesting exclusive access to some resource, for example to some input/output device or some communication area in memory. Information about necessary exclusive access should be taken into account during scheduling. Exploiting this information might, for example, be used to avoid the priority inversion problem (see page 188). Information concerning exclusive access to resources can be included in the graphs.

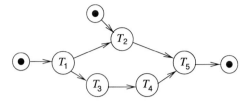

Figure 2.4. Graphs including I/O-nodes and edges

4 **Periodic schedules:** Many computations, especially in digital signal processing, are periodic. This means that we must distinguish more carefully
between a task and its execution (the latter is frequently called a **job** [Liu,
2000]). Task graphs for such schedules are infinite. Fig. 2.5 shows a task
graph including jobs J_{n-1} to J_{n+1} of a periodic task.

Figure 2.5. Graph including jobs

5 **Hierarchical graph nodes:** The complexity of the computations denoted
by graph nodes may be quite different. On one hand, specified computations may be quite involved and contain thousands of lines of program code.
On the other hand, programs can be split into small pieces of code so that in
the extreme case, each of the nodes corresponds only to a single operation.
The level of complexity of graph nodes is also called their **granularity**.
Which granularity should be used? There is no universal answer to this. For
some purposes, the granularity should be as large as possible. For example,
if we consider each of the nodes as one process to be scheduled by a real-
time operating system (RTOS), it may be wise to work with large nodes in
order to minimize context-switches between different processes. For other
purposes, it may be better to work with nodes modeling just a single operation. For example, nodes must be mapped to hardware or to software. If a
certain operation (such as the frequently used Discrete Cosine Transform,
or DCT) can be mapped to special purpose hardware, then it should not be
buried in a complex node that contains many other operations. It should
rather be modeled as its own node. In order to avoid frequent changes of
the granularity, hierarchical graph nodes are very useful. For example, at
a high hierarchical level, the nodes may denote complex tasks, at a lower

level basic blocks[4] and at an even lower level individual arithmetic operations. Fig. 2.6 shows a hierarchical version of the dependence graph in fig.
2.2, using a rectangle to denote a hierarchical node.

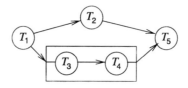

Figure 2.6.   Hierarchical task graph

As indicated above, MoCs can be classified according to the models of communication (reflected by edges in the task graphs) and the model of computations within the components (reflected by the nodes in the task graph). In the
following, we will explain prominent examples of such models:

- **Models of communication:**

  We distinguish between two communication paradigms: **shared memory**
  and **message passing**. Other communication paradigms exist (e.g. entan-
  gled states in quantum mechanics [Bouwmeester et al., 2000]), but are not
  considered in this book.

  - **Shared memory:**

    For shared memory, communication is carried out by accesses to the
    same memory from all components.

    Access to shared memory should be protected, unless access is totally
    restricted to reads. If writes are involved, exclusive access to the mem-
    ory must be guaranteed while components are accessing shared mem-
    ories. Segments of program code, for which exclusive access must
    be guaranteed, are called **critical sections**. Several mechanisms for
    guaranteeing exclusive access to resources have been proposed. These
    include semaphores, conditional critical regions and monitors. Refer
    to books on operating systems (e.g. Stallings [Stallings, 2009]) for a
    description of the different techniques. Shared memory-based commu-
    nication can be very fast, but is difficult to implement in multiprocessor
    systems if no common memory is physically available.

---

[4]Basic blocks are code blocks of maximum length not including any branch except possibly at their end and
not being branched into.

– **Message passing:** For message passing, messages are sent and received. Message passing can be implemented easily even if no common memory is available. However, message passing is generally slower than shared memory based communication. For this kind of communication, we can distinguish between the following three techniques:

* **asynchronous message passing**, also called **non-blocking communication**: In asynchronous message passing, components communicate by sending messages through channels which can buffer the messages. The sender does not need to wait for the recipient to be ready to receive the message. In real life, this corresponds to sending a letter or an e-mail. A potential problem is the fact that messages must be stored and that message buffers can overflow. There are several variations of this scheme, including communicating finite state machines (see page 54) and data flow models (see page 61).

* **synchronous message passing** or **blocking communication, *rendez-vous* based communication**: In synchronous message passing, available components communicate in atomic, instantaneous actions called *rendez-vous*. The component reaching the point of communication first has to wait until the partner has also reached its point of communication. In real life, this corresponds to physical meetings or phone calls. There is no risk of overflows, but the performance may suffer. Examples of languages following this model of computation include CSP (see page 102) and ADA (see page 102).

* **extended *rendez-vous*, remote invocation:** In this case, the sender is allowed to continue only after an acknowledgment has been received from the recipient. The recipient does not have to send this acknowledgment immediately after receiving the message, but can do some preliminary checking before actually sending the acknowledgment.

- **Organization of computations within the components:**

  – **Von-Neumann model:** This model is based on the sequential execution of sequences of primitive computations.

  – **Discrete event model:** In this model, there are events carrying a totally ordered time stamp, indicating the time at which the event occurs. Discrete event simulators typically contain a global event queue sorted by time. Entries from this queue are processed according to this order. The disadvantage is that this model relies on a global notion of event queues, making it difficult to map the semantic model onto parallel implementations. Examples include VHDL (see page 80), SystemC (see page 96), and Verilog (see page 98).

- **Finite state machines (FSMs):** This model is based on the notion of a finite set of states, inputs, outputs, and transitions between states. Several of these machines may need to communicate, forming so-called **communicating finite state machines (CFSMs)**.

- **Differential equations:** Differential equations are capable of modeling analog circuits and physical systems. Hence, they can find applications in cyber-physical system modeling.

- **Combined models:** Actual languages are typically combining a certain model of communication with an organization of computations within components. For example, StateCharts (see page 42) combines finite state machines with shared memories. SDL (see page 54) combines finite state machines with asynchronous message passing. ADA (see page 102) and CSP (see page 102) combine von-Neumann execution with synchronous message passing. Fig. 2.7 gives an overview of combined models which we will consider in this chapter. This figure also includes examples of languages for most of the MoCs.

Communication/ Organization of components	Shared memory	Message passing	
		synchronous	asynchronous
Undefined components	Plain text or graphics, use cases (Message) sequence charts		
Communicating finite state machines (CFSMs)	StateCharts		SDL
Data flow	(not useful)		Kahn networks SDF
Petri nets	C/E nets, P/T nets, ...		
Discrete event (DE) model[5]	VHDL, Verilog SystemC	(Only experimental systems) Distributed DE in Ptolemy	
Von-Neumann model	C, C++, Java	C, C++, Java, ... with libraries CSP, ADA	

Figure 2.7. Overview of MoCs and languages considered

---

[5]The classification of VHDL, Verilog and SystemC is based on the implementation of these languages in simulators. Message passing can be modeled in these languages "on top" of the simulation kernel.

Some MoCs have advantages in certain application areas, while others have advantages in others. Choosing the "best" MoC for a certain application may be difficult. Being able to mix MoCs (such as in the Ptolemy framework [Davis et al., 2001]) can be a way out of this dilemma. Also, models may be translated from one MoC into another one. Non-von-Neumann models are frequently translated into von-Neumann models. The distinction between the different models is blurring if the translation between them is easy.

Designs starting from non-von-Neumann models are frequently called **model-based designs**. The key idea of model-based design is to have some abstract model of the system under design (SUD). Properties of the system can then be studied at the level of this model, without having to care about software code. Software code is generated only after the behavior of the model has been studied in detail and this software is generated automatically. The term "model-based design" is not precisely defined. It is usually associated with models of control systems, comprising traditional control system elements such as integrators, differentiators etc. However, this view seems to be too restricted, since we could also start with abstract models of consumer systems.

In the following, we will present different MoCs, using existing languages as examples for demonstrating their features. A related (but shorter) survey is provided by Edwards [Edwards, 2006]. For a more comprehensive presentation see [Gomez and Fernandes, 2010].

2.3 Early design phases

The very first ideas about systems are frequently captured in a very informal way, possibly on paper. Frequently, only descriptions of the SUD in a natural language such as English or Japanese exist in the early phases of design projects. They are typically using a very informal style. These descriptions should be captured in some machine-readable document. They should be encoded in the format of some word processor and stored by a tool managing design documents. A good tool would allow links between the requirements, a dependence analysis as well as version management.

DOORS® [IBM, 2010b] exemplifies such a tool.

2.3.1 Use cases

For many applications, it is beneficial to envision potential usages of the SUD. Such usages are captured in **use cases**. Use cases describe possible applications of the SUD. Different notations for use cases could be used.

Support for a systematic approach to early specification phases is the goal of the so-called UML standardization effort [Object Management Group (OMG),

2010b], [Fowler and Scott, 1998], [Haugen and Moller-Pedersen, 2006]. UML stands for "Unified Modeling Language". UML was designed by leading software technology experts and is supported by commercial tools. UML primarily aims at the support of the software design process. UML provides a standardized form for use cases.

For use cases, there is neither a precisely specified model of the computations nor is there a precisely specified model of the communication. It is frequently argued that this is done intentionally in order to avoid caring about too many details during the early design phases.

For example, fig. 2.8 shows some use cases for an answering machine[6].

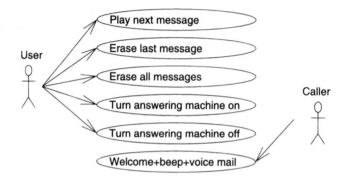

Figure 2.8. Use case example

Use cases identify different classes of users as well as the applications to be supported by the SUD. In this way, it is possible to capture expectations at a very high level.

2.3.2   (Message) Sequence Charts

At a slightly more detailed level, we might want to explicitly indicate the sequences of messages which must be exchanged between components in order to implement some use of the SUD. **Sequence charts** (SCs) -earlier called **message sequence charts** (MSCs)- provide a mechanism for this. Sequence charts use one dimension (usually the vertical dimension) of a 2-dimensional chart to denote sequences and the second dimension to reflect the different communication components. SCs describe partial orders between message transmissions. SCs display a possible behavior of a SUD.

---

[6]We assume that UML is covered in-depth in a software engineering course included in the curriculum. Therefore, UML is only briefly discussed in this book.

SCs are also standardized in UML. UML 2.0 has extended SCs with elements allowing a more detailed description than UML 1.0. Fig. 2.9 shows one of the use cases of the answering machine as an example.

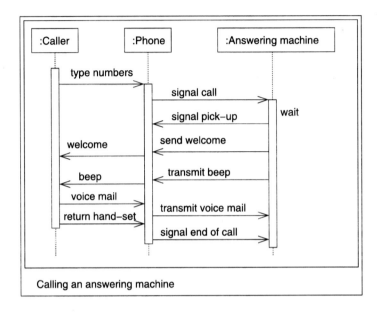

Figure 2.9. Answering machine in UML

Dashed lines are so-called "life-lines". Messages are assumed to be ordered according to their sequence along the life-line. We assume that, in this example, all information is sent in the form of messages. Arrows used in this diagram denote asynchronous messages. This means several messages can be sent by a sender without waiting for the receipt to be confirmed. Boxes on top of life-lines represent active control at the corresponding component. In the example, the answering machine is waiting for the user to pick up the phone within a certain amount of time. If he or she fails to do so, the machine signals a pick-up itself and sends a welcome message to the caller. The caller is then supposed to leave a voice-mail message. Alternative sequences (e.g. an early termination of the call by the caller or the callee picking up the phone) are not shown.

Complex control-dependent actions cannot be described by SCs. Other MoCs must be used for this. Frequently, certain preconditions must be met for a SC to apply. Such preconditions, a distinction between sequences which might happen and those which must happen, as well as other extensions are available in the so-called Live Sequence Charts [Damm and Harel, 2001].

Time/distance diagrams (TDDs) are a commonly used variant of SCs. In time/distance diagrams, the vertical dimension reflects real time, not just sequence. In some cases, the horizontal dimension also models the real distance between the components.

TDDs provide the right means for visualizing schedules of trains or buses. Fig. 2.10 is an example.

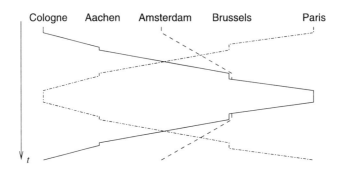

Figure 2.10. Time/distance diagram

This example refers to trains between Amsterdam, Cologne, Brussels and Paris. Trains can run from either Amsterdam or Cologne to Paris via Brussels. Aachen is included as an intermediate stop between Cologne and Brussels. Vertical segments correspond to times spent at stations. For one of the trains, there is a timing overlap between the trains coming from Cologne and Amsterdam at Brussels. There is a second train which travels between Paris and Cologne which is not related to an Amsterdam train.

This example and other examples can be simulated with the levi simulation software [Sirocic and Marwedel, 2007d]. A larger, more realistic example is shown in fig. 2.11. This example [Huerlimann, 2003] describes simulated Swiss railway traffic in the Lötschberg area. Slow and fast trains can be distinguished by their slope in the graph. This modeling technique is very frequently used in practice.

One of the key distinctions between the type of diagrams shown in figs. 2.9 and 2.11 is that fig. 2.9 does not include any reference to real time. UML was initially not designed for real-time applications. UML 2.0 includes **timing diagrams** as a special class of diagrams. Such diagrams enable referring to physical time. Also, certain UML "profiles" (see page 114) allow additional annotations to refer to time [Martin and Müller, 2005], [Müller, 2007].

TDDs are appropriate means for representing typical schedules. However, SCs and TDDs fail to provide information about necessary synchronization. For example, in the presented example of fig. 2.10 it is not known whether the tim-

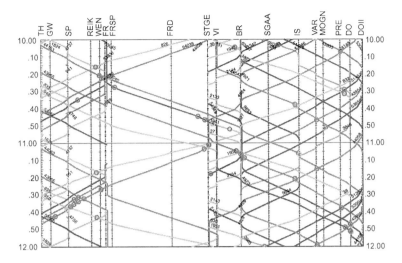

Figure 2.11. Railway traffic displayed by a TDD (courtesy H. Brändli, IVT, ETH Zürich), ©ETH Zürich

ing overlap at Brussels happens coincidentally or whether some real synchronization for connecting trains is required. Furthermore, permissible deviations from the presented schedule (min/max timing behavior) can hardly be included in these charts.

2.4 Communicating finite state machines (CFSMs)

If we start to represent our SUD at a more detailed level, we need more precise models. We mentioned at the beginning of this chapter that we need to describe state-oriented behavior. State diagrams are a classical means of doing this. Fig. 2.12 (the same as fig. 2.1) shows an example of a classical state diagram, representing a **finite state machine (FSM)**.

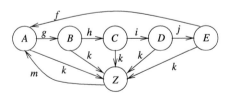

Figure 2.12. State diagram

Circles denote states. We will consider FSMs for which only one of their states is active. Such FSMs are called **deterministic** FSMs. Edges denote state transitions. Edge labels represent events. Let us assume that a certain state of the FSM is active, and that an event happens which corresponds to one of the out-going edges for the active state. Then, the FSM will change its state from the currently active state to the one indicated by the edge. FSMs may be implicitly clocked. Such FSMs are called **synchronous FSMs**. For synchronous FSMs, state changes will happen only at clock transitions. FSMs may also generate output (not shown in fig. 2.12). For more information about classical FSMs refer to, for example, Kohavi [Kohavi, 1987].

2.4.1   Timed automata

Classical FSMs do not provide information about time. In order to model time, classical automata have been extended to also include timing information. Timed automata are essentially automata extended with real-valued variables. *"The variables model the logical clocks in the system, that are initialized with zero when the system is started, and then increase synchronously with the same rate. Clock constraints, i.e. guards on edges, are used to restrict the behavior of the automaton. A transition represented by an edge can be taken when the clocks' values satisfy the guard labeled on the edge. Clocks may be reset to zero when a transition is taken"* [Bengtsson and Yi, 2004].

Fig. 2.13 shows an example.

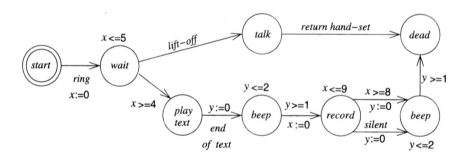

Figure 2.13.   Servicing an incoming line in an answering machine

The answering machine is usually in the initial state on the left. Whenever a ring signal is received, clock x is reset to 0 and a transition into a waiting state is made. If the called person lifts off the hand-set, talking can take place until the hand-set is returned. Otherwise, a transition to state *play text* can take place if time has reached a value of 4.

Once the transition took place, a recorded message is played and this phase is terminated with a beep. Clock y ensures that this beep lasts at least one time unit. After the beep, clock x is reset to 0 again and the answering machine is ready for recording. If time has reached a value of 8 or if the caller remains silent, the next beep is played. This second beep again lasts at least one time unit. After the second beep, a transition is made into the final state. In this example, transitions are either caused by inputs (such as *lift-off*) or by so-called **clock constraints**.

Clock constraints describe transitions which **can** take place, but they do not have to. In order to make sure that transitions actually take place, additional **location invariants** can be defined. Location invariants $x <= 5$, $x <= 9$ and $y <= 2$ are used in the example such that transitions will take place no later than one time unit after the enabling condition became true. Using two clocks is for demonstration purposes only; a single clock would be sufficient.

Formally speaking, timed automata can be defined as follows [Bengtsson and Yi, 2004]:

Let C be a set of real-valued, non-negative variables representing clocks. Let Σ be a finite alphabet of possible inputs.

Definition: A **clock constraint** is a conjunctive formula of atomic constraints of the form $x \circ n$ or $(x - y) \circ n$ for $x, y \in C, \circ \in \{\leq, <, =, >, \geq\}$ and $n \in \mathbb{N}$.

Note that constants n used in the constraints must be integers, even though clocks are real-valued. An extension to rational constants would be easy, since they could be turned into integers with simple multiplications. Let $B(C)$ be the set of clock constraints.

Definition [Bengtsson and Yi, 2004]: A **timed automaton** is a tuple (S, s_0, E, I) where:

- S is a finite set of states.

- s_0 is the initial state.

- $E \subseteq S \times B(C) \times \Sigma \times 2^C \times S$ is the set of edges. $B(C)$ models the conjunctive condition which must hold and Σ models the input which is required for a transition to be enabled. 2^C reflects the set of clock variables which are reset whenever the transition takes place.

- $I : S \to B(C)$ is the set of invariants for each of the states. $B(C)$ represents the invariant which must hold for a particular state S. This invariant is described as a conjunctive formula.

This first definition is usually extended to allow parallel compositions of timed automata. Timed automata having a large number of clocks tend to be difficult to understand. More details about timed automata can be found, for example, in papers by Dill et al. [Dill and Alur, 1994] and Bengtsson et al. [Bengtsson and Yi, 2004].

Timed automata extend classical automata with timing information. However, many of our requirements for specification techniques are not met by timed automata. In particular, in their standard form, they do no provide hierarchy and concurrency.

2.4.2   StateCharts: implicit shared memory communication

The StateCharts language is presented here as a very prominent example of a language based on automata and supporting hierarchical models as well as concurrency. It does include a limited way of specifying timing.

The StateCharts language was introduced by David Harel [Harel, 1987] in 1987 and later described more precisely in [Drusinsky and Harel, 1989]. According to Harel, the name was chosen since it was "*the only unused combination of flow or state with diagram or chart*".

2.4.2.1   Modeling of hierarchy

The StateCharts language describes extended FSMs. Due to this, they can be used for modeling state-oriented behavior. The key extension is **hierarchy**. Hierarchy is introduced by means of **super-states**.

Definitions:

- States comprising other states are called **super-states**.

- States included in super-states are called **sub-states** of the super-states.

Fig. 2.14 shows a StateCharts example. It is a hierarchical version of fig. 2.12.

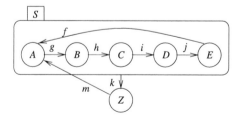

Figure 2.14.   Hierarchical state diagram

Super-state S includes states A, B, C, D and E. Suppose the FSM is in state Z (we will also call Z to be an **active state**). Now, if input m is applied to the FSM, then A and S will be the new active states. If the FSM is in S and input k is applied, then Z will be the new active state, regardless of whether the FSM is in sub-states A, B, C, D or E of S. In this example, all states contained in S are non-hierarchical states. In general, sub-states of S could again be super-states consisting of sub-states themselves. Also, **whenever a sub-state of some super-state is active, the super-state is active as well**.

Definitions:

- Each state which is not composed of other states is called a **basic state**.

- For each basic state s, the super states containing s are called **ancestor states**.

The FSM of fig. 2.14 can only be in one of the sub-states of super-state S at any time. Super states of this type are called **OR-super-states**[7].

In fig. 2.14, k might correspond to an exception for which state S has to be left. The example already shows that the hierarchy introduced in StateCharts enables a compact representation of exceptions.

StateCharts allows hierarchical descriptions of systems in which a system description comprises descriptions of subsystems which, in turn, may contain descriptions of subsystems. The **hierarchy** of the entire system can be represented by a **tree**. The root of the tree corresponds to the system as a whole, and all inner nodes correspond to hierarchical descriptions (in the case of State-Charts called super-nodes). The leaves of the hierarchy are non-hierarchical descriptions (in the case of StateCharts called basic states).

So far, we have used explicit, direct edges to basic states to indicate the next state. The disadvantage of that approach is that the internal structure of super-states cannot be hidden from the environment. However, in a true hierarchical environment, we should be able to hide the internal structure so that it can be described later or changed later without affecting the environment. This is possible with other mechanisms for describing the next state.

The first additional mechanism is the **default state mechanism**. It can be used in super-states to indicate the particular sub-states that will become active if the super-states become active. In diagrams, default states are identified by edges starting at small filled circles. Fig. 2.15 shows a state diagram using the

---

[7]More precisely, they should be called XOR-super-states, since the FSM is in **either** A, B, C, D or E. However, this name is not commonly used in the literature.

default state mechanism. It is equivalent to the diagram in fig. 2.14. Note that the filled circle does not constitute a state itself.

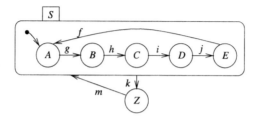

Figure 2.15.   State diagram using the default state mechanism

Another mechanism for specifying next states is the **history mechanism**. With this mechanism, it is possible to return to the last sub-state that was active before a super-state was left. The history mechanism is symbolized by a circle containing the letter H. In order to define the next state for the very initial transition into a super-state, the history mechanism is frequently combined with the default mechanism. Fig. 2.16 shows an example.

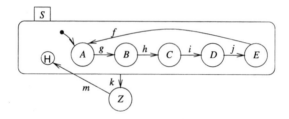

Figure 2.16.   State diagram using the history and the default state mechanism

The behavior of the FSM is now somewhat different. If we input m while the system is in Z, then the FSM will enter A if this is the very first time we enter S, and otherwise it will enter the last state that we were in before leaving S. This mechanism has many applications. For example, if k denotes an exception, we could use input m to return to the state we were in before the exception. States A, B, C, D and E could also call Z like a procedure. After completing "procedure" Z, we would return to the calling state.

Fig. 2.16 can also be redrawn as shown in fig. 2.17. In this case, the symbols for the default and the history mechanism are combined.

Specification techniques must also be able to describe concurrency conveniently. Towards this end, the StateCharts language provides a second class of super-states, so called **AND**-states.

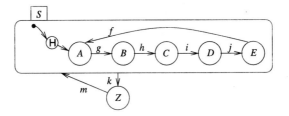

Figure 2.17. Combining the symbols for the history and the default state mechanism

Definition: Super-states *S* are called **AND-super-states** if the system containing *S* will be in all of the sub-states of *S* whenever it is in *S*.

An AND-super-state is included in the answering machine example shown in fig. 2.18.

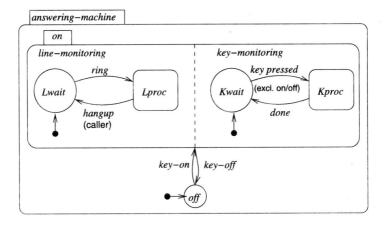

Figure 2.18. Answering machine

An answering machine normally performs two tasks concurrently: it is monitoring the line for incoming calls and the keys for user input. In fig. 2.18, the corresponding states are called *Lwait* and *Kwait*. Incoming calls are processed in state *Lproc* while the response to pressed keys is generated in state *Kproc*. For the time being, we assume that the on/off switch (generating events *key-off* and *key-on*) is decoded separately and pushing it does not result in entering *Kproc*. If this switch is pushed, the line monitoring state as well as the key monitoring state are left and re-entered only if the machine is switched on. At that time, default states *Lwait* and *Kwait* are entered. While switched on, the machine will always be in the line monitoring state as well as in the key monitoring state.

For AND-super-states, the sub-states entered as a result of some event can be defined independently. There can be any combination of history, default and explicit transitions. It is crucial to understand that **all** sub-states will always be entered, even if there is just one explicit transition to one of the sub-states. Accordingly, transitions out of an AND-super-state will always result in leaving **all** the sub-states.

For example, let us modify our answering machine such that the on/off switch, like all other switches, is decoded in state *Kproc* (see fig. 2.19).

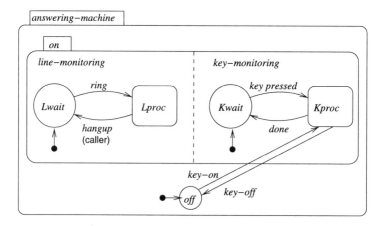

Figure 2.19. Answering machine with modified on/off switch processing

If pushing that key is detected in *Kwait*, transitions are assumed first into state *Kproc* and then into the *off* state. The second transition results in leaving the line-monitoring state as well. Switching the machine on again results in also entering the line-monitoring state.

AND-super-states provide the key mechanism for describing concurrency in StateCharts. Each sub-state can be considered a state machine by itself. These machines are communicating with each other, forming **communicating finite state machines** (CFSMs). This term has been used as the title of this section.

Summarizing, we can state the following: **States in StateCharts diagrams are either AND-states, OR-states or basic states.**

2.4.2.2   Timers

Due to the requirement to model time in embedded systems, StateCharts also provides timers. Timers are denoted by the symbol shown in fig. 2.20 (left).

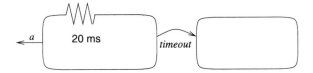

Figure 2.20. Timer in StateCharts

After the system has been in the state containing the timer for the specified period, a time-out will occur and the system will leave the specified state. Timers can also be used hierarchically.

Timers can be employed, for example, at the next lower level of the hierarchy of the answering machine in order to describe the behavior of state *Lproc*. Fig. 2.21 shows a possible behavior for that state. The timing specification is slightly different from the one in fig. 2.13.

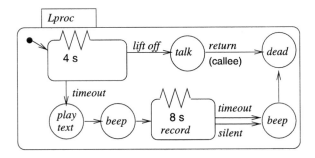

Figure 2.21. Servicing the incoming line in *Lproc*

Due to the exception-like transition for hangups by the caller in fig. 2.18, state *Lproc* is terminated whenever the caller hangs up. For hangups (returns) by the callee, the design of state *Lproc* results in an inconvenience: If the callee hangs up the phone first, the telephone will be dead (and quiet) until the caller has also hung up the phone.

The StateCharts language includes a number of other language elements. For a full description refer to Harel [Harel, 1987]. A more detailed description of the semantics of StateCharts is described by Drusinsky and Harel [Drusinsky and Harel, 1989].

2.4.2.3   Edge labels and StateMate semantics

Until now, we have not considered outputs generated by our extended FSMs. Generated outputs can be specified using edge labels. The general form of an edge label is "*event* [*condition*] / *reaction*". All three label parts are optional.

The *reaction*-part describes the reaction of the FSM to a state transition. Possible reactions include the generation of events and assignments to variables. The *condition*-part implies a test of the values of variables or a test of the current state of the system. The *event*-part refers to a test of current events. Events can be generated either internally or externally. Internal events are generated as a result of some transition and are described in *reaction*-parts. External events are usually described in the model environment.

Examples:

- *on-key* / *on*:=1 (Event-test and variable assignment),

- [*on*=1] (Condition test for a variable value),

- *off-key* [not in *Lproc*] / *on*:=0 (Event-test, condition test for a state, variable assignment. The assignment is performed if the event has occurred and the condition is true).

The semantics of edge labels can only be explained in the context of the semantics of StateMate [Drusinsky and Harel, 1989], a commercial implementation of StateCharts. StateMate assumes a step-based execution of StateMate-descriptions. Each step consists of three phases:

1 In the first phase, the impact of external changes on conditions and events is evaluated. This includes the evaluation of functions which depend on external events. This phase does not include any state changes. In our simple examples, this phase is not actually needed.

2 The next phase is to calculate the set of transitions that should be made in the current step. Variable assignments are evaluated, but the new values are only assigned to temporary variables.

3 In the third phase, state transitions become effective and variables obtain their new values.

The separation into phases 2 and 3 is especially important in order to guarantee a reproducible behavior of StateMate models. Consider the StateMate model of fig. 2.22.

In the second phase, new values for a and b are stored in temporary variables, say a' and b'. In the final phase, temporary variables are copied into the user-defined variables:

phase 2: $a':=b; b':=a;$

phase 3: $a:=a'; b:=b'$

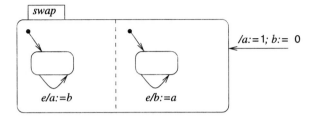

Figure 2.22. Mutually dependent assignments

As a result, the values of the two variables will be swapped each time an event *e* happens. This behavior corresponds to that of two cross-coupled registers (one for each variable) connected to the same clock (see fig. 2.23) and reflects the operation of a synchronous (clocked) finite state machine including those two registers[8].

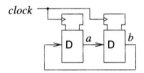

Figure 2.23. Cross-coupled D-type registers

Without the separation into phases, the same value would be assigned to both variables. The result would depend on the sequence in which the assignments were performed. The separation into (at least) two phases is quite typical for languages that try to reflect the operation of synchronous hardware. We will find the same separation in VHDL (see page 89). Due to the separation, the results do not depend on the order in which parts of the model are executed by the simulation. This property is extremely important. Otherwise, there could be simulation runs generating different results, all of which would be considered correct. This could be very confusing in all design procedures. This is not what we expect from the simulation of a real circuit with a fixed behavior.

There are different names for this property:

- Kahn [Kahn, 1974] calls this property **determinate**.

---

[8]We adopt IEEE standard schematic symbols [IEEE, 1991] for gates and registers for all the schematics in this book. The symbols in fig. 2.23 denote clocked D-type registers.

- In other papers, this property is called **deterministic**. However, this term is employed with different meanings:

  - This term is used to denote non-deterministic finite state machines, FSMs which can be in several states at the same time [Hopcroft et al., 2006].

  - Languages may have non-deterministic operators. For these operators, different behaviors are legal implementations.

  - Many authors consider systems to be non-deterministic if their behavior depends on some input not known before run-time.

  - In the sense Kahn uses the term "determinate".

In this book, we prefer to reduce possible confusion by following Kahn[9]. Note that StateMate models can be determinate only if there are no other reasons for an undefined behavior. For example, conflicts between transitions may be allowed (see fig. 2.24).

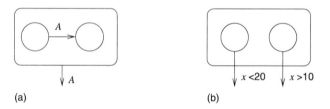

(a)   (b)

Figure 2.24. Conflicting StateCharts transitions

Consider fig. 2.24 (a). If event A takes place while the system is in the left state, we must figure out, which transition will take place. If these conflicts would be resolved arbitrarily, then we would have a non-determinate behavior. Typically, priorities are defined such that this type of a conflict is eliminated. Now, consider fig. 2.24 (b). There will be a conflict for x=15. Such conflicts are difficult to detect. Achieving a determinate behavior requires the absence of conflicts that are resolved in an arbitrary manner.

Note that there may be cases in which we would like to describe non-determinate behavior (e.g. if we have a choice to read from two inputs). In such a case, we would typically like to explicitly indicate that this choice can be taken at run-time (see the **select** statement of ADA on page 104).

Implementations of hierarchical state charts other than StateMate typically do not exhibit determinate behavior. These implementations correspond to a

---

[9]In earlier versions of the book, we used the term "deterministic" together with an additional explanation.

software-oriented view onto hierarchical state charts. In such implementations, choices are usually not explicitly described.

The three phases described on page 48 have to be repeatedly executed. Each execution is called a **step** (see fig. 2.25).

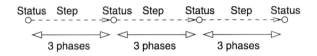

Figure 2.25. Steps during the execution of a StateMate model

Steps are assumed to be executed each time events or variables have changed. The set of all values of variables, together with the set of events generated (and the current time) is defined as the **status**[10] of a StateMate model. After executing the third phase, a new status is obtained. The notion of steps allows us to define the semantics of **events** more precisely. Events are generated, as mentioned, either internally or externally. **The visibility of events is limited to the step following the one in which they are generated.** Thus, events behave like single bit values which are stored in permanently enabled registers at one clock transition and have an effect on the values stored at the next clock transition. They do not live forever.

Variables, in contrast, retain their values, until they are reassigned. According to StateMate semantics, new values of variables are visible to all parts of the model from the step following the step in which the assignment was made onwards. That means, StateMate semantics implies that new values of variables are propagated to all parts of a model between two steps. StateMate implicitly assumes a **broadcast mechanism for updates on variables**. This means that StateCharts or StateMate can be implemented easily for shared memory-based platforms but are less appropriate for message passing and distributed systems. These languages essentially assume shared memory-based communication, even though this is not explicitly stated. For distributed systems, it will be very difficult to update all variables between two steps. Due to this broadcast mechanism, StateMate is not an appropriate language for modeling distributed systems.

---

[10]We would normally use the term "state" instead of "status". However, the term "state" has a different meaning in StateMate.

2.4.2.4   Evaluation and extensions

StateCharts' main application domain is that of local, control-dominated systems. The capability of nesting hierarchies at arbitrary levels, with a free choice of AND- and OR-states, is a key advantage of StateCharts. Another advantage is that the semantics of StateMate is defined at a sufficient level of detail [Drusinsky and Harel, 1989]. Furthermore, there are quite a number of commercial tools based on StateCharts. StateMate [IBM, 2010a] and StateFlow [MathWorks, 2010] are examples of commercial tools based on StateCharts. Many of them are capable of translating StateCharts into equivalent descriptions in C or VHDL (see page 80). From VHDL, hardware can be generated using synthesis tools. Therefore, StateCharts-based tools provide a complete path from StateCharts-based specifications down to hardware. Generated C programs can be compiled and executed. Hence, a path to software-based realizations exists as well.

Unfortunately, the efficiency of the automatic translation is sometimes a concern. For example, we could map sub-states of AND-states to UNIX-processes. This would hardly lead to efficient implementations on small processors. The productivity gain from object-oriented programming is not available in StateCharts, since it is not object-oriented. Furthermore, the broadcast mechanism makes it less appropriate for distributed systems. StateCharts do not comprise program constructs for describing complex computation and cannot describe hardware structures or non-functional behavior.

Commercial implementations of StateCharts typically provide some mechanisms for removing the limitations of the model. For example, C code can be used to represent program constructs and **module charts** of StateMate can represent hardware structures.

StateCharts allows timeouts. There is no straightforward way of specifying other timing requirements.

UML includes a variation of StateCharts and hence allows modeling state machines. In UML, these diagrams are called **state diagrams** in version 1 of UML and **state machine diagrams** from version 2.0 onwards. Unfortunately, the semantics of state machine diagrams in UML is different from StateMate: the three simulation phases are not included.

2.4.3   Synchronous languages

2.4.3.1   Motivation

Describing complex SUDs in terms of state machine diagrams is difficult. Such diagrams cannot express complex computations. Standard programming languages can express complex computations, but the sequence of executing several threads may be unpredictable. In a multi-threaded environment with preemptive scheduling there can be many different interleavings of the different computations. Understanding all possible behaviors of such concurrent systems is difficult. A key reason for this is that, in general, many different execution orders are feasible, i.e. the execution order is not specified. The order of execution may well affect the result. The resulting non-determinate behavior can have a number of negative consequences, such as, for example, problems with verifying a certain design. For distributed systems with independent clocks, determinate behavior is difficult to achieve. However, for non-distributed systems, we can try to avoid the problems of unnecessary non-determinate semantics.

For synchronous languages, finite state machines and programming languages are merged into one model. Synchronous languages can express complex computations, but the underlying execution model is that of finite automata. They describe concurrently operating automata. Determinate behavior is achieved by the following key feature: "... *when automata are composed in parallel, a transition of the product is made of the "simultaneous" transitions of all of them*" [Halbwachs, 1998]. This means: we do not have to consider all the different sequences of state changes of the automata that would be possible if each of them had its own clock. Instead, we can assume the presence of a single global clock. Each clock tick, all inputs are considered, new outputs and states are calculated and then the transitions are made. This requires a fast broadcast mechanism for all parts of the model. This idealistic view of concurrency has the advantage of guaranteeing **determinate behavior**. This is a restriction if compared to the general communicating finite state machines (CFSM) model, in which each FSM can have its own clock. Synchronous languages reflect the principles of operation in synchronous hardware and also the semantics found in control languages such as IEC 60848 [IEC, 2002] and STEP 7 [Siemens, 2010]. See Potop-Butucaru et al. [Potop-Butucaru et al., 2006] for a survey on synchronous languages.

2.4.3.2 Examples of synchronous languages: Esterel, Lustre and SCADE

Guaranteeing a determinate behavior for all language features has been a design goal for the synchronous languages Esterel [Esterel Technologies Inc., 2010], [Boussinot and de Simone, 1991] and Lustre [Halbwachs et al., 1991].

Esterel is a reactive language: when activated with an input event, Esterel models react by producing an output event. Esterel is a synchronous language: all reactions are assumed to be completed in zero time and it is sufficient to analyze the behavior at discrete moments in time. This idealized model avoids all discussions about overlapping time ranges and about events that arrive while the previous reaction has not been completed. Like other concurrent languages, Esterel has a parallelism operator, written ||. Similar to StateCharts, communication is based on a broadcast mechanism. In contrast to StateCharts, however, communication is instantaneous. Instantaneous in this context means "within the same clock cycle". This means that all signals generated in a particular clock cycle are also seen by the others parts of the model in the same clock cycle and these other parts, if sensitive to the generated signals, react in the same clock cycle. Several rounds of evaluations may be required until a stable state is reached. The propagation of values during the same macroscopic instant of time corresponds to the generation of a next status for the same moment in time in StateMate, except that the broadcast is now instantaneous and not delayed until the next round of evaluations like in StateMate. For more and updated information about Esterel, refer to the Esterel home page [Esterel Technologies Inc., 2010].

Esterel and Lustre use different syntactic techniques to denote CFSMs. Esterel appears as a kind of imperative language, whereas Lustre looks more like a data flow language (see page 61 for a description of data flow). SyncCharts is a graphical version of Esterel. In all three cases, semantics are explained by the closely-related underlying CFSMs. The commercial graphical language SCADE [Esterel Technologies, 2010] combines elements of all three languages. SCADE is used for a number of safety-critical software components, for example by Airbus.

Due to the three simulation phases in StateMate, StateMate has the key attributes of synchronous languages and it is determinate if conflicts are resolved. According to Halbwachs, *"StateMate is almost a synchronous language and the only feature missing in StateMate is the instantaneous broadcast"* [Halbwachs, 2008].

2.4.4   SDL: A case of message passing

2.4.4.1   Features of the language

StateCharts is not appropriate for modeling distributed communicating finite state machines. For distributed systems, message passing is the better communication paradigm. Therefore, we will now present a second example of a language based on communicating finite state machines, an example based on asynchronous message passing.

This language is called SDL (specification and description language). SDL was designed for distributed applications. It dates back to the 1970s. Formal semantics have been available since the 1980s. The language was standardized by the ITU (International Telecommunication Union). The first standards document is the Z.100 Recommendation published in 1980, with updates in 1984, 1988, 1992 (SDL-92), 1996 and 1999. Relevant versions of the standard include SDL-88, SDL-92 and SDL-2000 [SDL Forum Society, 2010].

Many users prefer graphical specification languages while others prefer textual ones. SDL pleases both types of users since it provides textual as well as graphical formats. Processes are the basic elements of SDL. Processes represent components modeled as extended finite state machines. Extensions include operations on data. Fig. 2.26 shows the graphical symbols used in the graphical representation of SDL.

Figure 2.26. Symbols used in the graphical form of SDL

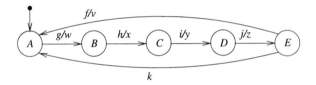

Figure 2.27. FSM to be described in SDL

As an example, we will consider how the state diagram in fig. 2.27 can be represented in SDL. Fig. 2.27 is the same as fig. 2.15, except that output has been added, state Z has been deleted, and the effect of signal *k* has been changed. Fig. 2.28 contains the corresponding graphical SDL representation.

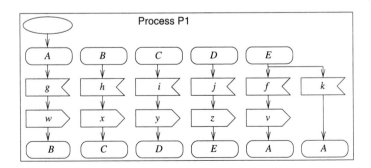

Figure 2.28. SDL-representation of fig. 2.27

Obviously, the representation in fig. 2.28 is equivalent to the state diagram of fig. 2.27.

As an extension to FSMs, SDL processes can perform operations on data. Variables can be declared locally for processes. Their type can either be pre-defined or defined in the SDL description itself. SDL supports abstract data types (ADTs). The syntax for declarations and operations is similar to that in other languages. Fig. 2.29 shows how declarations, assignments and decisions can be represented in SDL.

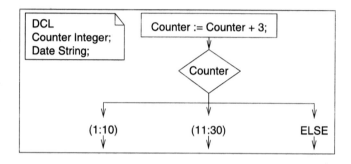

Figure 2.29. Declarations, assignments and decisions in SDL

SDL also contains programming language elements such as procedures. Procedure calls can also be represented graphically. Object-oriented features became available with version SDL-1992 of the language and were extended with SDL-2000.

Extended FSMs are just the basic elements of SDL descriptions. In general, SDL descriptions will consist of a set of interacting processes, or FSMs. Processes can send signals to other processes. Semantics of interprocess communication in SDL is based on asynchronous message passing and conceptually implemented through *first-in first-out* (FIFO)-*queues* associated with processes. There is exactly one queue per process. Signals sent to a particular process will be placed into the corresponding FIFO-queue (see fig. 2.30).

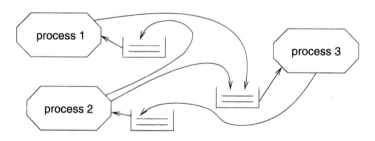

Figure 2.30. SDL interprocess communication

Each process is assumed to fetch the next available entry from the FIFO queue and check whether it matches one of the inputs described for the current state. If it does, the corresponding state transition takes place and output is generated. The entry from the FIFO-queue is ignored if it does not match any of the listed inputs (unless the so-called SAVE-mechanism is used). FIFO-queues are conceptually thought of as being of infinite length. This means: in the description of the semantics of SDL models, FIFO-overflow is never considered. In actual systems, however, infinite FIFO-queues cannot be implemented. They must be of finite length. This is one of the problems of SDL: in order to derive realizations from specifications, safe upper bounds on the length of the FIFO-queues must be proven.

Process interaction diagrams can be used for visualizing which of the processes are communicating with each other. Process interaction diagrams include **channels** used for sending and receiving signals. In the case of SDL, the term "signal" denotes inputs and outputs of modeled automata.

Example: Fig. 2.31 shows a process interaction diagram B1 with channels Sw1 and Sw2. Brackets include the names of signals propagated along a certain channel.

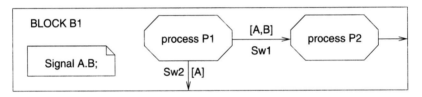

Figure 2.31. Process interaction diagram

There are three ways of indicating the recipient of signals:

1 **Through process identifiers:** by using identifiers of recipient processes in the graphical output symbol (see fig. 2.32 (left)).

Figure 2.32. Describing signal recipients

The number of processes does not need to be fixed at compile time, since processes can be generated dynamically at run-time. OFFSPRING represents identifiers of child processes generated dynamically by a process.

2 **Explicitly:** by indicating the channel name (see fig. 2.32 (right)). Sw1 is the name of a channel.

3 **Implicitly:** if signal names imply the channel names, those channels are used. Example: for fig. 2.31, signal B will implicitly always be communicated via channel Sw1.

No process can be defined within any other (processes cannot be nested). However, they can be grouped hierarchically into so-called **blocks**. Blocks at the highest hierarchy level are called **systems**. Process interaction diagrams are special cases of block diagrams. Process interaction diagrams are one level above the leaves of the hierarchical description. B1 can be used within intermediate level blocks (such as within B in fig. 2.33).

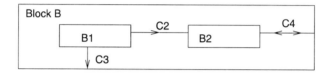

Figure 2.33. SDL block

At the highest level in the hierarchy, we have the system (see fig. 2.34). A system will not have any channels at its boundary if the environment is also modeled as a block.

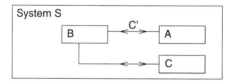

Figure 2.34. SDL system

Fig. 2.35 shows the hierarchy modeled by block diagrams 2.31, 2.33 and 2.34.

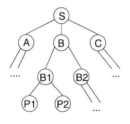

Figure 2.35. SDL hierarchy

Process interaction diagrams are next to the *leaves* of the hierarchical description, while system descriptions represent their *root*. Some of the restrictions of modeling hierarchy are removed in version SDL-2000 of the language. With SDL-2000, the descriptive power of blocks and processes is harmonized and replaced by a general *agent* concept.

In order to support the modeling of time, SDL includes **timers**. Timers can be declared locally for processes. They can be set and reset using SET and RESET primitives, respectively.

Fig. 2.36 shows the use of a timer T.

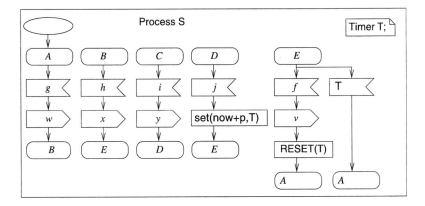

Figure 2.36. Using timer T

The diagram corresponds to that of fig. 2.28, with the exceptions that timer T is set to the current time plus p during the transition from state *D* to *E*. For the transition from *E* to *A* we now have a timeout of p time units. If these time units have elapsed before signal *f* has arrived, a transition to state *A* is taken without generating output signal *v*.

SDL can be used, for example, to describe protocol stacks found in computer networks. Fig. 2.37 shows three processors connected through a router. Communication between processors and the router is based on FIFOs.

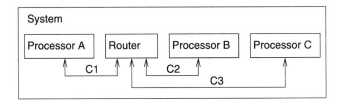

Figure 2.37. Small computer network described in SDL

The processors as well as the router implement layered protocols (see fig. 2.38).

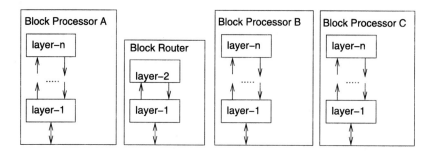

Figure 2.38. Protocol stacks represented in SDL

Each layer describes communication at a more abstract level. The behavior of each layer is typically modeled as a finite state machine. The detailed description of these FSMs depends on the network protocol and can be quite complex. Typically, this behavior includes checking and handling of error conditions, and sorting and forwarding of information packets.

Available tools for SDL include interfaces to UML (see page 113), and SCs (see page 36). A comprehensive list of tools is available from the SDL forum [SDL Forum Society, 2009].

Estelle [Budkowski and Dembinski, 1987] is another language which was designed to describe communication protocols. Similar to SDL, Estelle assumes communication via channels and FIFO-buffers. Attempts to unify Estelle and SDL failed.

2.4.4.2   Evaluation of SDL

SDL is excellent for distributed applications and has been used, for example, for specifying ISDN.

SDL is not necessarily determinate (the order, in which signals arriving at some FIFO at the same time are processed, is not specified).

Reliable implementations require the knowledge of a upper bound on the length of the FIFOs. This upper bound may be difficult to compute. The timer concept is sufficient for soft deadlines, but not for hard ones.

Hierarchies are not supported in the same way as in StateCharts.

There is no full programming support (but recent revisions of the standard have started to change this) and no description of non-functional properties.

It seems like the interest in SDL is decreasing, even though it is very useful as a reference model.

2.5 Data flow

2.5.1 Scope

Data flow is a very "natural" way of describing real life applications. Data flow models reflect the way in which data flows from component to component [Edwards, 2001]. Each component transforms the data in one way or the other. The following is a possible definition of data flow [Wikipedia, 2010]:

Definition: Data flow modeling "*is the process of identifying, modeling and documenting how data moves around an information system. Data flow modeling examines processes (activities that transform data from one form to another), data stores (the holding areas for data), external entities (what sends data into a system or receives data from a system), data flows (routes by which data can flow)*".

A **data flow program** is specified by a directed graph where the nodes (vertices), called **actors**, represent computations and the arcs represent communication channels. The computation performed by each actor is assumed to be functional, that is, based on the input values only. Each process in a data flow graph is decomposed into a sequence of firings, which are atomic actions. Each firing produces and consumes tokens.

For example, fig. 2.39 describes the flow of data in a video-on-demand system [Ko and Koo, 1996].

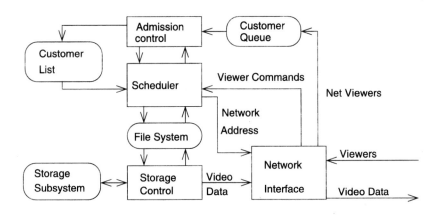

Figure 2.39. Video-on-demand system

For unrestricted data flow, it is difficult to prove requested system properties. Therefore, restricted models are commonly used.

2.5.2   Kahn process networks

Kahn process networks (KPN) [Kahn, 1974] are a special case of data flow models. Like other data flow models, KPNs consist of nodes and edges. Nodes correspond to computations performed by some program or task. KPN graphs, like all data flow graphs, show computations to be performed and their dependence, but not the order in which the computations must be performed (in contrast to specifications in von-Neumann languages such as C). Edges imply communication via channels containing potentially infinite FIFOs. Computation times and communication times may vary, but communication is guaranteed to happen within a finite amount of time. Writes are non-blocking, since the FIFOs are assumed to be as large as needed. Reads must specify a single channel to be read from. A node cannot check whether data is available before attempting a read. A process cannot wait for data on more than one port at a time. Read operations block whenever an attempt is made to read from an empty FIFO queue. Only a single process is allowed to read from a certain queue and only a single process is allowed to write into a queue. So, if output data has to be sent to more than a single process, duplication of data must be done inside processes. There is no other way for communication between processes except through FIFO-queues.

In the following example, p1 and p2 are incrementing and decrementing the value received from the partner:

```
process p1(in int u, out int v){
    int i;
    i = 0;
    for (;;) {
     send(i,v);    -- send i via channel v
     i = wait(u);  -- read i from channel u
     i = i-1;
    }}
process p2(in int v, out int u){
    int i;
    for (;;) {
     i = wait(v);
     i = i+1;
     send(i,u);
    }}
```

Fig. 2.40 shows a graphical representation of this KPN.

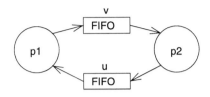

Figure 2.40. Graphical representation of KPN

Obviously, we do not really need the FIFOs in this example, since messages cannot accumulate in the channels. This example and other examples can be simulated with the levi simulation software [Sirocic and Marwedel, 2007b].

The restrictions are resulting in the **key beauty of KPNs**: the order in which a node is reading data from its channels is fixed by the sequence of read operations and does not depend on the order in which producers are transmitting data over the channels. This means that the sequence of operations is independent of the speed of the nodes producing data. **For a given set of input data, KPNs will always generate the same results, independently of the speed of the nodes.** This property is important, for example, for simulations: it does not matter how fast we are simulating the KPN, the result will always be the same. In particular, the result does not depend on using hardware accelerators for some of the nodes and a distributed execution will give the same result as a centralized one. This property has been called "determinate" and we are following this use. SDL-like conflicts at FIFOs do not exist. Due to this nice property, KPNs are frequently used as an internal representation within a design flow.

Sometimes, KPNs are extended with a "merge"-operator (corresponding to ADA's **select** statement, see page 104). This operation allows for queuing reads with a list of channels at the same time and waiting for channels to generate data. Such an operator introduces a non-determinate behavior: the order of processing inputs is not specified if both inputs arrive at the same time. This extension is useful in practice, but it destroys the key beauty of KPNs.

In general, Kahn processes require scheduling at run-time, since it is difficult to predict their precise behavior over time. These problems result from the fact that we do not make any assumptions regarding the speed of the channels and the nodes. The question of whether or not finite-length FIFOs are sufficient for an actual KPN model is undecidable in the general case. Nevertheless, execution times are actually unknown during early design phases and therefore this model is very adequate. Useful scheduling algorithms exist [Kienhuis et al., 2000]. For KPNs, the number of processes is fixed, i.e. it does not change at run-time.

2.5.3   Synchronous data flow

Scheduling becomes significantly easier and questions regarding buffer sizes can decidably be answered if we impose restrictions on the timing of nodes and channels. Synchronous data flow (SDF) [Lee and Messerschmitt, 1987] is such a model.

SDF can best be introduced by referring to its graphical notation. Fig. 2.41 (left) shows a synchronous data flow graph. The graph is a directed graph, nodes A and B denote computations * and +. Inputs to SDF graphs are assumed to consist of an infinite stream of samples. Nodes can start their computations when their inputs are available. Edges must be used whenever there is a data dependency between any two nodes.

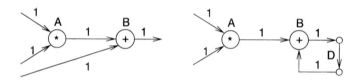

Figure 2.41.   Graphical representations of synchronous data flow

For each execution, the computation in a node is called a firing. For each firing, a number of tokens, representing data, is consumed and produced. In synchronous data flow, the number of tokens produced or consumed in one firing is constant. Constant edge labels denote the corresponding numbers of tokens. These constants facilitate the modeling of **multi-rate** signal processing applications, applications for which certain signals are generated at frequencies that are multiples of other frequencies. For example, in a TV set, some computations might be performed at a rate of 100 Hz while others are performed at a rate of 50 Hz. In general, the number of tokens sent to an edge must be equal to the number of tokens consumed. Let n_s be the number of tokens produced by some sender per firing, and let f_s be the corresponding rate. Let n_r be the corresponding number of tokens consumed per firing at the receiver, and let f_r be the corresponding rate. Then, we must have

$$n_s * f_s \; = \; n_r * f_r \tag{2.1}$$

This situation is also visualized in fig. 2.42. The FIFO is needed for buffering if $n_s \neq n_r$. In contrast to Kahn process networks, the size can be computed easily.

Figure 2.42. Multi-rate SDF model

The term **synchronous** data flow reflects the fact that tokens are consumed from the incoming arcs in a synchronous manner (all at the same instant in time). The term **asynchronous** message passing reflects the fact that tokens can be buffered using FIFOs. The property of producing and consuming a constant number of tokens makes it possible to determine execution order and memory requirements at compile time. Hence, complex run-time scheduling of executions is avoided. SDF graphs may include delays, denoted by the symbol D on an edge (see fig. 2.41 (right)). SDF graphs can be translated into periodic schedules for mono- as well as for multi-processor systems (see e.g. [Pino and Lee, 1995]). A legal schedule for the simple example of fig. 2.41 would consist of the sequence (A, B) (repeated forever). A sequence (A, A, B) (A executed twice as many times as B) would be illegal, since it would accumulate an infinite number of tokens on the implicit FIFO buffer between A and B.

SDF is very useful, for example, in modeling multimedia systems. In this case, each token would correspond to audio or video information, such as an audio sample or a video frame. The observer pattern, mentioned as a problem for modeling with von-Neumann languages on page 27, can be easily implemented correctly in SDF (see fig. 2.43). There is no risk of deadlocks. However, SDF does not allow adding new observers at run-time.

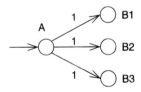

Figure 2.43. Observer pattern in SDF

SDF models are determinate, but they are not appropriate for modeling control flow, such as branches etc. Several extensions and variations of SDF models have been proposed (see, for example Stuijk [Stuijk, 2007]):

- For example, we can have **modes** corresponding to states of an associated finite state machine. For each of the modes, a different SDF graph could be relevant. Certain events could then cause transitions between these modes.

- Homogeneous synchronous data flow (HSDF) graphs are a special case of SDF graphs. For HSDF graphs, the number of tokens consumed and produced per firing is always 1.

- For cyclo-static data flow (CSDF), the number of tokens produced and consumed per firing can vary over time, but has to be periodic.

Complex SUDs including control flow must be modeled using more general computational graph structures.

2.5.4   Simulink

Computational graph structures are also frequently used in control engineering. For this domain, the Simulink toolbox of MATLAB [The MathWorks Inc., 2010], [Tewari, 2001] is very popular. MATLAB is a modeling and simulation tool based on mathematical models including, for example, partial differential equations. Fig. 2.44 shows an example of a Simulink model [Marian and Ma, 2007].

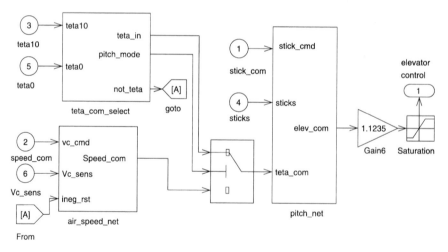

Figure 2.44.   Simulink model

The amplifier and the saturation component on the right demonstrate the inclusion of analog modeling. In the general case, the "schematic" could contain symbols denoting analog components such as integrators, differentiators. The switch in the center indicates that Simulink also allows some control flow modeling.

The graphical representation is intuitive and allows control engineers to focus on the control function, without caring about the code necessary to implement the function. The graphical symbols suggest that analog circuits are used as traditional components in control designs. A key goal is to synthesize software from such models. This approach is typically associated with the term **model-based design**, but there is no precise definition for this term.

Semantics of Simulink models reflect the simulation on a digital computer and the behavior may be similar to that of analog circuits, but possibly not quite the same. What is actually the semantics of a Simulink model? Marian and Ma [Marian and Ma, 2007] describe the semantics as follows: *"Simulink uses an idealized timing model for block (node) execution and communication. Both happen infinitely fast at exact points in simulated time. Thereafter, simulated time is advanced by exact time steps. All values on edges are constant in between time steps.* This means that we execute the model time step after time step. For each step, we compute the function of the nodes (in zero time) and propagate the new values to connected inputs. This explanation does not specify the distance between time steps. Also, it does not immediately tell us how to implement the system in software, since even slowly varying outputs may be recomputed frequently.

This approach is appropriate for modeling physical systems such as cars or trains at a high level and then simulating the behavior of these systems. Also, digital signal processing systems can be conveniently modeled with MATLAB and Simulink. In order to generate implementations, MATLAB/Simulink models first must be translated into a language supported by software or hardware design systems, such as C or VHDL.

Components in Simulink models provide a special case of **actors**. We can assume that actors are waiting for input and perform their operation once all required inputs have arrived. SDF is another case of actor-based languages. In **actor-based languages**, there is no need to pass control to these actors, like in von-Neumann languages.

2.6 Petri nets

2.6.1 Introduction

Very comprehensive descriptions of control flow are feasible with computational graphs known as Petri nets. Actually, Petri nets model **only** control and control dependencies. Modeling data as well requires extensions of Petri nets. Petri nets focus on the modeling of causal dependencies.

In 1962, Carl Adam Petri published his method for modeling causal dependencies, which became known as Petri nets [Petri, 1962]. Petri nets do not assume any global synchronization and are therefore especially suited for modeling distributed systems.

Conditions, **events** and a **flow relation** are the key elements of Petri nets. Conditions are either satisfied or not satisfied. Events can happen. The flow relation describes the conditions that must be met before events can happen and it also describes the conditions that become true if events happen.

Graphical notations for Petri nets typically use circles to denote conditions and boxes to denote events. Arrows represent flow relations. Fig. 2.45 shows a first example.

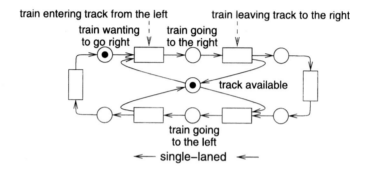

Figure 2.45. Single track railroad segment

This example describes mutual exclusion for trains at a railroad track that must be used in both directions. A token is used to prevent collisions of trains going into opposite directions. In the Petri net representation, that token is symbolized by a condition in the center of the model. A partially filled circle (a circle containing a second, filled circle) denotes the situation in which the condition is met (this means: the track is available). When a train wants to go to the right (also denoted by a partially filled circle in fig. 2.45), the two conditions that are necessary for the event "train entering track from the left" are met. We call these two conditions **preconditions**. If the preconditions of an event are met, it can happen. As a result of that event happening, the token is no longer available and there is no train waiting to enter the track. Hence, the preconditions are no longer met and the partially filled circles disappear (see fig. 2.46).

However, there is now a train going on that track from the left to the right and thus the corresponding condition is met (see fig. 2.46). A condition which is met after an event happened is called a **postcondition**. In general, an event can happen only if all its preconditions are true (or met). If it happens, the preconditions are no longer met and the postconditions become valid. Arrows identify those conditions which are preconditions of an event and those that

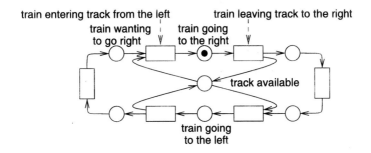

Figure 2.46. Using resource "track"

are postconditions of an event. Continuing with our example, we see that a train leaving the track will return the token to the condition at the center of the model (see fig. 2.47).

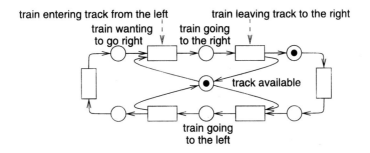

Figure 2.47. Freeing resource "track"

If there are two trains competing for the single-track segment (see fig. 2.48), only one of them can enter.

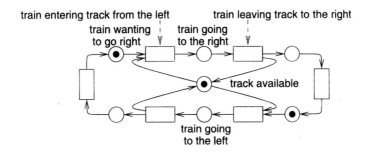

Figure 2.48. Conflict for resource "track"

In such situations, the next transition to be fired is non-deterministically cho-sen. Analyses of the net must consider all possible firing sequences. For Petri nets, we are intentionally modeling non-determinism.

A key advantage of Petri nets is that they can be the basis for formal proofs about system properties and that there are standardized ways of generating such proofs. In order to enable such proofs, we need a more formal definition of Petri nets. We will consider three classes of Petri nets: condition/event nets, place/transitions nets, and predicate transition nets.

2.6.2 Condition/event nets

Condition/event nets are the first class of Petri nets that we will define more formally.

Definition: $N = (C, E, F)$ is called a **net**, iff the following holds:

1 C and E are disjoint sets.

2 $F \subseteq (E \times C) \cup (C \times E)$ is a binary relation, called flow relation.

The set C is called conditions and the set E is called events.

Definition: Let N be a net and let $x \in (C \cup E)$. Then,

1 $^\bullet x := \{y | yFx, y \in (C \cup E)\}$ is called the **pre-set** of x. If x denotes an event, $^\bullet x$ is also called the set of **preconditions** of x.

2 $x^\bullet := \{y | xFy, y \in (C \cup E)\}$ is called the **post-set** of x. If x denotes an event, $x^\bullet$ is also called the set of **postconditions** of x.

The terms preconditions and postconditions are preferred if these sets actually denote conditions $\in C$, that is, if $x \in E$.

Definition: Let $(c, e) \in C \times E$.

1 (c, e) is called a **loop**, if $cFe \wedge eFc$.

2 N is called **pure**, if F does not contain any loops (see fig. 2.49, left).

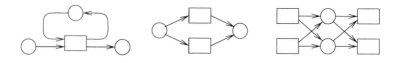

Figure 2.49.   Nets which are not pure (left) and not simple (center and right)

Definition: A net is called **simple** if no two transitions t_1 and t_2 have the same set of pre- and postconditions (see fig. 2.49 (center and right)).

Simple nets with no isolated elements meeting some additional restrictions are called **condition/event nets**. Condition/event nets are a special case of bipartite graphs (graphs with two disjoint sets of nodes). We will not discuss those additional restrictions in detail since we will consider more general classes of nets in the following.

2.6.3 Place/transition nets

For condition/event nets, there is at most one token per condition. For many applications, it is useful to remove this restriction and to allow more tokens per conditions. Nets allowing more than one token per condition are called place/transition nets. Places correspond to what we so far called conditions and transitions correspond to what we so far called events. The number of tokens per place is called a **marking**. Mathematically, a marking is a mapping from the set of places to the set of natural numbers extended by a special symbol ω denoting infinity.

Let N_0 denote the natural numbers including 0. Then, formally speaking, place/transition nets can be defined as follows:

Definition: (P, T, F, K, W, M_0) is called a place/transition net $\Longleftrightarrow$

1. $N = (P, T, F)$ is a net with places $p \in P$, transitions $t \in T$, and flow relation F.

2. Mapping $K : P \rightarrow (N_0 \cup \{\omega\}) \setminus \{0\}$ denotes the capacity of places (ω symbolizes infinite capacity).

3. Mapping $W : F \rightarrow (N_0 \setminus \{0\})$ denotes the weight of graph edges.

4. Mapping $M_0 : P \rightarrow N_0 \cup \{\omega\}$ represents the initial marking of places.

Edge weights affect the number of tokens that are required before transitions can happen and also identify the number of tokens that are generated if a certain transition takes place. Let $M(p)$ denote a current marking of place $p \in P$ and let $M'(p)$ denote a marking after some transition $t \in T$ took place. The weight of edges belonging to preconditions represents the number of tokens that are removed from places in the pre-set. Accordingly, the weight of edges belonging to the postconditions represents the number of tokens that are added to the places in the post-set. Formally, marking M' is computed as follows:

$$M'(p) = \begin{cases} M(p) - W(p,t), & \text{if } p \in {}^\bullet t \setminus t^\bullet \\ M(p) + W(t,p), & \text{if } p \in t^\bullet \setminus {}^\bullet t \\ M(p) - W(p,t) + W(t,p), & \text{if } p \in {}^\bullet t \cap t^\bullet \\ M(p) & \text{otherwise} \end{cases}$$

Fig. 2.50 shows an example of how transition t_j affects the current marking.

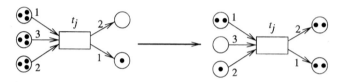

Figure 2.50.   Generation of a new marking

By default, unlabeled edges are considered to have a weight of 1 and unlabeled places are considered to have unlimited capacity ω.

We now need to explain the two conditions that must be met before a transition $t \in T$ can take place:

- for all places p in the pre-set, the number of tokens must at least be equal to the weight of the edge from p to t and

- for all places p in the post-set, the capacity must be large enough to accommodate the new tokens which t will generate.

Transitions meeting these two conditions are called **M-activated**. Formally, this can be defined as follows:

Definition: Transition $t \in T$ is said to be M-activated $\Longleftrightarrow$

$$(\forall p \in {}^\bullet t : M(p) \geq W(p,t)) \wedge (\forall p' \in t^\bullet : M(p') + W(t,p') \leq K(p'))$$

Activated transitions can happen, but they do not need to. If several transitions are activated, the sequence in which they happen is not deterministically defined.

The impact of a firing transition t on the number of tokens can be represented conveniently by a vector $\underline{t}$ associated with t. $\underline{t}$ is defined as follows:

$$\underline{t}(p) = \begin{cases} -W(p,t), & \text{if } p \in {}^\bullet t \setminus t^\bullet \\ +W(t,p), & \text{if } p \in t^\bullet \setminus {}^\bullet t \\ -W(p,t) + W(t,p), & \text{if } p \in {}^\bullet t \cap t^\bullet \\ 0 & \text{otherwise} \end{cases}$$

The new number M' of tokens, resulting from the firing of transition t, can be computed for all places p as follows:

$$M'(p) = M(p) + \underline{t}(p)$$

Using "+" to denote vector addition, we can rewrite this equation as follows:

$$M' = M + \underline{t}$$

The set of all vectors $\underline{t}$ form an incidence matrix $\underline{N}$. $\underline{N}$ contains vectors $\underline{t}$ as columns.

$$\underline{N} : P \times T \rightarrow \mathbb{Z}; \quad \forall t \in T : \underline{N}(p,t) = \underline{t}(p)$$

It is possible to formally prove system properties by using matrix $\underline{N}$. For example, we are able to compute sets of places, for which firing transitions will not change the overall number of tokens [Reisig, 1985]. Such sets are called **place invariants**. Let us initially consider a single transition t_j in order to find such invariants. Let us search for sets $R \subseteq P$ of places such that the total number of tokens does not change if t_j fires. The following must hold for such sets:

$$\sum_{p \in R} t_j(p) = 0 \tag{2.2}$$

Fig. 2.51 shows a transition for which the total number of tokens does not change if it fires.

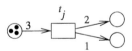

Figure 2.51. Transition with a constant number of tokens

We are now introducing the characteristic vector $\underline{c}_R$ of some set R of places:

$$\underline{c}_R(p) = \begin{cases} 1 & \text{iff } p \in R \\ 0 & \text{iff } p \notin R \end{cases}$$

With this definition, we can rewrite equation 2.2 as:

$$\sum_{p\in R} t_j(p) = \sum_{p\in P} t_j(p) * c_R(p) = t_j \cdot c_R = 0 \tag{2.3}$$

$\cdot$ denotes the scalar product. Now, we search for sets of places such that firings of **any** transition will not change the total number of tokens. This means that equation 2.3 must hold for all transitions t_j:

$$t_1 \cdot c_R = 0$$
$$t_2 \cdot c_R = 0 \tag{2.4}$$
$$\dots$$
$$t_n \cdot c_R = 0$$

Equations 2.4 can be combined into the following equation by using the transposed incidence matrix N^T:

$$N^T c_R = 0 \tag{2.5}$$

Equation 2.5 represents a system of linear, homogeneous equations. Matrix N represents edge weights of our Petri nets. We are looking for solution vectors c_R for this system of equations. Solutions must be characteristic vectors. Therefore, their components must be 1 or 0 (integer weights can be accepted if we use weighted sums of tokens). This is more complex than solving systems of linear equations with real-valued solution vectors. Nevertheless, it is possible to obtain information by solving equation 2.5. Using this proof technique, we can for example show that we are correctly implementing mutually exclusive access to shared resources.

Let us now consider a larger example: We are again considering the synchronization of trains. In particular, we are trying to model high-speed Thalys trains traveling between Amsterdam, Cologne, Brussels and Paris. Segments of the train run independently from Amsterdam and Cologne to Brussels. There, the segments get connected and then they run to Paris. On the way back from Paris, they get disconnected at Brussels again. We assume that Thalys trains must synchronize with some other train at Paris. The corresponding Petri net is shown in fig. 2.52.

Places 3 and 10 model trains waiting at Cologne and Amsterdam, respectively. Transitions 2 and 9 model trains driving from these cities to Brussels. After their arrival at Brussels, places 2 and 9 contain tokens. Transition 1 denotes connecting the two trains. The cup symbolizes the driver of one of the trains,

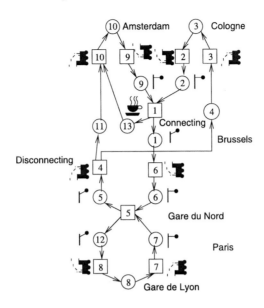

Figure 2.52. Model of Thalys trains running between Amsterdam, Cologne, Brussels, and Paris

who will have a break at Brussels while the other driver is continuing on to Paris. Transition 5 models synchronization with other trains at the Gare du Nord station of Paris. These other trains connect Gare du Nord with some other station (we have used Gare de Lyon as an example, even though the situation at Paris is somewhat more complex). Of course, Thalys trains do not use steam engines; they are just easier to visualize than modern high speed trains. Fig. 2.53 shows matrix N^T for this example.

	p_1	p_2	p_3	p_4	p_5	p_6	p_7	p_8	p_9	p_{10}	p_{11}	p_{12}	p_{13}
t_1	1	-1							-1				1
t_2		1	-1										
t_3			1	-1									
t_4				1	-1						1		
t_5					1	-1	-1					1	
t_6	-1					1							
t_7							1	-1					
t_8								1				-1	
t_9									1	-1			
t_{10}										1	-1		-1

Figure 2.53. N^T for the Thalys example

For example, row 2 indicates that firing t_2 will increase the number of tokens on p_2 by 1 and decrease the number of tokens on p_3 by 1. Using techniques from linear algebra, we are able to show that the following four vectors are solutions for this system of linear equations:

$c_{R,1} = (1,1,1,1,1,1,0,0,0,0,0,0,0)$

$c_{R,2} = (1,0,0,0,1,1,0,0,1,1,1,0,0)$

$c_{R,3} = (0,0,0,0,0,0,0,0,1,1,0,0,1)$

$c_{R,4} = (0,0,0,0,0,0,1,1,0,0,0,1,0)$

These vectors correspond to the places along the track for trains from Cologne, to the places along the track for trains from Amsterdam, to the places along the path for drivers of trains from Amsterdam, and to the places along the track within Paris, respectively. Therefore, we are able to show that the number of trains and drivers along these tracks is constant (something which we actually expect). This example demonstrates that place invariants provide us with a standardized technique for proving properties about systems.

2.6.4   Predicate/transition nets

Condition/event nets as well as place/transition nets can quickly become very large for large examples. A reduction of the size of the nets is frequently possible with predicate/transition nets. We will demonstrate this, using the so-called "dining philosophers problem" as an example. The problem is based on the assumption that a set of philosophers is dining at a round table. In front of each philosopher, there is a plate containing spaghetti (see fig. 2.54).

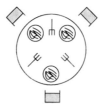

Figure 2.54.   The dining philosophers problem

Between each of the plates, there is just one fork. Each philosopher is either eating or thinking. Eating philosophers need their two adjacent forks for that, so they can only eat if their neighbors are not eating.

This situation can be modeled as a condition/event net, as shown in fig. 2.55.

Conditions t_j correspond to the thinking states, conditions e_j correspond to the eating states, and conditions f_j represent available forks. Considering the

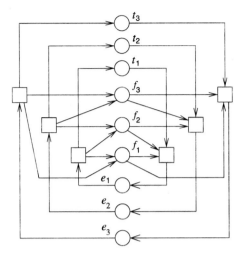

Figure 2.55. Place/transition net model of the dining philosophers problem

small size of the problem, this net is already very large. The size of this net can be reduced by using predicate/transition nets. Fig. 2.56 is a model of the same problem as a predicate/transition net.

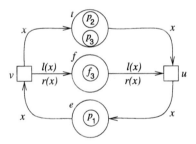

Figure 2.56. Predicate/transition net model of the dining philosophers problem

With predicate/transition nets, tokens have an identity and can be distinguished from each other[11]. We use this in fig. 2.56 in order to distinguish between the three different philosophers p_1 to p_3 and to identify fork f_3. Furthermore, edges can be labeled with variables and functions. In the example, we use variables to represent the identity of philosophers and functions $l(x)$ and $r(x)$ to denote the left and right forks of philosopher x, respectively. These two forks

---

[11]We could also think of adding a **color** to each of the tokens.

are required as a precondition for transition u and returned as a postcondition by transition v. Note that this model can be easily extended to the case of $n > 3$ philosophers. We just need to add more tokens. In contrast to the net in fig. 2.55, the structure of the net does not have to be changed.

2.6.5 Evaluation

The key advantage of Petri nets is their power for modeling causal dependencies. Standard Petri nets have no notion of time and all decisions can be taken locally, by just analyzing transitions and their pre- and postconditions. Therefore, they can be used for modeling geographically distributed systems. Furthermore, there is a strong theoretical foundation for Petri nets, simplifying formal proofs of system properties. Petri nets are not necessarily determinate: different firing sequences can lead to different results. The descriptive power of Petri nets encompasses that of other MoCs, including finite state machines.

In certain contexts, their strength is also their weakness. If time is to be explicitly modeled, standard Petri nets cannot be used. Furthermore, standard Petri nets have no notion of hierarchy and no programming language elements, let alone object oriented features. In general, it is difficult to represent data.

There are extended versions of Petri nets avoiding the mentioned weaknesses. However, there is no universal extended version of Petri nets meeting all requirements mentioned at the beginning of this chapter. Nevertheless, due to the increasing amount of distributed computing, Petri nets became more popular than they were initially.

UML includes extended Petri nets called **activity diagrams**. Extensions include symbols denoting decisions (just like in ordinary flow charts). The placement of symbols is somewhat similar to SDL. Fig. 2.57 shows an example.

The example shows the procedure to be followed during a standardization process. Forks and joins of control correspond to transitions in Petri nets and they use the symbols (horizontal bars) that were initially used for Petri nets as well. The diamond at the bottom shows the symbol used for decisions. Activities can be organized into "swim-lanes" (areas between vertical dotted lines) such that the different responsibilities and the documents exchanged can be visualized. It is interesting to note how a technique like Petri nets was initially certainly not a mainstream technique. Decades after its invention, it has become a frequently applied technique due to its inclusion in UML.

2.7   Discrete event based languages

The discrete event-based model of computation is based on the idea of simulating the generation of events and the processing of events over time. In

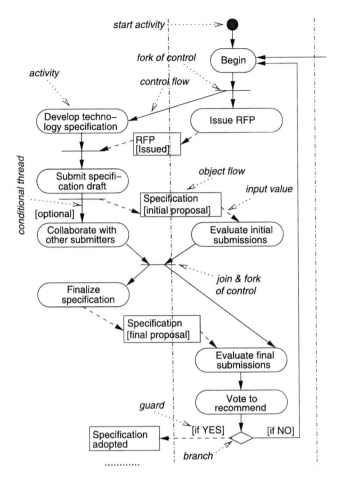

Figure 2.57. Activity diagram [Kobryn, 2001]

this model, we are using a queue of future events. These events are sorted by the time at which they should be processed. Semantics is defined by removing the events concerning the current time from the queue, performing the corresponding actions, possibly entering new events into the queue. Time is advanced whenever no action exists, which should be performed at the current time.

Hardware description languages (HDLs) are designed to model hardware. They are typically based on the discrete event model. We will use HDLs as a prominent example of discrete event modeling. The focus will be on the hardware description language VHDL, and we will briefly cover other HDLs as well.

A key distinction between common software languages and hardware description languages is the need to model time in HDLs. Another distinction comes
from the requirement to describe concurrency among different hardware components.

2.7.1   VHDL

2.7.1.1   Introduction

VHDL is a prominent example of HDLs. VHDL uses **processes** for modeling
concurrency. Each process models one component of the potentially concurrent hardware. For simple hardware components, a single process may be
sufficient. More complex components may need several processes for modeling their operations. Processes communicate through **signals**. Signals roughly
correspond to physical connections (wires).

The origin of VHDL can be traced back to the 1980s. At that time, most design
systems used graphical HDLs. The most common building block was the gate.
However, in addition to using graphical HDLs, we can also use textual HDLs.
The strength of textual languages is that they can easily represent complex
computations including variables, loops, function parameters and recursion.
Accordingly, when digital systems became more complex in the 1980s, textual
HDLs almost completely replaced graphical HDLs. Textual HDLs were initially a research topic at universities. See Mermet et al. [Mermet et al., 1998]
for a survey of languages designed in Europe at that time. MIMOLA was one
of these languages and the author of this book contributed to its design and
applications [Marwedel and Schenk, 1993], [Marwedel, 2008b]. Textual languages became popular when VHDL and its competitor Verilog (see page 98)
were introduced.

VHDL was designed in the context of the VHSIC program of the Department
of Defense (DoD) in the US. VHSIC stands for *very high speed integrated circuits*[12]. Initially, the design of VHDL (VHSIC hardware description language)
was done by three companies: IBM, Intermetrics and Texas Instruments. A
first version of VHDL was published in 1984. Later, VHDL became an IEEE
standard, called IEEE 1076. The first IEEE version was standardized in 1987;
updates were designed in 1992, in 1997, in 2002 and in 2006 [Lewis et al.,
2007]. VHDL-AMS allows modeling analog and mixed-signal systems by including differential equations in the language. The design of VHDL used ADA
(see page 102) as the starting point, since both languages were designed for the
DoD. Since ADA is based on PASCAL, VHDL has some of the syntactical fla-

---

[12]The design of the Internet was also part of the VHSIC program.

vor of PASCAL. However, the syntax of VHDL is much more complex and it is necessary not to get distracted by the syntax. In the current book, we will just focus on some concepts of VHDL which are useful also in other languages. A full description of VHDL is beyond the scope of this book. The standard is available from IEEE (see, for example, [IEEE, 2002]).

2.7.1.2 Entities and architectures

VHDL, like all other HDLs, includes the necessary support for modeling concurrent operation of hardware components. Hardware components are modeled by so-called **design entities** or **VHDL entities**. Entities contain **processes** used to model concurrency. According to the VHDL grammar, design entities are composed of two types of ingredients: an **entity declaration** and one (or several) **architectures** (see fig. 2.58).

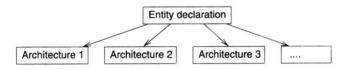

Figure 2.58. An entity consists of an entity declaration and architectures

For each entity, the most recently analyzed architecture will be used by default. Using other architectures can be specified. Architectures may contain several processes.

We will discuss a full adder as an example. Full adders have three input ports and two output ports (see fig. 2.59).

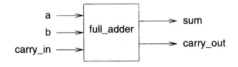

Figure 2.59. Full-adder and its interface signals

An entity declaration corresponding to fig. 2.59 is the following:

```
entity full_adder is              -- entity declaration
   port (a, b, carry_in: in Bit;   -- input ports
         sum, carry_out: out Bit); -- output ports
end full_adder;
```

Two hyphens (--) are starting comments. They extend until the the end of the line. Architectures consist of architecture headers and architectural bodies.

We can distinguish between different styles of bodies, in particular between structural and behavioral bodies. We will show how the two are different using the full adder as an example. Behavioral bodies include just enough information to compute output signals from input signals and the local state (if any), including the timing behavior of the outputs. The following is an example of this ($<=$ denotes assignments to signals):

> **architecture** behavior **of** full_adder **is** -- architecture
>
> **begin**
>
>> sum $<=$ (a xor b) xor carry_in **after** 10 ns;
>>
>> carry_out $<=$ (a and b) or (a and carry_in) or
>>
>>> (b and carry_in) **after** 10 ns;
>
> **end** behavior;

VHDL-based simulators are capable of displaying output signal waveforms resulting from stimuli applied to the inputs of the full adder described above.

In contrast, structural bodies describe the way entities are composed of simpler entities. For example, the full adder can be modeled as an entity consisting of three components (see fig. 2.60). These components are called i1 to i3 and are of type half_adder or or_gate.

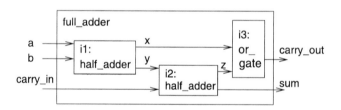

Figure 2.60.   Schematic describing structural body of the full adder

In the 1987 version of VHDL, these components must be declared in a so-called component declaration. This declaration is very similar to (and it serves the same purpose) as forward declarations in other languages. This declaration provides the necessary information about the component even if the full description of that component is not yet stored in the VHDL database (this may happen in the case of so-called top-down designs). From the 1992 version of VHDL onwards, such declarations are not required if the relevant components are already stored in the component database.

Connections between local component and entity ports are described in **port maps**. The following VHDL code represents the structural body shown in fig. 2.60:

```
architecture structure of full_adder is  -- architecture head
    component half_adder
        port (in1, in2: in Bit; carry: out Bit; sum: out Bit);
    end component;
    component or_gate
        port (in1, in2: in Bit; o: out Bit);
    end component;
    signal x, y, z: Bit;          -- local signals
    begin                         -- port map section
    i1: half_adder               -- introduction of half_adder i1
        port map (a, b, x, y);   -- connections between ports
    i2: half_adder port map (y, carry_in, z, sum);
    i3: or_gate    port map (x, z, carry_out);
    end structure;
```

2.7.1.3 VHDL processes and assignments

VHDL treats components described above as processes. The syntax used above is just a shorthand for processes. The general syntax for processes is as follows:

label : -- optional

process

declarations -- optional

begin

statements-- optional

end process ;

Assignments are special cases of statements. In VHDL, there are two kinds of assignments:

- **Variable assignments**: The syntax of variable assignments is

  variable := expression

  Whenever control reaches such an assignment, the expression is computed and assigned to the variable. Such assignments behave like assignments in common programming languages.

- **Signal assignments**: Signals and signal assignments are introduced in an attempt to model electrical signals in real hardware systems. Signals associate values with instances in time. In VHDL, such a mapping from time to values is represented by **waveforms**. Waveforms are computed from signal assignments. The syntax of signal assignments is

  signal $<=$ *expression*;

  signal $<=$ **transport** *expression* **after** *delay*;

  signal $<=$ *expression* **after** *delay*;

  signal $<=$ **reject** *time* **inertial** *expression* **after** *delay*;

Whenever control reaches such an assignment, the expression is computed and used to extend predicted future values of the waveform. In order to compute future values, **simulators are assumed to include a queue of events to happen later than the current simulated time**. This queue is sorted by the time, at which future events (e.g. updates of signals) should happen. Executing a signal assignment results in the creation of entries in this queue. Each entry contains a time for executing the event, the affected signal and the value to be assigned. For signal assignments not containing any **after** clause (first syntactical form), the entry will contain the current simulation time as the time at which this assignment has to be performed. In this case, the change will take place after an infinitesimally small amount of time, called δ-delay (see below). This allows us to update signals without changing macroscopic time.

For signal assignments containing a **transport** prefix (second syntactical form), the update of the signal will be delayed by the specified amount. This form of the assignment is following the so-called **transport delay model**. This model is based on the behavior of simple wires: wires are (as a first order of approximation) delaying signals. Even short pulses propagate along wires. The transport delay model can be used for logic circuits, even though its main application is to model wires. Suppose that we model a simple or-gate using a transport delay signal assignment:

  c $<=$ **transport** a or b **after** 10 ns;

Such a model would propagate even short pulses (see fig. 2.61).

Transport delay signal assignments will delete all entries in the queue corresponding to the time of the computed update or later times (if we first execute an assignment with a rather large delay and then execute an assignment with a smaller delay, then the entry resulting from the first assignment will be deleted).

For signal assignments containing an after clause, but no transport clause, **inertial delay** is assumed. The inertial delay model reflects the fact that

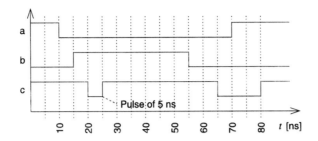

Figure 2.61. Gate modeled with transport delay

real circuits come with some "inertia". This means that short spikes will be suppressed. For the third syntactical form of the signal assignment, all signals changes which are shorter than the specified delay are suppressed. For the fourth form, all signal changes which are shorter than the indicated amount are removed from the predicted waveform. Suppose that we model a simple or-gate using inertial delay:

c <= a or b **after** 10 ns;

For such a model, short spikes would be suppressed (see fig. 2.62).

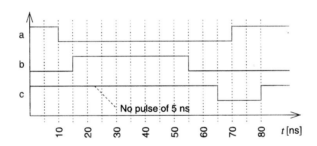

Figure 2.62. Gate modeled with inertial delay

The implementation of inertial delay relies on the removal of entries in the predicted waveform. The subtle rules for removals are not repeated here.

In addition to assignments, processes may contain **wait** statements. Such statements can be used to suspend a process. There are the following kinds of **wait** statements:

- **wait on** *signal list*; suspend until one of the signals in the list changes;

- **wait until** *condition*; suspend until *condition* is met, e.g. a = '1';

- **wait for** *duration*; suspend for a specified period of time;

- **wait**; suspend indefinitely.

As an alternative to explicit **wait** statements, a list of signals can be added to the process header. In that case, the process is activated whenever one of the signals in that list changes its value. Example: The following model of an and-gate will execute its body once and will restart from the beginning every time one of the inputs changes its value:

> **process**(x, y) **begin**
>
> > prod $<=$ x and y ;
>
> **end process**;

This model is equivalent to

> **process begin**
>
> > prod $<=$ x and y ;
>
> **wait on** x,y;
>
> **end process**;

2.7.1.4   The VHDL simulation cycle

According to the original standards document [IEEE, 1997], the execution of a VHDL model is described as follows: *"The execution of a model consists of an **initialization phase** followed by the **repetitive execution of process statements** in the description of that model. Each such repetition is said to be a **simulation cycle**. In each cycle, the values of all signals in the description are computed. If as a result of this computation an event occurs on a given signal, process statements that are sensitive to that signal will resume and will be executed as part of the simulation cycle."*

The initialization phase takes signal initializations into account and executes each process once. It is described in the standards as follows[13]:

"At the beginning of initialization, the current time, T_c is assumed to be 0 ns. The initialization phase consists of the following steps:[14]

- *The driving value and the effective value of each explicitly declared signal are computed, and the current value of the signal is set to the effective*

---

[13] We leave out the discussion of implicitly declared signals and so-called postponed processes introduced in the 1997 version of VHDL.

[14] In order not to get lost in the amount of details provided by the standard, some of its sections (indicated by "...") are omitted in the citation.

value. This value is assumed to have been the value of the signal for an infinite length of time prior to the start of the simulation. ...

- *Each ... process in the model is executed until it suspends. ...*

- *The time of the next simulation cycle (which in this case is the first simulation cycle), T_n is calculated according to the rules of step e of the simulation cycle, below."*

Each simulation cycle starts with setting the current time to the next time at which changes must be considered. This time T_n was either computed during the initialization or during the last execution of the simulation cycle. Simulation terminates when the current time reaches its maximum, $TIME'HIGH$. According to the original document, the simulation cycle is described as follows: "*A simulation cycle consists of the following steps:*

a) *The current time, T_c is set equal to T_n. Simulation is complete when $T_n = TIME'HIGH$ and there are no active drivers or process resumptions at T_n.*

b) *Each active explicit signal in the model is updated. (Events may occur as a result.)" ...*

   In the cycle preceding the current cycle, new future values for some of the signals have been computed. If T_c corresponds to the time at which these values become valid, they are now assigned. New values of signals are never immediately assigned while executing a simulation cycle: they are not assigned before the next simulation cycle, at the earliest. Signals that change their value generate so-called events which, in-turn, may enable the execution of processes that are sensitive to that signal.

c) "*For each process P, if P is currently sensitive to a signal S and if an event has occurred on S in this simulation cycle, then P resumes.*

d) *Each ... process that has resumed in the current simulation cycle is executed until it suspends.*

e) *The time of the next simulation cycle, T_n is determined by setting it to the earliest of*

   1 *$TIME'HIGH$* (This is the end of simulation time).

   2 *The next time at which a driver becomes active* (this is the next instance in time, at which a driver specifies a new value), *or*

   3 *The next time at which a process resumes* (this time is determined by **wait for** statements).

   If $T_n = T_c$, then the next simulation cycle (if any) will be a delta cycle."

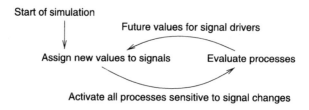

Figure 2.63.   VHDL simulation cycles

The iterative nature of simulation cycles is shown in fig. 2.63.

Delta (δ) simulation cycles have been the source of many discussions. Their purpose is to introduce a infinitesimally small delay even in cases in which the user did not specify any. As an example, we will show the effect of these cycles using a flip-flop as an example. Fig. 2.64 shows the schematic of the flip-flop.

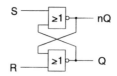

Figure 2.64.   RS-Flipflop

The flip-flop is modeled in VHDL as follows:

```
entity RS_Flipflop is
   port (R: in   BIT;      -- reset
      S: in   BIT;     -- set
      Q: inout BIT;   -- output
      nQ: inout BIT; ); -- Q-bar
   end RS_Flipflop;
   architecture one of RS_Flipflop is
     begin
     process: (R,S,Q,nQ)
       begin
         Q  <= R nor nQ;  nQ <= S nor Q;
       end process;
     end one;
```

Ports Q and nQ must be of mode **inout** since they are also read internally, which would not be possible if they were of mode **out**. Fig. 2.65 shows the simulation times at which signals are updated for this model. During each cycle, updates are propagated through one of the gates. Simulation terminates after three δ cycles. The last cycle does not change anything, since Q is already '0'.

	< 0ns	0ns	0ns+δ	0ns+2*δ	0ns+3*δ
R	0	1	1	1	1
S	0	0	0	0	0
Q	1	1	0	0	0
nQ	0	0	0	1	1

Figure 2.65. δ cycles for RS-flip-flop

δ cycles correspond to an infinitesimally small unit of time, which will always exist in reality. δ cycles ensure that simulation respects causality.

The results do not depend on the order in which parts of the model are executed by the simulation. This feature is enabled by the separation between the computation of new values for signals and their actual assignment. In a model containing the lines

    a <= b;

    b <= a;

signals a and b will always be swapped. If the assignments were performed immediately, the result would depend on the order in which we execute the assignments (see also page 48). **VHDL models are therefore determinate**. This is what we expect from the simulation of a real circuit with a fixed behavior.

There can be arbitrarily many δ cycles before the current time T_c is advanced. This possibility of infinite loops can be confusing. One of the options of avoiding this possibility would be to disallow zero delays, which we used in our model of the flip-flop.

The propagation of values using signals also allows an easy implementation of the observer pattern (see page 27). In contrast to SDF, the number of observers can vary, depending on the number of processes waiting for changes on a signal.

What is the communication model behind VHDL? The description of the semantics of VHDL relies heavily on a **single, centralized** queue of future events, storing values of all signals in the future. The purpose of this queue is **not** to implement asynchronous message passing. Rather, this queue is supposed to be accessed by the simulation kernel, one entry at a time, in a non-distributed fashion. Attempts to perform distributed VHDL simulations are typically suffering from a poor performance. All modeled components can access values of signals and variables which are in their scope without any message-based communication. Therefore, we tend towards associating VHDL with a shared memory based implementation of the communication. However, FIFO-based message passing could be implemented in VHDL on top of the VHDL simulator as well.

2.7.1.5   Multi-valued logic and IEEE 1164

In this book, we are restricting ourselves to embedded systems implemented with binary logic. Nevertheless, it may be advisable or necessary to use more than two values for modeling such systems. For example, our systems might contain electrical signals of different strengths and it may be necessary to compute the strength and the logic level resulting from a connection of two or more sources of electrical signals. In the following, we will therefore distinguish between the **level** and the **strength** of a **signal**. While the former is an abstraction of the signal voltage, the latter is an abstraction of the impedance (resistance) of the voltage source. We will be using discrete sets of signal values representing the signal level and the strength. **Using discrete sets of strengths avoids the problems of having to solve Kirchhoff's equations and enables us to avoid analog models used in electrical engineering.** We will also model unknown electrical signals by special signal values.

In practice, electronic design systems use a variety of value sets. Some systems allow only two, while others allow 9 or 46. The overall goal of developing discrete value sets is to avoid the problems of solving network equations (e.g. Kirchoff's laws) and still model existing systems with sufficient precision. In the following, we will present a systematic technique for building up value sets and for relating these to each other. We will use the strength of electrical signals as the key parameter for distinguishing between various value sets. A systematic way of building up value sets, called CSA-theory, was presented by Hayes [Hayes, 1982]. CSA stands for "connector, switch, attenuator". These three elements are key elements of this theory. We will later show how the standard value set used for most cases of VHDL-based modeling can be derived as a special case.

1 signal strength (Two logic values)

In the simplest case, we will start with just two logic values, called '0' and '1'. These two values are considered to be of the same strength. This means: if two wires **connect** values '0' and '1', we will not know anything about the resulting signal level.

A single signal strength may be sufficient if no two wires carrying values '0' and '1' are connected and no signals of different strength meet at a particular node of electronic circuits.

2 signal strengths (Three and four logic values)

In many circuits, there may be instances in which a certain electrical signal is not actively driven by any output. This may be the case, when a certain wire is not connected to ground, the supply voltage or any circuit node.

For example, systems may contain open-collector outputs (see fig. 2.66, left). If the "pull-down" transistor PD is non-conducting, the output is effectively disconnected. For the tristate outputs (see fig. 2.66, right), an enable signal of '0' will generate a '0' at the outputs of the and-gates (denoted by &), and will make both transistors non-conducting. As a result, output A will be discon-nected[15]. Hence, using appropriate input signals, such outputs can be effec-tively disconnected from a wire.

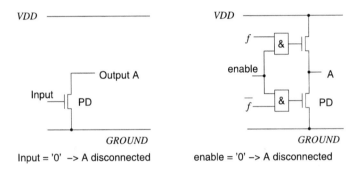

Figure 2.66. Outputs that can be effectively disconnected from a wire

Obviously, the signal strength of disconnected outputs is the smallest strength that we can think of. In particular, the signal strength of Z is smaller than that of '0' and '1'. Furthermore, the signal level of such an output is unknown. This combination of signal strength and signal value is represented by a logic value called 'Z'. If a signal of value 'Z' is connected to another signal, that other signal will always dominate. For example, if two tristate outputs are connected to the same bus and if one output contributes a value of 'Z', the resulting value on the bus will always be the value contributed by the second output (see fig. 2.67).

In VHDL, each output is associated with a so-called signal **driver**. Computing the value resulting from the contributions of multiple drivers to the same sig-

---

[15] In practice, pull-up transistors may be depletion transistors and the tri-state outputs may be inverting.

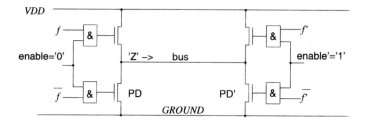

Figure 2.67.   Right output dominates bus

nal is called **resolution** and resulting values are computed by functions called **resolution functions**.

In most cases, three-valued logic sets {'0','1','Z'} are extended by a fourth value called 'X'. 'X' represents an unknown signal level of the same strength as '0' or '1'. More precisely, we are using 'X' to represent unknown values of signals that can be either '0' or '1' or some voltage representing neither '0' nor '1'[16].

The resolution that is required if multiple drivers get connected can be computed very easily, if we make use of a partial order among the four signal values '0', '1', 'Z', and 'X'. The partial order is depicted in the **Hasse diagram** in fig. 2.68.

Figure 2.68.   Partial order for value set {'0', '1', 'Z', 'X'}

Edges in this figure reflect the domination of signal values. Edges define a relation $>$. If $a > b$, then a dominates b. '0' and '1' dominate 'Z'. 'X' dominates all other signal values. Based on the relation $>$, we define a relation $\geq$. $a \geq b$ holds iff $a > b$ or $a = b$.

We define an operation *sup* on two signals, which returns the **supremum** of the two signal values. The supremum c of the two values a and b is the weakest value for which $c \geq a$ and $c \geq b$ holds. For example, *sup* ('Z', '0')='0', *sup*('Z','1')='1' etc. **The interesting observation is that resolution functions should compute the *sup* function according to the above definition.** The supremum corresponds to the **connect** element of the CSA theory.

---

[16]There are other interpretations of 'X', but the one presented above is the most useful one in our context.

3 signal strengths (Seven signal values)

In many circuits, two signal strengths are not sufficient. A common case that requires more values is the use of depletion transistors (see fig. 2.69).

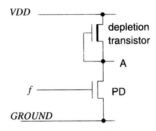

Figure 2.69. Output using depletion transistor

The effect of the depletion transistor is similar to that of a resistor providing a low conductance path to the supply voltage *VDD*. The depletion transistor as well as the "pull-down transistor" PD act as drivers for node A of the circuit and the signal value at node A can be computed using resolution. The pull-down transistor provides a driver value of '0' or 'Z', depending upon the input to PD. The depletion transistor provides a signal value, which is weaker than '0' and '1'. Its signal level corresponds to the signal level of '1'. We represent the value contributed by the depletion transistor by 'H', and we call it a "weak logic one". Similarity, there can be weak logic zeros, represented by 'L'. The value resulting from the possible connection between 'H' and 'L' is called a "weak logic undefined", denoted as 'W'. As a result, we have three signal strengths and seven logic values {'0', '1', 'L', 'H', 'W', 'X', 'Z'}. Resolution can again be based on a partial order among these seven values. The corresponding partial order is shown in fig. 2.70.

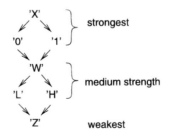

Figure 2.70. Partial order for value set {'0', '1', 'L', 'H', 'W', 'X', 'Z'}

This order also defines an operation *sup* returning the weakest value at least as strong as the two arguments. For example, $sup('H','0') = '0'$, $sup('H','Z') = 'H'$, $sup('H','L') = 'W'$.

'0' and 'L' represent the same signal levels, but a different strength. The same holds for the pairs '1' and 'H'. Devices increasing the signal strength are called **amplifiers**, devices reducing the signal strength are called **attenuators**.

Ten signal values (4 signal strengths)

In some cases, three signal strengths are not sufficient. For example, there are circuits using charges stored on wires. Such wires are charged to levels corresponding to '0' or '1' during some phases of the operation of the electronic circuit. This stored charge can control the (high impedance) inputs of some transistors. However, if these wires get connected to even the weakest signal source (except 'Z'), they lose their charge and the signal value from that source dominates.

For example, in fig. 2.71, we are driving a bus from a specialized output.

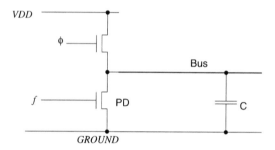

Figure 2.71. Pre-charging a bus

The bus has a high capacitive load C. While function f is still '0', we set ϕ to '1', charging capacitor C. Then we set ϕ to '0'. If the real value of function f becomes known and it turns out to be '1', we discharge the bus. The key reason for using pre-charging is that charging a bus using an output such as the one shown in fig. 2.69 is a slow process, since the resistance of depletion transistors is large. Discharging through regular pull-down transistors PD is a much faster process.

In order to model such cases, we need signal values which are weaker than 'H' and 'L', but stronger than 'Z'. We call such values "very weak signal values" and denote them by 'h' and 'l'. The corresponding very weak unknown value is denoted by 'w'. As a result, we obtain ten signal values {'0', '1', 'L', 'H', 'l', 'h', 'X', 'W', 'w', 'Z'}. Using the signal strength, we can again define a partial order among these values (see fig. 2.72).

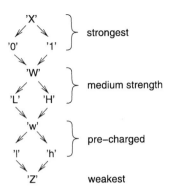

Figure 2.72. Partial order for value set {'0', '1', 'Z', 'X', 'H', 'L', 'W', 'h', 'l', 'w'}

Five signal strengths

So far, we have ignored power supply signals. These are stronger than the strongest signals we have considered so far. Signal value sets taking power supply signals into account have resulted in the definition of 46-valued value sets [Coelho, 1989]. However, such models are not very popular.

IEEE 1164

In VHDL, there is no predefined number of signal values, except for some basic support for two-valued logic. Instead, the used value sets can be defined in VHDL itself and different VHDL models can use different value sets.

However, portability of models would suffer in a very severe manner if this capability of VHDL was applied in this way. In order to simplify exchanging VHDL models, a standard value set was defined and standardized by the IEEE. This standard is called IEEE 1164 and is employed in many system models. IEEE 1164 has nine values: {'0', '1', 'L', 'H', 'X', 'W', 'Z', 'U', '-'}. The first seven values correspond to the seven signal values described above. 'U' denotes an uninitialized value. It is used by simulators for signals that have not been explicitly initialized.

'-' denotes the **input don't care**. This value needs some explanation. Frequently, hardware description languages are used for describing Boolean functions. The VHDL **select** statement is a very convenient means for doing that. The **select** statement corresponds to **switch** and **case** statements found in other languages and its meaning is different from the **select** statement in ADA (see page 104).

Example: Suppose that we would like to represent the Boolean function

$$f(a,b,c) = a\overline{b} + bc$$

Furthermore, suppose that f should be undefined for the case of $a = b = c = '0'$. A very convenient way of specifying this function would be the following:

    f <= **select** a & b & c -- & denotes concatenation

        '1' **when** "10-" -- corresponds to first term

        '1' **when** "-11" -- corresponds to second term

        'X' **when** "000"

This way, functions given above could be easily translated into VHDL. Unfortunately, the **select** statement denotes something completely different. Since IEEE 1164 is just one of a large number of possible value sets, it does not include any knowledge about the "meaning" of '-'. Whenever VHDL tools evaluate select statements such as the one above, they check if the selecting expression (a & b & c in the case above) is equal to the values in the when clauses. In particular, they check if e.g. a & b & c is equal to "10-". In this context, '-' behaves like any other value: VHDL systems check if c has a value of '-'. Since '-' is never assigned to any of the variables, these tests will never be true. Therefore, '-' is of limited benefit. The non-availability of convenient input don't care values is the price that one has to pay for the flexibility of defining value sets in VHDL itself[17].

The nice property of the general discussion on pages 90 to 95 is the following: it allows us to immediately draw conclusions about the modeling power of IEEE 1164. The IEEE standard is based on the 7-valued value set described on page 93 and, therefore, is capable of modeling circuits containing depletion transistors. It is, however, not capable of modeling charge storage[18].

2.7.2   SystemC

Due to the trend of implementing more and more functionality in software, a growing number of embedded systems includes a mixture of hardware and software. Most of the embedded system software is specified in C. For example, embedded systems implement standards such as MPEG 1/2/4 or decoders for mobile phone standards such as GSM or UMTS. The standards are frequently available in the form of "reference implementations", consisting of C programs not optimized for speed but providing the required functionality. The disadvantage of design methodologies based on VHDL or Verilog is the fact that these standards must be rewritten in order to generate hardware.

---

[17]This problem was corrected in VHDL 2006 [Lewis et al., 2007].

[18]As an exception, if the capability of modeling depletion transistors or pull-up resistors is not needed, one could interpret weak values as stored charges. This is, however, not very practical since pull-up resistors are found in most actual systems.

Furthermore, simulating hardware and software together requires interfacing software and hardware simulators. Typically, this involves a loss of simulation efficiency and inconsistent user interfaces. Also, designers must learn several languages.

Therefore, there has been a search for techniques for representing hardware structures in software languages. Some fundamental problems must be solved before hardware can be modeled with software languages:

- **Concurrency**, as it is found in hardware, has to be modeled in software.

- There has to be a representation of simulated **time**.

- **Multiple-valued logic** and **resolution** as described earlier must be supported.

- The **determinate behavior** of almost all useful hardware circuits must be guaranteed.

SystemC^{TM} [SystemC, 2010], [Open SystemC Initiative, 2005] is a C++ class library designed to solve these problems. With SystemC, specifications can be written in C or C++, making appropriate references to the class libraries.

SystemC comprises a notion of processes executed concurrently. Simulation semantics are similar to VHDL, including the presence of delta cycles. The execution of these processes is controlled via sensitivity lists and calls to **wait** primitives. The sensitivity list concept includes dynamic sensitivity lists.

SystemC includes a model of time. Earlier SystemC 1.0 used floating point numbers to denote time. In the current standard, an integer model of time is preferred. SystemC also supports physical units such as picoseconds, nanoseconds, microseconds etc.

SystemC data types include all common hardware types: four-valued logic ('0', '1', 'X' and 'Z') and bitvectors of different lengths are supported. Writing digital signal processing applications is simplified due to the availability of fixed-point data types.

Determinate behavior (see page 49) is not guaranteed in general, unless a certain modeling style is used. Using a command line option, the simulator can be directed to run processes in different orders. This way, the user can check if the simulation results depend on the sequence in which the processes are executed. However, for models of realistic complexity, only the presence of non-determinate behavior can be shown, not its absence.

Reusing hardware components in different contexts is simplified by the separation of computation and communication. SystemC provides channels, ports and interfaces as abstract components for communication. The introduction of these mechanisms facilitate so-called transaction-level modeling, as defined by Grötker et al. [Grötker et al., 2002]:

Definition: *"**Transaction-level modeling** (TLM) is a high-level approach to modeling digital systems where details of communication among modules are separated from the details of the implementation of functional units or of the communication architecture. Communication mechanisms such as buses or FIFOs are modeled as channels, and are presented to modules using SystemC interface classes. Transaction requests take place by calling interface functions of these channel models, which encapsulate low-level details of the information exchange. At the transaction level, the emphasis is more on the functionality of the data transfers - what data are transferred to and from what locations - and less on their actual implementation, that is, on the actual protocol used for data transfer. This approach makes it easier for the system-level designer to experiment, for example, with different bus architectures (all supporting a common abstract interface) without having to recode models that interact with any of the buses, provided these models interact with the bus through the common interface."*

SystemC has the potential for replacing existing VHDL-based design flows. Hardware synthesis starting from SystemC has become available [Herrera et al., 2003a], [Herrera et al., 2003b]. There are also commercial offerings. Methodology and applications for SystemC-based design are described in a book on that topic [Müller et al., 2003]. SystemC has been standardized as IEEE standard 1666-2005 [Open SystemC Initiative, 2005].

2.7.3 Verilog and SystemVerilog

Verilog is another hardware description language. Initially it was a proprietary language, but it was later standardized as IEEE standard 1364, with versions called IEEE standard 1364-1995 (Verilog version 1.0) and IEEE standard 1364-2001 (Verilog 2.0). Some features of Verilog are quite similar to VHDL. Just like in VHDL, designs are described as a set of connected design entities, and design entities can be described behaviorally. Also, processes are used to model concurrency of hardware components. Just like in VHDL, bitvectors and time units are supported. There are, however, some areas in which Verilog is less flexible and focuses more on comfortable built-in features. For example, standard Verilog does not include the flexible mechanisms for defining enumerated types such as the ones defined in the IEEE 1164 standard. However, support for four-valued logic is built into the Verilog language, and the standard IEEE 1364 also provides multiple valued logic with 8 different signal strengths. Multiple-valued logic is more tightly integrated into Verilog than into VHDL. The Verilog logic system also provides more features for transistor-level descriptions. However, VHDL is more flexible. For example, VHDL allows hardware entities to be instantiated in loops. This can be used to generate a structural description for, e.g. *n*-bit adders without having to specify *n* adders and their interconnections manually.

Verilog has a similar number of users as VHDL. While VHDL is more popular in Europe, Verilog is more popular in the US.

Verilog versions 3.0 and 3.1 are also known as SystemVerilog. They include numerous extensions to Verilog 2.0. These extensions include [Accellera Inc., 2003], [Sutherland, 2003]:

- additional language elements for modeling behavior,

- C data types such as int and type definition facilities such as typedef and struct,

- definition of interfaces of hardware components as separate entities,

- standardized mechanism for calling C/C++ functions and, to some extent, to call built-in Verilog functions from C,

- significantly enhanced features for describing an environment (called test-bench) for the hardware circuit under design (called CUD), and for using the testbench to validate the CUD by simulation,

- classes known from object-oriented programming for use within testbenches,

- dynamic process creation,

- standardized interprocess communication and synchronization, including semaphores,

- automatic memory allocation and deallocation,

- language features that provide a standardized interface to formal verification (see page 203).

Due to the capability of interfacing with C and C++, interfacing to SystemC models is also possible. Improved facilities for simulation- as well as for formal verification-based design validation and the possible interfacing to SystemC will potentially create a very good acceptance. Recently, Verilog and SystemVerilog have been merged into one standard, IEEE 1800-2009 [IEEE, 2009].

2.7.4   SpecC

The SpecC language [Gajski et al., 2000] is based on the clear separation between communication and computation that should be used for modeling embedded systems. This separation paves the way for re-using components in different contexts and enables *plug-and-play* for system components. SpecC models systems as hierarchical networks of behaviors communicating through channels. SpecC descriptions consist of behaviors, channels and interfaces. Behaviors include ports, locally instantiated components, private variables and functions and a public main function. Channels encapsulate communication. They include variables and functions, which are used for the definition of a communication protocol. Interfaces are linking behaviors and channels together. They declare the communication protocols which are defined in a channel.

SpecC can model hierarchies with nested behaviors. Fig. 2.73 [Gajski et al., 2000] shows a component B including sub-components b1 and b2.

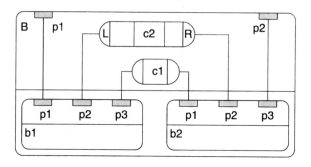

Figure 2.73.   Structural hierarchy of SpecC example

The sub-components are communicating through integer c1 and through channel c2. The structural hierarchy includes b1 and b2 as the leaves. b1 and b2 are executed concurrently, denoted by the keyword **par** in SpecC. This structural hierarchy is described in the following SpecC model.

```
interface L {void Write(int x); };
interface R {int Read(void); };
channel C implements L,R
    {int Data; bool Valid;
    void Write(int x) {Data=x; Valid=true;}
    int Read (void)
        {while (!Valid) waitfor (10); return (Data);} }
```

```
    behavior B1(in int p1, L p2, in int p3)
      {void main (void) {/* ...*/ p2.Write(p1);} };
    behavior B2 (out int p1, R p2, out int p3)
      {void main(void) {/*...*/ p3=p2.Read(); } };
  behavior B(in int p1, out int p2)
    {int c1;  C c2;  B1 b1(p1, c2, c1);  B2 b2(c1, c2, p2);
    void main (void)
      {par {b1.main(); b2.main();}}
    };
```

Note that the interface protocol implemented in channel C, consisting of methods for read and write operations, can be changed without changing behaviors B1 and B2. For example, communication can be bit-serial or parallel and the choice does not affect the models of B1 and B2. This is a necessary feature for IP-reuse.

In order to simplify designs containing software and hardware components, the syntax of SpecC is based on C and C++. In fact, SpecC models are translated into C++ for simulation.

At the specification level, SpecC can model any kind of communication and typically uses message passing. The implementation of simulators is nevertheless typically based on a non-distributed system. The communication model of SpecC has inspired communication in SystemC 2.0.

2.8   Von-Neumann languages

The sequential execution of von-Neumann languages is their common characteristic. Also, such languages allow an almost unrestricted access to global variables. Model-based design using CFSMs and computational graphs is very appropriate for embedded system design. Nevertheless, the use of standard von-Neumann languages is still widespread. Therefore, we cannot ignore these languages.

However, the distinction between KPNs and properly restricted von-Neumann languages is blurring. For KPNs, we do also have sequential execution of the code for each of the nodes. We are still keeping the distinction between KPN and von-Neumann languages since for KPNs, the emphasis of modeling is on the communication and details of the execution within the nodes are irrelevant. For the first two languages covered in this section, communication is built into the languages. For the remaining languages, focus is on the computations and communication can be replaced by selecting different libraries.

2.8.1   CSP

CSP (*communicating sequential processes*) [Hoare, 1985] is one of the first languages comprising mechanisms for interprocess communication. Communication is based on channels.

Example:

process A	**process** B
.....	......
var a ..	**var** b ...
a := 3;	...
c!a; −− output to channel c	c?b; −− input from channel c
end;	**end**;

Both processes will wait for the other process to arrive at the input or output statement. This is a case of *rendez-vous*-based, **blocking** or **synchronous message passing**.

CSP is determinate, since it relies on the commitment to wait for input from a particular channel, like in Kahn process networks.

CSP has laid the foundation for the OCCAM language that was proposed as a programming language of the **transputer** [Thiébaut, 1995]. The focus on communication channels has been picked up again in the design of the XS1 processor [XMOS Ltd., 2010].

2.8.2   ADA

During the 1980s, the Department of Defense (DoD) in the United States realized that the dependability and maintainability of the software in its military equipment could soon become a major source of problems, unless some strict policy was enforced. It was decided that all software should be written in the same real-time language. Requirements for such a language were formulated.

No existing language met the requirements and, consequently, the design of a new one was started. The language which was finally accepted was based on PASCAL. It was called ADA (after Ada Lovelace, who can be considered being the first (female) programmer). ADA'95 [Kempe, 1995], [Burns and Wellings, 2001] is an object-oriented extension of the original standard.

One of the interesting features of ADA is the ability to have nested declarations of processes (called tasks in ADA). Tasks are started whenever control passes into the scope in which they are declared.

The following is an example (according to Burns et al. [Burns and Wellings, 1990]):

```
procedure example1 is

    task a;

    task b;

    task body a is

        -- local declarations for a

        begin

            -- statements for a

        end a;

    task body b is

        -- local declarations for b

        begin

            -- statements for b

        end b;

    begin

        -- Tasks a and b will start before the 1st statement of example1

        -- statements for example1

    end;
```

The communication concept of ADA is another key concept. It is based on the *rendez-vous* paradigm. Whenever two tasks want to exchange information, the task reaching the "meeting point" first has to wait until its partner has also reached a corresponding point of control. Syntactically, procedures are used for describing communication. Procedures which can be called from other tasks must be identified by the keyword **entry**.

Example [Burns and Wellings, 1990]:

```
task screen_out is

    entry call (val : character; x, y : integer);

end screen_out;
```

Task screen_out includes a procedure named call which can be called from other processes. Some other task can call this procedure by prefixing it with the name of the task:

```
screen_out.call('Z',10,20);
```

The calling task has to wait until the called task has reached a point of control, at which it accepts calls from other tasks. This point of control is indicated by the keyword **accept**:

```
task body screen_out is

    ...

    begin

        ...

        accept call (val : character; x, y : integer) do

        ...

        end call;

    ...

    end screen_out;
```

Obviously, task screen_out may be waiting for several calls at the same time. The ADA **select**-statement provides this capability.

Example:

```
task screen_output is
    entry call_ch(val:character; x, y: integer);
    entry call_int(z, x, y: integer);
end screen_out;
task body screen_output is

    ...

    select
      accept call_ch ...  do...
      end call_ch;
    or
      accept call_int ... do ..
      end call_int;
    end select;

    ...
```

In this case, task screen_out will be waiting until either call_ch or call_int are called.

Due to the presence of the **select**-statement, ADA is not determinate. ADA has been the preferred language for military equipment produced in the Western hemisphere for some time. Recently produced information about ADA is available from a web sites (see, for example [Kempe Software Capital Enterprises (KSCE), 2010]).

2.8.3 Java

For Java, communication can be selected by choosing between different libraries. Computation is strictly sequential.

Java was designed as a platform-independent language. It can be executed on any machine for which an interpreter of the internal byte-code representation of Java-programs is available. This byte-code representation is a very compact representation, which requires less memory space than a standard binary machine code representation. Obviously, this is a potential advantage in system-on-a-chip applications, where memory space is limited.

Also, Java was designed as a safe language. Many potentially dangerous features of C or C++ (like pointer arithmetic) are not available in Java. Hence, Java meets the safety requirements for specification languages for embedded systems. Java supports exception handling, simplifying recovery in case of run-time errors. There is no danger of memory leakages due to missing memory deallocation, since Java provides automatic garbage collection. This feature avoids potential problems in applications that must run for months or even years without ever being restarted. Java also meets the requirement to support concurrency since it includes threads (light-weight processes).

In addition, Java applications can be implemented quite fast, since Java supports object orientation and since Java development systems come with powerful libraries.

However, standard Java is not really designed for real-time and embedded systems. A number of characteristics which would make it a real-time and embedded programming language are missing:

- The size of Java run-time libraries has to be added to the size of the application itself. These run-time libraries can be quite large. Consequently, only really large applications benefit from the compact representation of the application itself.

- For many embedded applications, direct control over I/O devices is necessary (see page 26). For safety reasons, no direct control over I/O devices is available in standard Java.

- Automatic garbage collection requires some computing time. In standard Java, the instance in time at which automatic garbage collection is started cannot be predicted. Hence, the worst case execution time is very difficult to predict. Only extremely conservative estimates can be made.

- Java does not specify the order in which threads are executed if several threads are ready to run. As a result, worst-case execution time estimates must be even more conservative.

- Java programs are typically less efficient than C programs. Hence, Java is less recommended for resource constrained systems.

Proposals for solving the problems were made by Nilsen [Nilsen, 1998]. Proposals include hardware-supported garbage-collection, replacement of the run-time scheduler and tagging of some of the memory segments.

Currently (in 2010) relevant Java programming environments include the Java Enterprise Edition (J2EE), the Java Standard Edition (J2SE), the Java Micro Edition (J2ME), and CardJava [Sun, 2010]. CardJava is a stripped-down version of Java with emphasis on security for SmartCard applications. J2ME is the relevant Java environment for all other types of embedded systems. Two library profiles have been defined for J2ME: CDC and CLDC. CLDC is used for mobile phones, using the so-called MIDP 1.0/2.0 as its standard for the application programming interface (API). CDC is used, for example, for TV sets and powerful mobile phones. Currently relevant sources for Java real-time programming include book by Wellings [Wellings, 2004], Dibble [Dibble, 2008] and Bruno [Bruno and Bollella, 2009] as well as web sites [Java Community Process, 2002] and [Anonymous, 2010b].

2.8.4   Pearl and Chill

Pearl [Deutsches Institut für Normung, 1997] was designed for industrial control applications. It does include a large repertoire of language elements for controlling processes and referring to time. It requires an underlying real-time operating system. Pearl has been very popular in Europe and a large number of industrial control projects has been implemented in Pearl. Pearl supports semaphores which can be used to protect communication based on shared buffers.

Chill [Winkler, 2002] was designed for telephone exchange stations. It was standardized by the CCITT and used in telecommunication equipment. Chill is a kind of extended PASCAL.

2.8.5   Communication libraries

Standard von-Neumann languages do not come with built-in communication primitives. However, communication can be provided by libraries. There is a trend towards supporting communication within some local system as well as communication over longer distances. The use of internet protocols is becoming more popular. Libraries will be described in more detail in the section on system software (see page 195).

2.9   Levels of hardware modeling

In practice, designers start design cycles at various levels of abstraction. In some cases, these are high levels describing the overall behavior of the system to be designed. In other cases, the design process starts with the specification of electrical circuits at lower levels of abstraction. For each of the levels, a variety of languages exists, and some languages cover various levels. In the following, we will describe a set of possible levels. Some lower end levels are presented here for context reasons. Specifications should not start at those levels. The following is a list of frequently used names and attributes of levels:

- **System level models:** The term system level is not clearly defined. It is used here to denote the entire embedded system and the system into which information processing is embedded ("the product"), and possibly also the environment (the physical input to the system, reflecting e.g. the roads, weather conditions etc.). Obviously, such models include mechanical as well as information processing aspects and it may be difficult to find appropriate simulators. Possible solutions include VHDL-AMS (the analog extension to VHDL), SystemC or MATLAB. MATLAB and VHDL-AMS support modeling partial differential equations, which is a key requirement for modeling mechanical systems. It is a challenge to model information processing parts of the system in such a way that the simulation model can also be used for the synthesis of the embedded system. If this is not possible, error-prone manual translations between different models may be needed.

- **Algorithmic level:** At this level, we are simulating the algorithms that we intend to use within the embedded system. For example, we might be simulating MPEG video encoding algorithms in order to evaluate the resulting video quality. For such simulations, no reference is made to processors or instruction sets.

  Data types may still allow a higher precision than the final implementation. For example, MPEG standards use double precision floating point numbers. The final embedded system will hardly include such data types. If data types have been selected such that every bit corresponds to exactly one bit in the final implementation, the model is said to be **bit-true**. Translating non-bit-true into bit-true models should be done with tool support (see page 286).

  Models at this level may consist of single processes or of sets of cooperating processes.

- **Instruction set level:** In this case, algorithms have already been compiled for the instruction set of the processor(s) to be used. Simulations at this level allow counting the executed number of instructions. There are several variations of the instruction set level:

- In a coarse-grained model, only the effect of the instructions is simulated and their timing is not considered. The information available in assembly reference manuals (instruction set architecture (ISA)) is sufficient for defining such models.

- **Transaction level modeling:** In transaction level modeling (see also page 98), transactions, such as bus reads and writes, and communication between different components is modeled. Transaction level modeling includes less details than cycle-true modeling (see below), enabling significantly superior simulation speeds [Clouard et al., 2003].

- In a more fine-grained model, we might have **cycle-true instruction set simulation**. In this case, the exact number of clock cycles required to run an application can be computed. Defining cycle-true models requires a detailed knowledge about processor hardware in order to correctly model, for example, pipeline stalls, resource hazards and memory wait cycles.

- **Register-transfer level (RTL):** At this level, we model all the components at the register-transfer level, including arithmetic/logic units (ALUs), registers, memories, muxes and decoders. Models at this level are always cycle-true. Automatic synthesis from such models is not a major challenge.

- **Gate-level models:** In this case, models contain gates as the basic components. Gate-level models provide accurate information about signal transition probabilities and can therefore also be used for power estimations. Also delay calculations can be more precise than for the RTL. However, typically no information about the length of wires and hence no information about capacitances is available. Hence, delay and power consumption calculations are still estimates.

  The term "gate-level model" is sometimes also employed in situations in which gates are only used to denote Boolean functions. Gates in such a model do not necessarily represent physical gates; we are only considering the behavior of the gates, not the fact that they also represent physical components. More precisely, such models should be called "Boolean function models"[19], but this term is not frequently used.

- **Switch-level models:** Switch level models use switches (transistors) as their basic components. Switch level models use digital values models

---

[19] These models could be represented with binary decision diagrams (BDDs) [Wegener, 2000].

(refer to page 90 for a description of possible value sets). In contrast to gate-level models, switch level models are capable of reflecting bidirectional transfer of information.

- **Circuit-level models:** Circuit theory and its components (current and voltage sources, resistors, capacitances, inductances, and frequently possible macro-models of semiconductors) form the basis of simulations at this level. Simulations involve partial differential equations. These equations are linear if and only if the behavior of semiconductors is linearized (approximated). The most frequently used simulator at this level is SPICE [Vladimirescu, 1987] and its variants.

- **Layout models:** Layout models reflect the actual circuit layout. Such models include **geometric** information. Layout models cannot be simulated directly, since the geometric information does not directly provide information about the behavior. Behavior can be deduced by correlating the layout model with a behavioral description at a higher level or by extracting circuits from the layout, using knowledge about the representation of circuit components at the layout level. In a typical design flow, the length of wires and the corresponding capacitances are extracted from the layout and **back-annotated** to descriptions at higher levels. This way, more precision can be gained for delay and power estimations.

- **Process and device models:** At even lower levels, we can model fabrication processes. Using information from such models, we can compute parameters (gains, capacitances etc) for devices (transistors).

2.10 Comparison of models of computation

2.10.1 Criteria

Models of computation can be compared according to several criteria. For example, Stuijk [Stuijk, 2007] compares MoCs according to the following criteria:

- **Expressiveness** and **succinctness** indicate, which systems can be modeled and how compact they are.

- **Analyzability** relates to the availability of scheduling algorithms and the need for run-time support.

- The **implementation efficiency** is influenced by the required scheduling policy and the code size.

Fig. 2.74 classifies data flow models according to these criteria.

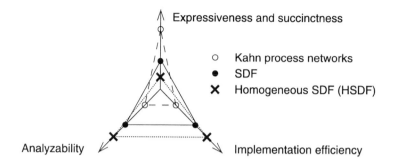

Figure 2.74. Comparison between data flow models

This figure reflects the fact that Kahn process networks are expressive: they are Turing-complete, meaning that any problem which can be computed on a Turing machine can also be computed in a KPN. Turing machines are used as the standard model of universal computers [Herken, 1995]. However, termination properties and upper bounds on buffer sizes of KPNs are difficult to analyze. SDF graphs, on the other hand, are not Turing-complete. The underlying reason is that they cannot model control flow. However, deadlock properties and upper bounds on buffer sizes of SDF graphs are easier to analyze. Homogeneous SDF (HSDF) graphs (graphs for which all rates are equal to one) are even less expressive, but also easier to analyze.

We could compare MoCs also with respect to the type of processes supported:

- The **number of processes** can be either **static** or **dynamic**. A static number of processes simplifies the implementation and is sufficient if each process models a piece of hardware and if we do not consider "hot-plugging" (dynamically changing the hardware architecture). Otherwise, dynamic process creation (and termination) should be supported.

- Processes can either be statically **nested** or all declared at the same level. For example, StateCharts allows nested process declarations while SDL (see page 54) does not. Nesting provides encapsulation of concerns.

- Different techniques for **process creation** exist. Process creation can result from an elaboration of the process declaration in the source code, through the fork and join mechanism (supported for example in Unix), and also through explicit process creation calls.

The expressiveness of different data flow oriented models of computation is also shown in fig. 2.75 [Basten, 2008]. MoCs not discussed in this book are indicated by dashed lines.

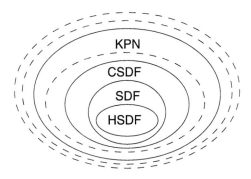

Figure 2.75. Expressiveness of data flow models

None of the MoCs and languages presented so far meets all the requirements for specification languages for embedded systems. Fig. 2.76 presents an overview over some of the key properties of some of the languages.

Language	Behavioral Hierarchy	Structural Hierarchy	Programming Language Elements	Exceptions Supported	Dynamic Process Creation
StateCharts	+	-	-	+	-
VHDL	+	+	+	-	-
SpecCharts	+	-	+	+	-
SDL	+-	+-	+-	-	+
Petri nets	-	-	-	-	+
Java	+	-	+	+	+
SpecC	+	+	+	+	+
SystemC	+	+	+	+	+
ADA	+	-	+	+	+

Figure 2.76. Language comparison

Interestingly, SpecC and SystemC meet all listed requirements. However, some other requirements (like a precise specification of deadlines, etc.) is not included. It is not very likely that a single MoC or language will ever meet all requirements, since some of the requirements are essentially conflicting. A language supporting hard real-time requirements well may be inconvenient to use for less strict real-time requirements. A language appropriate for distributed control-dominated applications may be poor for local data-flow dominated applications. Hence, we can expect that we will have to live with compromises and possibly with mixed models.

Which compromises are actually used in practice? In practice, assembly language programming was very common in the early years of embedded systems

programming. Programs were small enough to handle the complexity of problems in assembly languages. The next step is the use of C or derivatives of C.
Due to the ever increasing complexity of embedded system software (see page
xiv), higher level languages are to follow the introduction of C. Object oriented
languages and SDL are languages which provide the next level of abstraction.
Also, languages like UML are required to capture specifications at an early
design stage. In practice, these languages can be used like shown in fig. 2.77.

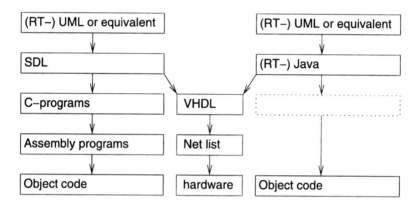

Figure 2.77.   Using various languages in combination

According to fig. 2.77, languages like SDL or StateCharts can be translated
into C. These C descriptions are then compiled. Starting with SDL or State-
Charts also opens the way to implementing the functionality in hardware, if
translators from these languages to VHDL are provided. Both C and VHDL
will certainly survive as intermediate languages for many years. Java does not
need intermediate steps but does also benefit from good translation concepts to
assembly languages. In a similar way, translations between various graphs are
feasible. For example, SDF graphs can be translated into a subclass of Petri
nets [Stuijk, 2007]. Also, they correspond to a subclass of the **computation
graph model** proposed by Karp and Miller [Karp and Miller, 1966]. Linking
the various models of computation is facilitated by formal techniques [Chen
et al., 2007].

Several languages for embedded system design are covered in a book edited
by M. Radetzki [Radetzki, 2009]. Popovici et al. [Popovici et al., 2010] use a
combination of Simulink and SystemC.

2.10.2   UML

UML^{TM} is a language including diagrams reflecting several MoCs. Fig. 2.78 classifies the UML diagrams mentioned so far with respect to our table of MoCs.

Communication/ Components	Shared memory	Message passing	
		synchronous	asynchronous
Undefined components		Use cases	
		Sequence charts, timing diagrams	
Finite state machines	State diagrams	-	-
Data flow	(not useful)	Data flow diagrams	
Petri nets	(not useful)	Activity charts	
Distributed event model	-	-	
Von-Neumann model	-	-	

Figure 2.78.   Models of computation available in UML

This figure shows how UML covers several models of computation, with a focus on early design phases. Semantics of communication is typically imprecisely defined. Therefore, our classification cannot be precise in this respect. In addition to the diagrams already mentioned, the following diagrams can be modeled:

- **Deployment diagrams**: These diagrams are important for embedded systems: they describe the "execution architecture" of systems (hardware or software nodes).

- **Package diagrams**: Package diagrams represent the partitioning of software into software packages. They are similar to module charts in State-Mate.

- **Class diagrams**: These diagrams describe inheritance relations of object classes.

- **Communication diagrams** (called **Collaboration diagrams** in UML 1.x): These graphs represent classes, relations between classes, and messages that are exchanged between them.

- **Component diagrams:** They represent the components used in applications or systems.

- **Object diagrams, interaction overview diagrams, composite structure diagrams**: This list consists of three types of diagrams which are less frequently used. Some of them may actually be special cases of other types of diagrams.

Available tools provide some consistency checking between the different diagram types. Complete checking, however, seems to be impossible. One reason for this is that the semantics of UML initially was left undefined. It has been argued that this was done intentionally, since one does not like to bother about the precise semantics during the early phases of the design. As a consequence, precise, executable specifications can only be obtained if UML is combined with some other, executable language. Available design tools have combined UML with SDL [IBM, 2009] and C++. There are, however, also some first attempts to define the semantics of UML.

Version 1.4 of UML was not designed for embedded systems. Therefore, it lacks a number of features required for modeling embedded systems (see page 21). In particular, the following features are missing [McLaughlin and Moore, 1998]:

- the partitioning of software into tasks and processes cannot be modeled,

- timing behavior cannot be described at all,

- the presence of essential hardware components cannot be described.

Due to the increasing amount of software in embedded systems, UML is gaining importance for embedded systems as well. Hence, several proposals for UML extensions to support real-time applications have been made [McLaughlin and Moore, 1998], [Douglass, 2000]. These extensions have been considered during the design of UML 2.0. UML 2.0 includes 13 diagram types (up from nine in UML 1.4) [Ambler, 2003]. Special profiles are taking the requirements of real-time systems into account [Martin and Müller, 2005], [Müller, 2007]. Profiles include class diagrams with constraints, icons, diagram symbols, and some (partial) semantics. There are UML profiles for [Müller, 2007]:

- Schedulability, Performance, and Time Specification (SPT) [Object Management Group (OMG), 2005b],

- Testing [Object Management Group (OMG), 2010a],

- Quality of Service (QoS) and Fault Tolerance [Object Management Group (OMG), 2010a],

- a Systems Modeling Language called SysML [Object Management Group (OMG), 2008],

- Modeling and Analysis of Real-Time Embedded Systems (MARTE), [Object Management Group (OMG), 2009]

- UML and SystemC interoperability [Riccobene et al., 2005],

- The SPRINT profile for reuse of intellectual property (IP) [Sprint Consortium, 2008].

Using such profiles, we can -for example- attach timing information to sequence charts. However, profiles may be incompatible. Also, UML has been designed for modeling and frequently leaves too many semantical issues open to allow automatic synthesis of implementations [Müller, 2007].

2.10.3 Ptolemy II

The Ptolemy project [Davis et al., 2001] focuses on modeling, simulation, and design of heterogeneous systems. Emphasis is on embedded systems that mix different technologies and, accordingly, also MoCs. For example, analog and digital electronics, hardware and software, and electrical and mechanical devices can be described. Ptolemy supports different types of applications, including signal processing, control applications, sequential decision making, and user interfaces. Special attention is paid to the generation of embedded software. The idea is to generate this software from the MoC which is most appropriate for a certain application. Version 2 of Ptolemy (Ptolemy II) supports the following MoCs and corresponding domains (see also page 33):

1 Communicating sequential processes (CSP)

2 Continuous time (CT): This model is appropriate for mechanical systems and analog circuits. It is supported through a set of extensible differential equation solvers.

3 Discrete event model (DE): this is the model used by many simulators, e.g. VHDL simulators.

4 Distributed discrete events (DDE). Discrete event systems are difficult to simulate in parallel, due to the inherent centralized queue of future events. Attempts to distribute this data structure have not been very successful so far. Therefore, this special (experimental) domain is introduced. Semantics can be defined such that distributed simulation becomes more efficient than in the DE model.

5 Finite state machines (FSM)

6 Process networks (PN), using Kahn process networks (see page 62).

7 Synchronous dataflow (SDF)

8 Synchronous/reactive (SR) MoC: This model uses discrete time, but signals do not need to have a value at every clock tick. Esterel (see page 53) is a language following this style of modeling.

This list clearly shows the focus on different models of computation in the Ptolemy project.

2.11   Assignments

1 Prepare a list of up to 6 requirements for specification languages for embedded systems!

2 Simulate trains between Paris, Brussels, Amsterdam and Cologne, using the levi simulation software [Sirocic and Marwedel, 2007d]! Modify the examples included with the software such that two independent tracks exist between any two stations and demonstrate an (arbitrary) schedule involving 10 trains!

3 Suppose the StateCharts in fig. 2.79 model is given.

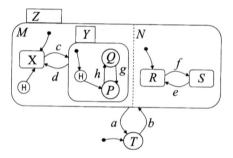

Figure 2.79.   StateCharts example

Also, suppose that we have the following sequence of input events: *b c f h g h e a b c*. In the diagram in fig. 2.80, mark all the states the StateCharts model will be in after a particular input has been applied! Note that H denotes the history mechanism.

4 Are StateCharts determinate models if we follow the StateMate semantics? Please explain your answer!

5 Which three types of Petri nets did we discuss in this book?

6 One of the types of Petri nets allows several non-distinguishable tokens per place. Which components are used in a mathematical model of such nets? Hint: $N=(P, \dots \dots)$

7 How does a compact model of the dining philosopher's problem look like?

8 CSA theory leads to 2, 3 and 4 logic strengths, corresponding to 4, 7 and 10 logic values. How many strengths and values are we using in IEEE 1164?

	M	N	P	Q	R	S	T	X	Y	Z
(Reset)							v			
b										
c										
f										
h										
g										
h										
e										
a										
b										
c										

Figure 2.80. States of the StateCharts example

Please show the partial order among the values of IEEE 1164 in a diagram! Which of the values of IEEE 1164 are not included in the partial order and what is the meaning of these values?

9 Which of the following circuits can be modeled with IEEE 1164: complementary CMOS outputs, outputs with a depletion transistor, open collector outputs, tristate outputs, pre-charging on buses (if depletion transistors are used as well)?

10 Suppose that a bus as shown in fig. 2.81 is given. Rectangles containing an &-sign denote AND-gates.

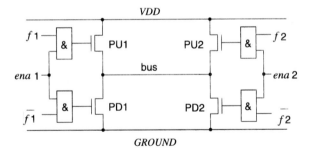

Figure 2.81. Bus driven by tri-state outputs

Which of the IEEE 1164 values will be on the bus if both enable inputs are set to '0' (*ena*1 = *ena*2 = '0')? Which of the IEEE 1164 values will be on the bus if *ena*1 = '0', *ena*2 = '1' and $f2$ = '1'?

11 Simulate a Kahn process network computing Fibonacci numbers, using the levi simulation software [Sirocic and Marwedel, 2007b].

12 Which of the following languages are using asynchronous message passing communication: StateCharts, SDL, VHDL, CSP, Petri nets?

13 Which of the following languages use a broadcast mechanism for updating variables: StateCharts, SDL, Petri nets?

14 Which of the following diagram types are supported by UML: Sequence charts, record charts, Y-charts, use cases, activity diagrams, circuit diagrams?

Chapter 3

EMBEDDED SYSTEM HARDWARE

3.1 Introduction

It is one of the characteristics of embedded and cyber-physical systems that both hardware and software must be taken into account. The reuse of available hard- and software components is at the heart of the **platform-based design methodology** (see also page 236). Consistent with the need to consider available hardware components and with the design information flow shown in fig. 3.1, we are now going to describe some of the essentials of embedded system hardware.

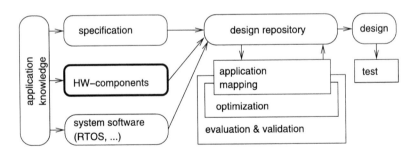

Figure 3.1. Simplified design flow

Hardware for embedded systems is much less standardized than hardware for personal computers. Due to the huge variety of embedded system hardware, it is impossible to provide a comprehensive overview of all types of hardware components. Nevertheless, we will try to provide a survey of some of the essential components which can be found in most systems.

P. Marwedel, *Embedded System Design*, Embedded Systems,
DOI 10.1007/978-94-007-0257-8_3, © Springer Science+Business Media B.V. 2011

In many of the cyber-physical systems, especially in control systems, hardware is used in a loop (see fig. 3.2).

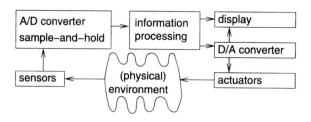

Figure 3.2.   Hardware in the loop

In this loop, information about the physical environment is made available through **sensors**. Typically, sensors generate continuous sequences of analog values. In this book, we will restrict ourselves to information processing where digital computers process discrete sequences of values. Appropriate conversions are performed by two kinds of circuits: sample-and-hold-circuits and analog-to-digital (A/D) converters. After such conversion, information can be processed digitally. Generated results can be displayed and also be used to control the physical environment through actuators. Since most actuators are analog actuators, conversion from digital to analog signals is also needed.

This model is obviously appropriate for control applications. For other applications, it can be employed as a first order approximation. In the following, we will describe essential hardware components of cyber-physical systems following the loop structure of fig. 3.2.

3.2   Input

3.2.1   Sensors

We start with a brief discussion of sensors. Sensors can be designed for virtually every physical quantity. There are sensors for weight, velocity, acceleration, electrical current, voltage, temperature, etc. A wide variety of physical effects can be exploited in the construction of sensors [Elsevier B.V., 2010a]. Examples include the law of induction (generation of voltages in an electric field), and photoelectric effects. There are also sensors for chemical substances [Elsevier B.V., 2010b].

Recent years have seen the design of a huge range of sensors, and much of the progress in designing smart systems can be attributed to modern sensor technology. It is impossible to cover this subset of cyber-physical hardware technology comprehensively and we can only give characteristic examples:

- **Acceleration sensors:** Fig. 3.3 shows a small sensor manufactured using microsystem technology. The sensor contains a small mass in its center. When accelerated, the mass will be displaced from its standard position, thereby changing the resistance of the tiny wires connected to the mass.

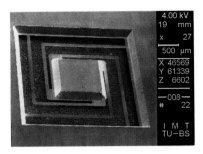

Figure 3.3.   Acceleration sensor (courtesy S. Bütgenbach, IMT, TU Braunschweig), ©TU Braunschweig, Germany

- **Rain sensors:** In order to remove distraction from drivers, some cars contain rain sensors. Using these, the speed of the wipers can be automatically adjusted to the amount of rain.

- **Image sensors:** There are essentially two kinds of image sensors: charge-coupled devices (CCDs) and CMOS sensors. In both cases, arrays of light sensors are used. The architecture of CMOS sensor arrays is similar to that of standard memories: individual pixels can be randomly addressed and read out. CMOS sensors use standard CMOS technology for integrated circuits [Dierickx, 2000]. Due to this, sensors and logic circuits can be integrated on the same chip. This allows some preprocessing to be done already on the sensor chip, leading to so-called smart sensors. CMOS sensors require only a single standard supply voltage and interfacing in general is easy. Therefore, CMOS-based sensors can be cheap.

  In contrast, CCD technology is optimized for optical applications. In CCD technology, charges must be transferred from one pixel to the next until they can finally be read out at an array boundary. This sequential charge transfer also gave CCDs their name. For CCD sensors, interfacing is more complex.

  Selecting the most appropriate image sensor is not so obvious. The image quality of CMOS sensors has been significantly improved over the recent years. Therefore, achieving a good image quality is feasible with CCD and with CMOS sensors. However, CMOS sensors are in general less power efficient than CCD sensors. Hence, if a very small power consumption is a target, CCD sensors are preferred. If minimum cost is an issue, CMOS sensors are preferred. Also, CMOS sensors are preferred if smart sensors are to be designed. Due to their smaller power consumption, compact cameras with live view displays typically use CMOS sensors [Belbachir, 2010]. For other cameras, the situation is less clear.

- **Biometric sensors:** Demands for higher security standards as well as the need to protect mobile and removable equipment have led to an increased interest in authentication. Due to the limitations of password based security (e.g. stolen and lost passwords), smartcards, biometric sensors and biomedical authentication receive significant attention. Biometric authentication tries to identify whether or not a certain person is actually the person she or he claims to be. Methods for biometric authentication include iris scans, finger print sensors and face recognition. Finger print sensors are typically fabricated using the same CMOS technology [Weste et al., 2000] which is used for manufacturing integrated circuits. Possible applications include notebooks which grant access only if the user's finger print is recognized [IBM, 2002]. CCD and CMOS image sensors described above are used for face recognition. False accepts as well as false rejects are an inherent problem of biometric authentication. In contrast to password based authentication, exact matches are not possible.

- **Artificial eyes**: Artificial eye projects have received significant attention. While some projects attempt to actually affect the eye, others try to provide vision in an indirect way.

  For example, the Dobelle Institute experimented with a setup in which a little camera was attached to glasses. This camera was connected to a computer translating these patterns into electrical pulses. These pulses were then sent directly to the brain, using a direct contact through an electrode. The resolution was in the order of 128 by 128 pixels, enabling blind persons to drive a car in controlled areas [The Dobelle Institute, 2003].

  More recently, the translation of images into audio has been preferred. Obviously, it is less invasive.

- **Radio frequency identification (RFID)**: RFID technology is based on the response of a **tag** to radio frequency signals [Hunt et al., 2007]. The tag consists of an integrated circuit and an antenna. The tag provides its identification to **RFID readers**. The maximum distance between tags and readers depends on the type of the tag. The technology can be applied wherever objects, animals or people should be identified.

- **Other sensors:** Other common sensors include: pressure sensors, proximity sensors, engine control sensors, Hall effect sensors, and many more.

Sensors are generating **signals**. Mathematically, the following definition applies:

Definition: A **signal** σ is a mapping from the time domain D_T to a value domain D_V:

$$\sigma : D_T \rightarrow D_V$$

Signals may be defined over a continuous or a discrete time domain as well as over a continuous or a discrete value domain.

3.2.2 Discretization of time: Sample-and-hold circuits

All known digital computers work in a **discrete** time domain D_T. This means that they can process discrete sequences or **streams** of values. Hence, incoming signals over the continuous time domain must be converted to signals over the discrete time domain. This is the purpose of **sample-and-hold circuits**. Fig. 3.4 (left) shows a simple sample-and-hold circuit.

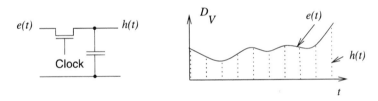

Figure 3.4. Sample-and-hold-circuit

In essence, the circuit consists of a clocked transistor and a capacitor. The transistor operates like a switch. Each time the switch is closed by the clock signal, the capacitor is charged so that its voltage $h(t)$ is practically the same as the incoming voltage $e(t)$. After opening the switch again, this voltage will remain essentially unchanged until the switch is closed again. Each of the values stored on the capacitor can be considered as an element of a discrete sequence of values $h(t)$, generated from a continuous function $e(t)$ (see fig. 3.4, right). If we sample $e(t)$ at times $\{t_s\}$, then $h(t)$ will be defined only at those times.

An ideal sample-and-hold circuit would be able to change the voltage at the capacitor in an arbitrarily short amount of time. This way, the input voltage at a particular instance in time could be transfered to the capacitor and each element in the discrete sequence would correspond to the input voltage at a particular point in time. In practice, however, the transistor has to be kept closed for a short time window in order to really charge or discharge the capacitor. The voltage stored on the capacitor will then correspond to a voltage reflecting that short time window.

An interesting question is this one: would we be able to reconstruct the original signal $e(t)$ from the sampled signal $h(t)$? At this time, we revert to the fact that arbitrary signals can be approximated by summing (possibly phase-shifted) sine functions of different frequencies (Fourier approximation)[1]. For example, fig. 3.5 and fig. 3.6 demonstrate how even a square wave can be approximated by sine waves of increasing frequencies.

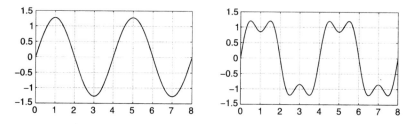

Figure 3.5. Approximation of a square wave by sine waves for $K=1$ (left) and $K=3$ (right)

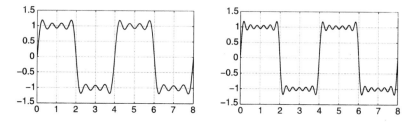

Figure 3.6. Approximation of a square wave by sine waves for $K=7$ (left) and $K=11$ (right)

These graphs display a graphical representation of equation 3.1 [Oppenheim et al., 2009], where p is the period:

$$e'_K(t) \quad = \quad \sum_{k=1,3,5,7,9,...}^{K} \left(\frac{4}{\pi k} \sin\left(\frac{2\pi k t}{p}\right) \right) \tag{3.1}$$

---

[1] The presentation in this book is based on the assumption that a full presentation of the theory of Fourier approximations cannot be included in a course on embedded systems. Therefore, only the impact of this theory is demonstrated by using examples. Students would benefit from knowing the theory behind these examples.

A data processing transformation Tr is said to be **linear**, if for signals $e_1(t)$ and $e_2(t)$ we have:

$$Tr(e_1 + e_2) = Tr(e_1) + Tr(e_2) \tag{3.2}$$

In the following, we restrict ourselves to linear systems. Then, in order to answer the question raised above, we study the effect of sampling on each of the sine waves independently.

Suppose that our input signal corresponds to either of the two functions e_3 or e_4:

$$e_3(t) = \sin(\frac{2\pi t}{8}) + 0.5\sin(\frac{2\pi t}{4}) \tag{3.3}$$

$$e_4(t) = \sin(\frac{2\pi t}{8}) + 0.5\sin(\frac{2\pi t}{4}) + 0.5\sin(\frac{2\pi t}{1}) \tag{3.4}$$

The sine waves used in these functions have periods of $p = 8, 4$, and 1, respectively (this can be seen easily by comparing these sine waves with those used in eq. 3.1). A graphical representation of these functions is shown in fig. 3.7.

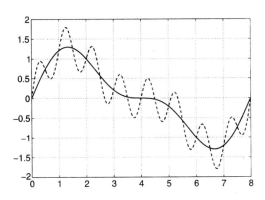

Figure 3.7. Visualization of functions $e_3(t)$ (solid) and $e_4(t)$ (dotted)

Suppose that we will be sampling these signals at integer times. It then so happens that both signals have the same value whenever they are sampled. Obviously, it is not possible to distinguish between $e_3(t)$ and $e_4(t)$ if we sample at these instances in time as shown and if only the sampled signal is available.

In general, sampled signals will not allow us to distinguish between some slow signal $e_3(t)$ and some other faster varying signal $e_4(t)$ if $e_3(t)$ and $e_4(t)$ are identical each time we are sampling the signals. The fact that two or more unsampled signals can have the same sampled representation is called **aliasing**. We are not sampling $e_4(t)$ frequently enough to notice, for example, that it has slope changes between integer times. So, from this counterexample we can conclude that **reconstruction of the original unsampled signal is not feasible unless we have additional knowledge about the frequencies or the waveforms present in the input signal**.

How frequently do we have to sample signals to be able to distinguish between different sine waves?

Let us assume that we are sampling the input signal at constant time intervals, such that p_s is the **sampling period**[2]:

$$\forall s \quad : \quad p_s = t_{s+1} - t_s \tag{3.5}$$

Let

$$f_s \;\; = \;\; \frac{1}{p_s} \tag{3.6}$$

be the sampling rate or sampling frequency.

According to the theory of sampling [Oppenheim et al., 2009], **aliasing is avoided if we restrict the frequencies of the incoming signal to less than half of the sampling frequency** f_s:

$$p_s \;\; < \;\; \frac{p_N}{2} \text{ where } p_N \text{ is the period of the "fastest" sine wave, or} \tag{3.7}$$

$$f_s \;\; > \;\; 2 f_N \text{ where } f_N \text{ is the frequency of the "fastest" sine wave} \tag{3.8}$$

Definition: f_N is called the Nyquist frequency, f_s is the sampling rate.

The condition in equation 3.8 is called **sampling criterion**, and sometimes the **Nyquist sampling criterion**.

Therefore, reconstruction of input signals $e(t)$ from discrete samples $h(t)$ can be successful only if we make sure that higher frequency components such

---

[2] In order to be consistent with the notation in scheduling theory, we denote the period by p_s instead of by T_s. The latter notation is frequently used in digital signal processing.

as the one in $e_4(t)$ are removed. This is the purpose of anti-aliasing filters. Anti-aliasing filters are placed in front of the sample-and-hold circuit (see fig. 3.8).

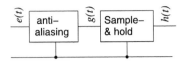

Figure 3.8.   Anti-aliasing placed in front of the sample-and-hold circuit

Fig. 3.9 demonstrates the ratio between the amplitudes of the output and the input waves as a function of the frequency for this filter.

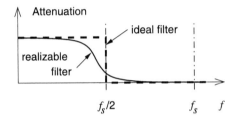

Figure 3.9.   Ideal and realizable anti-aliasing filters (low-pass filters)

Ideally, such a filter would remove all frequencies at and above half the sampling frequency and keep all other components unchanged. This way, it would convert signal $e_4(t)$ into signal $e_3(t)$. In practice, such ideal filters do not exist. Realizable filters will already start attenuating frequencies smaller than $f_s/2$ and will still not eliminate all frequencies larger than $f_s/2$ (see fig. 3.9). Attenuated high frequency components will exist even after filtering. For frequencies smaller than $f_s/2$, there may also be some "overshooting", i.e. frequencies for which there is some amplification of the input signal. The design of good anti-aliasing filters is an art by itself.

3.2.3   Discretization of values: A/D-converters

Since we are restricting ourselves to digital computers, we must also replace signals that map time to a continuous value domain D_V by signals that map time to a discrete value domain D'_V. This conversion from analog to digital values is done by analog-to-digital (A/D) converters. There is a large range of A/D converters with varying speed/precision characteristics. In this book, we will present two extreme cases:

- **Flash A/D converter**: This type of A/D converters uses a large number of comparators. Each comparator has two inputs, denoted as + and -. If the voltage at input + exceeds that at input -, the output corresponds to a logical '1' and it corresponds to a logical '0' otherwise[3].

In the A/D-converter, all - inputs are connected to a voltage divider. If input voltage $h(t)$ exceeds $\frac{3}{4}V_{ref}$, the comparator at the top of fig. 3.10 (a) will generate a '1'. The encoder at the output of the comparators will try to identify the most significant '1' and will encode this case as the largest output value. The case $h(t) > V_{ref}$ should normally be avoided since V_{ref} is typically close to the supply voltage of the circuit and input voltages exceeding the supply voltage can lead to electrical problems. In our case, input voltages larger than V_{ref} generate the largest digital value as long as the converter does not fail due to the high input voltage.

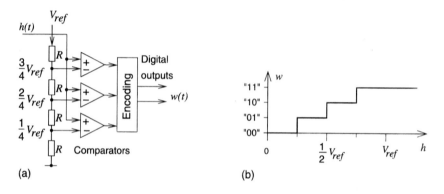

Figure 3.10. (a) Flash A/D converter   (b) w as a function of h

Now if input voltage $h(t)$ is less than $\frac{3}{4}V_{ref}$, but still larger than $\frac{2}{4}V_{ref}$, the comparator at the top of fig. 3.10 will generate a '0', while the next comparator will still signal a '1'. The encoder will encode this as the second-largest value.

Similar arguments hold for cases $\frac{1}{4}V_{ref} < h(t) < \frac{2}{4}V_{ref}$, and $0 < h(t) < \frac{1}{4}V_{ref}$, which will be encoded as the third-largest and the smallest value, respectively. Fig. 3.10 (b) shows the relation between input voltages and generated digital values.

The outputs of the comparators encode numbers in a special way: if a certain comparator output is equal to '1', then all the less significant outputs

---

[3]In practice, the case of equal voltages is not relevant, as the actual behavior for very small differences between the voltages at the two inputs depends on many factors (like temperatures, manufacturing processes etc.) anyway.

are all equal to '1'. The encoder transforms this representation of numbers into the usual representation of natural numbers. The encoder is actually a so-called "priority encoder", encoding the most significant input number carrying a '1' in binary [4].

The circuit can convert positive analog input voltages into digital values. Converting both positive and negative voltages and generating two's complement numbers requires some extensions.

A/D-converters are characterized by their **resolution**. This term has several different but related meanings [Analog Devices Inc. Eng., 2004]. The resolution (measured in bits) is the number of bits produced by an A/D-converter. For example, A/D-converters with a resolution of 16 bits are needed for many audio applications. However, the resolution is also measured in volts, and in this case it denotes the difference between two input voltages causing the output to be incremented by 1:

$$Q = \frac{V_{FSR}}{n}$$

Where:

V_{FSR} : is the difference between the largest and the smallest voltage,
Q : is the resolution in volts per step, and
n : is the number of voltage intervals (**not** the number of bits).

Example: For the A/D-converter of fig. 3.10, the resolution is 2 bits or $\frac{1}{4}V_{ref}$ volts, if we assume V_{ref} as the largest voltage.

The key advantage of the flash A/D-converter is its speed. It does not need any clock. The delay between the input and the output is very small and the circuit can be used easily, for example, for high-speed video applications. The disadvantage is its hardware complexity: we need $n-1$ comparators in order to distinguish between n values. Imagine using this circuit in generating digital audio signals for CD recorders. We would need $2^{16}-1$ comparators! High-resolution A/D-converters must be built in a different way.

- **Successive approximation:** Distinguishing between a large number of digital values is possible with A/D converters using successive approximation. The circuit is shown in fig. 3.11.

---

[4] Such encoders are also useful for finding the most significant '1' in the mantissa of floating point numbers.

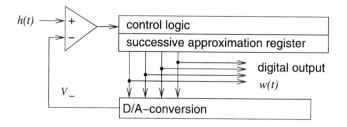

Figure 3.11.   Circuit using successive approximation

The key idea of this circuit is to use binary search. Initially, the most significant output bit of the successive approximation register is set to '1', all other bits are set to '0'. This digital value is then converted to an analog value, corresponding to $0.5\times$ the maximum input voltage[5]. If $h(t)$ exceeds the generated analog value, the most significant bit is kept at '1', otherwise it is reset to '0'.

This process is repeated with the next bit. It will remain set to '1' if the input value is either within the second or the fourth quarter of the input value range. The same procedure is repeated for all the other bits. Fig. 3.12 shows an example.

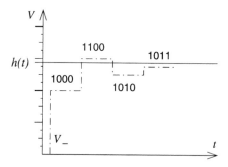

Figure 3.12.   Steps used during successive approximation

In fig. 3.12, initially the most significant bit is set to '1'. This value is kept, since the resulting V_- is less than $h(t)$. Then, the second-most significant bit is set to '1'. It is reset to '0', since the resulting V_- is exceeding $h(t)$. Next, the third-most significant bit is tried, and so on. Obviously, $h(t)$ must be constant during the conversion. This requirement is met if we employ

---

[5]Fortunately, the conversion from digital to analog values (D/A-conversion) can be implemented very efficiently and can be very fast (see page 164).

a sample-and-hold circuit as shown above. The resulting digital signal is called $w(t)$.

The key advantage of the successive approximation technique is its hardware efficiency. In order to distinguish between n digital values, we need $log_2(n)$ bits in the successive approximation register and the D/A converter. The disadvantage is its speed, since it needs $O(log_2(n))$ steps. These converters can therefore be used for high-resolution applications, where moderate speeds are required. Examples include audio applications.

Fig. 3.13 highlights the behavior an A/D-converter when the input signal is that of equation 3.3. Only the behavior for a positive input signal is shown.

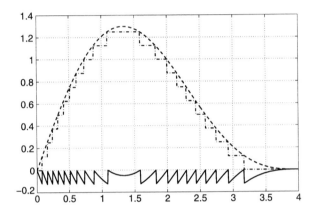

Figure 3.13. $h(t)$ (dashed), step function $w(t)$ (dash-dotted), $w(t) - h(t)$ (solid)

The figure includes the voltage corresponding to the digital value, the original voltage, and the difference between the two. Obviously, the converter is "truncating" the digital representation of the analog signal to the number of available bits (i.e. the digital value is always less than or equal to the analog value). This is a consequence of the way in which we are doing comparisons. "Rounding" converters would need an internal correction by "half a bit".

Effectively, the digital signal encodes values corresponding to the sum of the original analog values and the difference $w(t) - h(t)$. This means, it appears **as if the difference between the two signals had been added to the original signal**. This difference is a signal called **quantization noise**:

$$\text{quantization noise}(t) \quad = \quad w(t) - h(t) \tag{3.9}$$

$$|\text{quantization noise}(t)| \; < \; Q \qquad\qquad (3.10)$$

Obviously, it is possible to decrease quantization noise by increasing the resolution (in bits) of the A/D-converters. The impact of quantization noise is frequently captured in the definition of the signal-to-noise ratio (SNR). The SNR is measured in decibels (tens of a Bel, named after Alexander G. Bell):

$$\text{SNR (in decibels)} \; = \; 10 \cdot log \frac{\text{power of the ``useful'' signal}}{\text{power of the noise signal}} \qquad (3.11)$$

$$= \; 20 \cdot log \frac{\text{voltage of the ``useful'' signal}}{\text{voltage of the noise signal}} \qquad (3.12)$$

In this case, we have used the fact that, for any given impedance R, the power of a signal is equal to the square of the voltage. Decibels are no physical units, since the signal-to-noise ratio is dimension-less.

For any signal $h(t)$, the power of the quantization noise is equal to $\alpha \cdot Q$, where $\alpha \leq 1$ depends on the waveform of $h(t)$. If $h(t)$ can always be represented exactly by a digital value, $\alpha = 0$. If $h(t)$ is always "just a little" below the next value that can be represented, α may be close to 1.

For example (for $\alpha \sim 1$), the SNR of 16-bit CD audio is in the order of:

$$20 \cdot log(2^{16}) = 96 \text{ decibels}(dB)$$

For high-quality 24-bit CDs we would obtain an SNR of about 144 dB. Values of $\alpha < 1$ and imperfections of A/D-converters may change these numbers a bit.

There are several other types of A/D-converters. They differ by their speed and their precision [O'Neill, 2006]. Techniques for automatically selecting the most appropriate converter exist [Vogels and Gielen, 2003].

3.3   Processing Units

3.3.1   Overview

Currently available embedded systems require electrical energy to operate. The amount of electrical energy used is frequently called "consumed energy". Strictly speaking, this term is not correct, since this electrical energy is converted to other forms of energy, typically thermal energy. For embedded systems, energy availability is a deciding factor. This was already observed in a Dutch road mapping effort: "*Power is considered as the most important constraint in embedded systems*" [Eggermont, 2002]. The importance of power and energy efficiency was initially recognized for embedded systems. The focus on these objectives was later taken up for general purpose computing as well and led to initiatives such as the **green computing initiative**. For information processing in embedded systems, we will consider ASICs (application-specific integrated circuits) using hardwired multiplexed designs, reconfigurable logic, and programmable processors. These three technologies are quite different as far as their energy efficiency is concerned. Fig. 3.14 repeats the information already provided on page 6.

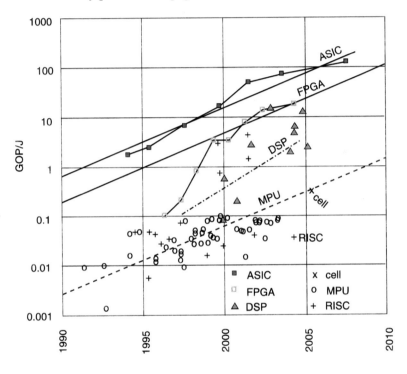

Figure 3.14. Hardware efficiency (©De Man and Philips)

Fig. 3.14 reflects the efficiency/flexibility conflict of currently available hardware technologies: if we want to aim at very power- and energy-efficient designs, we should use ASICs instead of flexible designs based on processors or re-programmable logic. If we go for excellent flexibility, we cannot be power-efficient.

The energy E for a certain application is closely related to the power P required per operation, since

$$E = \int P dt \tag{3.13}$$

Let us assume that we start with some design having a power consumption of $P_0(t)$, leading to an energy consumption of

$$E_0 = \int_0^{t_0} P_0(t)dt$$

after t_0 units of execution time. Suppose that a modified design finishing computations already at time t_1 comes with a power consumption of $P_1(t)$ and an energy consumption of

$$E_1 = \int_0^{t_1} P_1(t)dt$$

If $P_1(t)$ is not too much larger than $P_0(t)$, then a reduction of the execution time also reduces the energy consumption. However, in general this is not necessarily always true. The situation is also shown in fig. 3.15: E_1 may be smaller than E_0, but E_1 can also be larger than E_0.

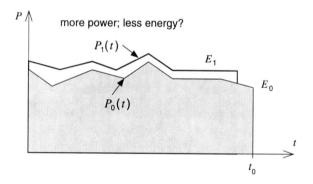

Figure 3.15.   Comparison of energies E_0 and E_1

Minimization of power and energy consumption are both important. Power consumption has an effect on the size of the power supply, the design of the voltage regulators, the dimensioning of the interconnect, and short term cooling. Minimizing the energy consumption is required especially for mobile applications, since battery technology is only slowly improving [ITRS Organization, 2009], and since the cost of energy may be quite high. Also, a reduced energy consumption decreases cooling requirements and improves the reliability (since the lifetime of electronic circuits decreases for high temperatures).

We will consider ASICs first.

3.3.2 Application-Specific Circuits (ASICs)

For high-performance applications and for large markets, application-specific integrated circuits (ASICs) can be designed. However, the cost of designing and manufacturing such chips is quite high. For example, the cost of the mask set which is used for transferring geometrical patterns onto the chip can amount to about 10^5-10^6 Euros or dollars. In fact, the cost for mask sets has grown exponentially over the recent years. Also, this approach suffers from long design times and the lack of flexibility: correcting design errors typically requires a new mask set and a new fabrication run. Therefore, ASICs are appropriate only if either maximum energy efficiency is needed and if the market accepts the costs or if a large number of such systems can be sold. Consequently, the design of ASICs is not covered in this book.

3.3.3 Processors

The key advantage of processors is their flexibility. With processors, the overall behavior of embedded systems can be changed by just changing the software running on those processors. Changes of the behavior may be required in order to correct design errors, to update the system to a new or changed standard or in order to add features to the previous system. Because of this, processors have become very popular.

Embedded processors must be efficient and they do not need to be instruction set compatible with commonly used personal computers (PCs). Therefore, their architectures may be different from those processors found in PCs. Efficiency has a number of different aspects (see page 5):

- **Energy-efficiency**: Architectures must be optimized for their energy-efficiency and we must make sure that we are not losing efficiency in the software generation process. For example, compilers generating 50% overhead in terms of the number of cycles will take us further away from the efficiency of ASICs, possibly by even more than 50%, if the supply voltage and the clock frequency must be increased in order to meet timing deadlines.

  There is a large amount of techniques available that can make processors energy efficient and energy efficiency should be considered at various levels of abstraction, from the design of the instruction set down to the design of the chip manufacturing process [Burd and Brodersen, 2003]. Gated clocking is an example of such a technique. With gated clocking, parts of the processor are disconnected from the clock during idle periods. For example, no clock is applied to direct memory access (DMA) hardware or bus bridges if they are not needed. Also, there are attempts, to get rid of the clock for major parts of the processor altogether. There are two contrasting approaches: globally synchronous, locally asynchronous processors and globally asynchronous, locally synchronous processors (GALS) [Iyer and Marculescu, 2002]. Further information about low power design techniques is available in a book by E. Macii [Macii, 2004] and in the PATMOS proceedings (see [Monteiro and van Leuken, 2010] for a recent issue).

Two techniques can be applied at a rather high level of abstraction:

– **Dynamic power management (DPM):** With this approach, processors have several power saving states in addition to the standard operating state. Each power saving state has a different power consumption and a different time for transitions into the operating state. Fig. 3.16 shows the three states for the StrongArm SA 1100 processor.

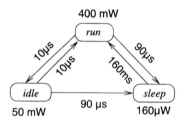

Figure 3.16. Dynamic power management states of the StrongArm Processor SA 1100 [Benini et al., 2000]

The processor is fully operational in the *run* state. In the *idle* state, it is just monitoring the interrupt inputs. In the *sleep* state, on-chip activity is shutdown, the processor is reset and the chip's power supply is shut-off [Wolf, 2001]. A separate I/O-power supply provides power to power manager hardware. The processor can be restarted by the power manager hardware by a preprogrammed wake-up event. Note the large difference in the power consumption between the *sleep* state and the other states, and note also the large delay for transitions from the *sleep* to the *run* state.

– **Dynamic voltage scaling (DVS):** This approach exploits the fact that the energy consumption of CMOS processors increases quadratically with the supply voltage V_{dd}. The power consumption P of CMOS circuits is given by [Chandrakasan et al., 1992]:

$$P = \alpha\, C_L\, V_{dd}^2\, f \tag{3.14}$$

where α is the switching activity, C_L is the load capacitance, V_{dd} is the supply voltage and f is the clock frequency. The delay of

CMOS circuits can be approximated as [Chandrakasan et al., 1992], [Chandrakasan et al., 1995]:

$$\tau \;=\; k \cdot C_L \cdot \frac{V_{dd}}{(V_{dd} - V_t)^2} \qquad (3.15)$$

where k is a constant, and V_t is the threshold voltage. V_t has an impact on the transistor input voltage required to switch the transistor on. For example, for a maximum supply voltage of $V_{dd,max}$=3.3 volts, V_t may be in the order of 0.8 volts. Consequently, the maximum clock frequency is a function of the supply voltage. However, decreasing the supply voltage reduces the power quadratically, while the run-time of algorithms is only linearly increased (ignoring the effects of the memory system). This can be exploited in a technique called **dynamic voltage scaling (DVS)**. For example, the Crusoe^{TM} processor by Transmeta [Klaiber, 2000] provided 32 voltage levels between 1.1 and 1.6 volts, and the clock could be varied between 200 MHz and 700 MHz in increments of 33 MHz. Transitions from one voltage/frequency pair to the next took about 20 ms. Design issues for DVS-capable processors are described in a paper by Burd and Brodersen [Burd and Brodersen, 2000]. According to the same paper, potential power savings will exist even for future technologies with a decreased maximum V_{dd}, since the threshold voltages will also be decreased (unfortunately, this will lead to increased leakage currents, increasing the standby power consumption). In 2004, six different speed/voltage pairs were provided with the Intel® SpeedStep^{TM} technology for the Pentium® M [Intel, 2004].

- **Code-size efficiency**: Minimizing the code size is very important for embedded systems, since hard disk drives are typically not available and since the capacity of memory is typically also very limited[6]. This is even more pronounced for **systems on a chip** (SoCs). For SoCs, the memory and processors are implemented on the same chip. In this particular case, memory is called **embedded memory**. Embedded memory may be more expensive to fabricate than separate memory chips, since the fabrication processes for memories and processors must be compatible. Nevertheless, a large percentage of the total chip area may be consumed by the memory. There are several techniques for improving the code-size efficiency:

  - **CISC machines**: Standard RISC processors have been designed for speed, not for code-size efficiency. Earlier Complex Instruction Set

---

[6]The availability of large flash memories makes memory size constraints less tight.

Processors (CISC machines) were actually designed for code-size efficiency, since they had to be connected to slow memories. Caches were not frequently used. Therefore, "old-fashioned" CISC processors are finding applications in embedded systems. ColdFire processors [Freescale semiconductor, 2005], which are based on the Motorola 68000 family of CISC processors are an example.

– **Compression techniques**: In order to reduce the amount of silicon needed for storing instructions as well as in order to reduce the energy needed for fetching these instructions, instructions are frequently stored in the memory in compressed form. This reduces both the area as well as the energy necessary for fetching instructions. Due to the reduced bandwidth requirements, fetching can also be faster. A (hopefully small and fast) decoder is placed between the processor and the (instruction) memory in order to generate the original instructions on the fly (see fig. 3.17, right)[7]. Instead of using a potentially large memory of uncompressed instructions, we are storing the instructions in a compressed format.

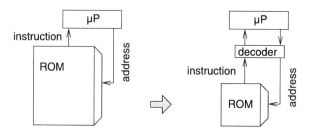

Figure 3.17.   Decompression of compressed instructions

The goals of compression can be summarized as follows:

* We would like to save ROM and RAM areas, since these may be more expensive than the processors themselves.

* We would like to use some encoding technique for instructions and possibly also for data with the following properties:

  · There should be little or no run-time penalty for these techniques.

---

[7]We continue denoting multiplexers, arithmetic units and memories by shape symbols, due to their widespread use in technical documentation. For memories, we adopt shape symbols including an explicit address decoder (included in the shape symbols for the ROMs on the right). These decoders identify the address input.

- Decoding should work from a limited context (it is, for example, impossible to read the entire program to find the destination of a branch instruction).

- Word-sizes of the memory, of instructions and addresses must be taken into account.

- Branch instructions branching to arbitrary destination addresses must be supported.

- Fast encoding is only required if writable data is encoded. Otherwise, fast decoding is sufficient.

There are several variations of this scheme:

∗ For some processors, there is a **second instruction set**. This second instruction set has a narrower instruction format. An example of this is the ARM processor family. The ARM instruction set is a 32 bit instruction set. The ARM instruction set includes predicated execution. This means an instruction is executed if and only if a certain condition is met (see page 148). This condition is encoded in the first four bits of the instruction format. Most ARM processors also provide a second instruction set, with 16 bit wide instructions, called THUMB instructions. THUMB instructions are shorter, since they do not support predication, use shorter and less register fields and use shorter immediate fields (see fig. 3.18).

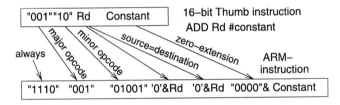

Figure 3.18. Re-encoding THUMB into ARM instructions

THUMB instructions are dynamically converted into ARM instructions while programs are decoded. THUMB instructions can use only half the registers in arithmetic instructions. Therefore, register-fields of THUMB instructions are concatenated with a '0'-bit[8]. In the THUMB instruction set, source and destination registers are identical and the length of constants that can be used, is reduced by 4 bits. During decoding, pipelining is used to keep the run-time penalty low.

---

[8]Using VHDL-notation (see page 80), concatenation is denoted by an &-sign and constants are enclosed in quotes in fig. 3.18.

Similar techniques also exist for other processors. The disadvantage of this approach is that the tools (compilers, assemblers, debuggers etc.) must be extended to support a second instruction set. Therefore, this approach can be quite expensive in terms of software development cost.

* A second approach is the use of **dictionaries**. With this approach, each instruction pattern is stored only once. For each value of the program counter, a look-up table then provides a pointer to the corresponding instruction in the instruction table, the dictionary (see fig. 3.19).

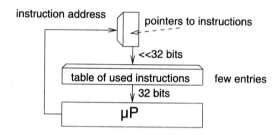

Figure 3.19. Dictionary approach for instruction compression

This approach relies on the idea that only very few different instruction patterns are used. Therefore, only few entries are required for the instruction table. Correspondingly, the bit width of the pointers can be quite small. Many variations of this scheme exist. Some are called *two-level control store* [Dasgupta, 1979], *nanoprogramming* [Stritter and Gunter, 1979], or *procedure ex-lining* [Vahid, 1995].

Beszedes [Beszedes, 2003] and Latendresse [Latendresse, 2004] provide overviews of known compression techniques.

■ **Run-time efficiency**: In order to meet time constraints without having to use high clock frequencies, architectures can be customized to certain application domains, such as digital signal processing (DSP). One can even go one step further and design application-specific instruction set processors (ASIPs). As an example of domain-specific processors, we will consider processors for DSP. In digital signal processing, digital filtering is a very frequent operation. Let us assume that we are extending the processing pipeline of fig. 3.4 by such filtering. Naming conventions for the involved signals are shown in fig. 3.20.

Equation 3.16 describes a digital filter generating an output signal $x(t)$ from an input signal $w(t)$. Both signals are defined over the (usually unbounded)

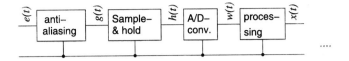

Figure 3.20. Naming conventions for signals

domain $\{t_s\}$ of sampling instances. For brevity, we write x_s instead of $x(t_s)$ and w_s instead of $w(t_s)$:

$$x_s = \sum_{k=0}^{n-1} w_{s-k} * a_k \tag{3.16}$$

A certain output element x_s corresponds to a weighted average over the last n signal elements of w and can be computed iteratively, adding one product at a time. Processors for DSP are designed such that each iteration can be encoded as a single instruction. Let us consider an example. Fig. 3.21 shows the internal architecture of an ADSP 2100 DSP processor.

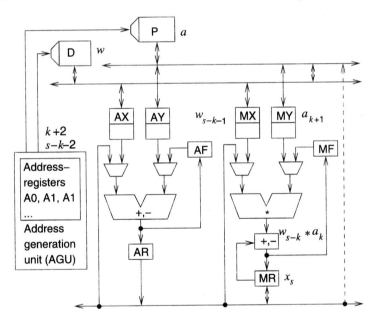

Figure 3.21. Internal architecture of the ADSP 2100 processor

The processor has two memories, called D and P. A special address generating unit (AGU) can be used to provide the pointers for accessing these memories. There are separate units for additions and multiplications, each with their own argument registers AX, AY, AF, MX, MY and MF. The multiplier is connected to a second adder in order to compute the series of multiplications and additions quickly.

For this processor, one iteration is essentially performed in a single cycle. For this purpose, the two memories are allocated to hold the two arrays w and a and address registers are allocated such that relevant pointers can be easily updated in the AGU. Partial sums are stored in MR. The pipelined computation involves registers A1, A2, MX, and MY.

```
-- outer loop over sampling times t_s
{ MR:=0;  A1:=1;  A2:=s-1;  MX:=w[s];  MY:=a[0];
  for (k=0;  k <= (n − 1); k++)
    {MR:=MR + MX * MY;  MX:=w[A2];  MY:=a[A1];
     A1++; A2--; }
  x[s]:=MR;}
```

The outer loop corresponds to the progressing time. A single instruction encodes the inner loop body, comprising the following operations:

- reading of two arguments from argument registers MX and MY, multiplying them and adding the product to register MR storing partial sums,
- fetching the next elements of arrays a and w from memories P and D and storing them in argument registers MX and MY,
- updating pointers to the next arguments, stored in address registers A1 and A2,
- testing for the end of the loop.

This way, each iteration of the inner loop requires just a single instruction. In order to achieve this, several operations are performed in parallel. For given computational requirements, this (limited) form of parallelism leads to relatively low clock frequencies. Furthermore, the registers in this architecture perform different functions. They are said to be **heterogeneous**. Heterogeneous register files are a common characteristic for DSP processors. In order to avoid extra cycles for testing for the end of the loop, **zero-overhead loop instructions** are frequently provided in DSP processors. With such instructions, a single or a small number of instructions can be executed a fixed number of times. Processors not optimized for DSP would probably need several instructions per iteration and would therefore require a higher clock frequency, if available.

The approach in its presented form would require arrays w and x of unlimited size if $\{t_s\}$ is unbounded. The size of these arrays can be constrained since we need to access only the n most recent values. Reuse of space in these arrays is possible with modulo addressing (see below).

3.3.3.1   Digital Signal Processing (DSP)

In addition to allowing single instruction realizations of loop bodies for filtering, DSP processors provide a number of other application-domain oriented features:

- **Specialized addressing modes:** In the filter application described above, only the last n elements of w need to be available. Ring buffers can be used for that. These can be implemented easily with modulo addressing. In modulo addressing, addresses can be incremented and decremented until the first or last element of the buffer is reached. Additional increments or decrements will result in addresses pointing to the other end of the buffer.

- **Separate address generation units:** Address generation units (AGUs) are typically directly connected to the address input of the data memory (see fig. 3.22).

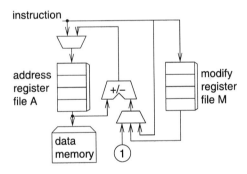

Figure 3.22.   AGU using special address registers

Addresses which are available in address registers can be used in register-indirect addressing modes. This saves machine instructions, cycles and energy. In order to increase the usefulness of address registers, instruction sets typically contain auto-increment and -decrement options for most instructions using address registers.

- **Saturating arithmetic:** Saturating arithmetic changes the way overflows and underflows are handled. In standard binary arithmetic, wrap-around is used for the values returned after an overflow or underflow. Fig. 3.23 shows an example in which two unsigned four-bit numbers are added. A carry is generated which cannot be returned in any of the standard registers. The result register will contain a pattern of all zeros. No result could be further away from the true result than this one.

	0111
+	1001
Standard *wrap-around* arithmetic	10000
saturating arithmetic	1111

Figure 3.23. Wrap-around vs. saturating arithmetic for unsigned integers

In saturating arithmetic, we try to return a result which is as close as possible to the true result. For saturating arithmetic, the largest value is returned in the case of an overflow and the smallest value is returned in the case of an underflow. This approach makes sense especially for video and audio applications: the user will hardly recognize the difference between the true result value and the largest value that can be represented. Also, it would be useless to raise exceptions if overflows occur, since it is difficult to handle exceptions in real-time. Note that we need to know whether we are dealing with signed or unsigned add instructions in order to return the right value.

- **Fixed-point arithmetic:** Floating-point hardware increases the cost and power-consumption of processors. Consequently, it has been estimated that 80 % of the DSP processors do not include floating-point hardware [Aamodt and Chow, 2000]. However, in addition to supporting integers, many such processors do support fixed-point numbers. Fixed-point data types can be specified by a 3-tuple $(wl, iwl, sign)$, where wl is the total word-length, iwl is the integer word-length (the number of bits left of the binary point), and sign $s \in \{s, u\}$ denotes whether we are dealing with unsigned or signed numbers. See also fig. 3.24. Furthermore, there may be different rounding modes (e.g. truncation) and overflow modes (e.g. saturating and wrap-around arithmetic).

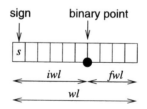

Figure 3.24. Parameters of a fixed-point number system

For fixed-point numbers, the position of the binary point is maintained after multiplications (some low order bits are truncated or rounded). For fixed-point processors, this operation is supported by hardware.

- **Real-time capability:** Some of the features of modern processors used in PCs are designed to improve the average execution time of programs. In many cases, it is difficult if not impossible to formally verify that they improve the worst case execution time. In such cases, it may be better not to implement these features. For example, it is difficult (though not impossible [Absint, 2002]) to guarantee a certain speedup resulting from the use of caches. Therefore, many embedded processors do not have caches. Also, virtual addressing and demand paging are normally not found in embedded systems. Techniques for computing worst case execution times will be presented in section 5.2.2.

- **Multiple memory banks or memories:** The usefulness of multiple memory banks was demonstrated in the ADSP 2100 example: the two memories D and P allow fetching both arguments at the same time. Several DSP processors come with two memory banks.

- **Heterogeneous register files:** Heterogeneous register files were already mentioned for the filter application.

- **Multiply/accumulate instructions:** These instructions perform multiplications followed by additions. They were also already used in the filter application.

3.3.3.2 Multimedia processors/instruction sets

Registers and arithmetic units of many modern architectures are at least 64 bits wide. Therefore, two 32 bit data types ("double words"), four 16 bit data types ("words") or eight 8 bit data types ("bytes") can be packed into a single register (see fig. 3.25).

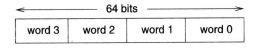

Figure 3.25. Using 64 bit registers for packed words

Arithmetic units can be designed such that they suppress carry bits at double word, word or byte boundaries. Multimedia instruction sets exploit this fact by supporting operations on packed data types. Such instructions are sometimes called single-instruction, multiple-data (SIMD) instructions, since a single instruction encodes operations on several data elements. With bytes packed into

64-bit registers, speed-ups of up to about eight over non-packed data types are possible. Data types are typically stored in packed form in memory. Unpacking and packing are avoided if arithmetic operations on packed data types are used. Furthermore, multimedia instructions can usually be combined with saturating arithmetic and therefore provide a more efficient form of overflow handling than standard instructions. Hence, the overall speed-up achieved with multimedia instructions can be significantly larger than the factor of eight enabled by operations on packed data types. Due to the advantages of operations on packed data types, new instructions have been added to several processors. For example, so-called **streaming SIMD extensions** (SSE) have been added to Intel's family of Pentium®-compatible processors [Intel, 2008]. New instructions have also been called **short vector instructions**. Currently (in 2010), Intel® Advanced Vector Extensions (AVX) are being introduced [Intel, 2010a].

3.3.3.3   Very long instruction word (VLIW) processors

Computational demands for embedded systems are increasing, especially when multimedia applications, advanced coding techniques or cryptography are involved. Performance improvement techniques used in high-performance microprocessors are not appropriate for embedded systems: driven by the need for instruction set compatibility, processors found, for example, in PCs spend a huge amount of resources and energy on automatically finding parallelism in application programs. Still, their performance is frequently not sufficient. For embedded systems, we can exploit the fact that instruction set compatibility with PCs is not required. Therefore, we can use instructions which explicitly identify operations to be performed in parallel. This is possible with **explicit parallelism instruction set computers** (EPICs). With EPICs, detection of parallelism is moved from the processor to the compiler. This avoids spending silicon and energy on the detection of parallelism at runtime. As a special case, we consider very long instruction word (VLIW) processors. For VLIW processors, several operations or instructions are encoded in a long instruction word (sometimes called **instruction packet**) and are assumed to be executed in parallel. Each operation/instruction is encoded in a separate field of the instruction packet. Each field controls certain hardware units. Four such fields are used in fig. 3.26, each one controlling one of the hardware units.

For VLIW architectures, the compiler has to generate instruction packets. This requires that the compiler is aware of the available hardware units and to schedule their use.

Instruction fields must be present, regardless of whether or not the corresponding functional unit is actually used in a certain instruction cycle. As a result,

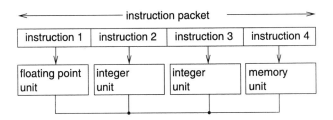

Figure 3.26. VLIW architecture (example)

the code density of VLIW architectures may be low, if insufficient parallelism is detected to keep all functional units busy. The problem can be avoided if more flexibility is added. For example, the Texas Instruments TMS 320C6xx family of processors implements a variable instruction packet size of up to 256 bits. In each instruction field, one bit is reserved to indicate whether or not the operation encoded in the next field is still assumed to be executed in parallel (see fig. 3.27). No instruction bits are wasted for unused functional units.

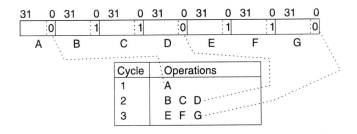

Figure 3.27. Instruction packets for TMS 320C6xx

Due to its variable length instruction packets, TMS 320C6xx processors do not quite correspond to the classical model of VLIW processors. Due to their explicit description of parallelism, they are EPIC processors, though.

Partitioned Register Files

Implementing register files for VLIW and EPIC processors is far from trivial. Due to the large number of operations that can be performed in parallel, a large number of register accesses has to be provided in parallel. Therefore, a large number of ports is required. However, the delay, size and energy consumption of register files increases with their number of ports. Hence, register files with very large numbers of ports are inefficient. As a consequence, many VLIW/EPIC architectures use partitioned register files. Functional units are then only connected to a subset of the register files. As an example, fig. 3.28 shows the internal structure of the TMS 320C6xx processors. These processors have two register files and each of them is connected to half of the functional units. During each clock cycle, only a single path from one register file to the functional units connected to the other register file is available.

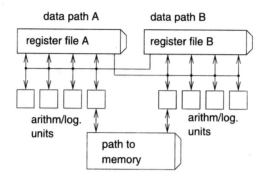

Figure 3.28.   Partitioned register files for TMS 320C6xx

Alternative partitionings are considered by Lapinskii et al. [Lapinskii et al., 2001].

Many DSP processors are actually VLIW processors. As an example, we are considering the M3-DSP processor [Fettweis et al., 1998]. The M3-DSP processor is a VLIW processor containing (up to) 16 parallel data paths. These data paths are connected to a group memory, providing the necessary arguments in parallel (see fig. 3.29).

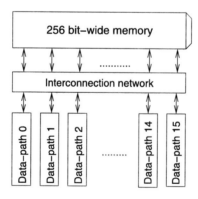

Figure 3.29.   M3-DSP (simplified)

Predicated Execution

A potential problem of VLIW and EPIC architectures is their possibly large **delay penalty**: This delay penalty might originate from branch instructions found in some instruction packets. Instruction packets normally must pass through pipelines. Each stage of these pipelines implements only part of the operations to be performed by the instructions executed. The fact that branch instructions exist cannot be detected in the first stage of the pipeline. When the execution of the branch instruction is finally completed, additional instructions have already entered the pipeline (see fig. 3.30).

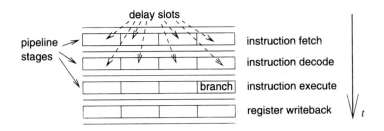

Figure 3.30. Branch instruction and delay slots

There are essentially two ways to deal with these additional instructions:

1 They are executed as if no branch had been present. This case is called **delayed branch**. Instruction packet slots that are still executed after a branch are called **branch delay slots**. These branch delay slots can be filled with instructions which would be executed before the branch if there were no delay slots. However, it is normally difficult to fill all delay slots with useful instructions and some must be filled with no-operation instructions (NOPs). The term **branch delay penalty** denotes the loss of performance resulting from these NOPs.

2 The pipeline is stalled until instructions from the branch target address have been fetched. There are no branch delay slots in this case. In this organization the branch delay penalty is caused by the stall.

Branch delay penalties can be significant. For example, the TMS 320C6xx family of processors has up to 40 delay slots. Therefore, efficiency can be improved by avoiding branches, if possible. In order to avoid branches originating from if-statements, **predicated instructions** have been introduced. For each predicated instruction, there is a predicate. This predicate is encoded in a few bits and evaluated at run-time. If the result is true, the instruction is executed. Otherwise, it is effectively turned into a NOP. Predication can also be found in RISC machines such as the ARM processor. Example: ARM instructions, as introduced on page 139, include a four-bit field. These four bits encode various expressions involving the condition code registers. Values stored in these registers are checked at run-time. They determine whether or not a certain instruction has an effect.

Predication can be used to implement small if-statements efficiently: the condition is stored in one of the condition registers and if-statement-bodies are implemented as predicated instructions which depend on this condition. This way, if-statement bodies can be evaluated in parallel with other operations and no delay penalty is incurred.

The Crusoe processor is a (commercially finally unsuccessful) example of an EPIC processor designed for PCs [Klaiber, 2000]. Efforts for making EPIC instruction sets available in the PC sector resulted in Intel's IA-64 instruction set [Intel, 2010b] and its implementation in the Itanium® processor. Due to legacy problems, the main application is in the server market. Many MPSoCs (see page 151) are based on VLIW and EPIC processors.

3.3.3.4   Micro-controllers

A large number of the processors in embedded systems are in fact micro-controllers. Micro-controllers are typically not very complex and can be used easily. Due to their relevance for designing control systems, we introduce one of the most frequently used processors: the Intel 8051. This processor has the following characteristics:

- 8 bit CPU, optimized for control applications,

- large set of operations on Boolean data types,

- program address space of 64 k bytes,

- separate data address space of 64 k bytes,

- 4 k bytes of program memory on chip, 128 bytes of data memory on chip,

- 32 I/O lines, each of which can be addressed individually,

- 2 counters on the chip,

- universal asynchronous receiver/transmitter for serial lines available on the chip,

- clock generation on the chip,

- many variations commercially available.

All these characteristics are quite typical for micro-controllers.

3.3.3.5 Multiprocessor systems-on-a-chip (MPSoCs)

Further increase of clock rates of processors has recently come to a stand-still. The large energy consumption of processors using multi-gigahertz clock speeds is a key reason for this. In order to still improve the overall perfor-mance, several processors must be employed. This led to the design of chips comprising multiple processors as well as additional components such as pe-ripheral devices and memories. Systems implemented in that way are called **MPSoCs** (MultiProcessor System-on-a-Chip). For general purpose computing and PCs, multi-processor systems are typically **homogeneous** (all processors are of the same type). The term **multi-core** system is usually associated with such systems. For embedded systems, energy efficiency has top priority. En-ergy efficiency is typically obtained with highly specialized processors. For example, there may be specialized processors for mobile communication or image processing. Fig. 3.31 contains a simplified version of the floor-plan of the SH-MobileG1 chip [Hattori, 2007].

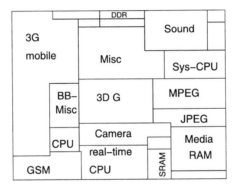

Figure 3.31.   Floor-plan of the SH-MobileG1 chip

The chip demonstrates the fact, that highly specialized processors are being used: there are special processors for MPEG- and JPEG-encoding, for GSM- and 3G mobile communication etc. In order to save energy, unused areas are typically powered-down. Using such multi-processor-based systems from ap-plications written in a sequential language is a challenge, which will be ad-dressed in Chapter 6. Mapping techniques for such processors are important, since examples demonstrate that a power efficiency close to that of ASICs can be achieved. For example, for IMEC's ADRES processor, an efficiency of 55×10^9 operations per Watt (about 50% of the power efficiency of ASICs) has been predicted [Man, 2007], [IMEC, 2010].

3.3.4   Reconfigurable Logic

In many cases, full-custom hardware chips (ASICs) are too expensive and software-based solutions are too slow or too energy consuming. Reconfigurable logic provides a solution if algorithms can be efficiently implemented in custom hardware. It can be almost as fast as special-purpose hardware, but in contrast to special-purpose hardware, the performed function can be changed by using configuration data. Due to these properties, reconfigurable logic finds applications in the following areas:

- **Fast prototyping:** modern ASICs can be very complex and the design effort can be large and takes a long time. It is therefore frequently desirable to generate a prototype, which can be used for experimenting with a system which behaves "almost" like the final system. The prototype can be more costly and larger than the final system. Also, its power consumption can be larger than the final system, some timing constraints can be relaxed, and only the essential functions need to be available. Such a system can then be used for checking the fundamental behavior of the future system.

- **Low volume applications:** If the expected market volume is too small to justify the development of special-purpose ASICs, reconfigurable logic can be the right hardware technology for applications, for which software would be too slow or too inefficient.

- **Real-time systems**: The timing of FPGA-based designs is typically known very precisely. Therefore, FPGAs can be used to implement timing-predictable systems.

Reconfigurable hardware frequently includes random access memory (RAM) to store configurations during normal operation of the hardware. Such RAM is normally **volatile** (the information is stored only while power is applied). Therefore, the configuration data must be copied into the configuration RAM at power-up. **Persistent** storage technology such as read-only memories (ROMs) and Flash memories will then provide the configuration data.

Field programmable gate arrays (FPGAs) are the most common form of reconfigurable hardware. As the name indicates, such devices are programmable "in the field" (after fabrication). Furthermore, they consist of arrays of processing elements. As an example, fig. 3.32 shows the array structure of Xilinx Virtex-II arrays [Xilinx, 2007].

The more recent Virtex-5 arrays contain up to 240×108 **configurable logic blocks** (CLBs) [Xilinx, 2009]. These can be connected using a programmable interconnect structure. Arrays also contain up to 1200 user input/output connections. In addition, there are up to 1056 DSP blocks comprising 25×18

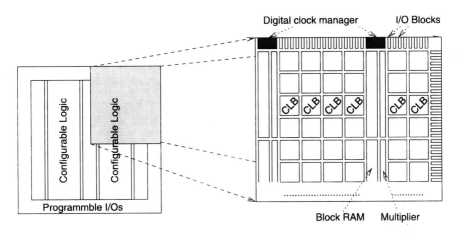

Figure 3.32. Floor-plan of Virtex-II FPGAs

bit multipliers and 16416 kbits of RAM (Block RAM). Each CLB consists of 2 so-called slices (see fig. 3.33).

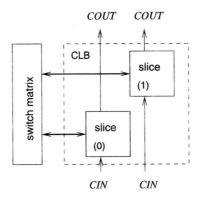

Figure 3.33. Virtex-5 CLB

Each slice contains four memories. Each memory can be used as a look-up table (LUT) for implementing a single 6-input logic function or two 5-input logic functions. All 2^{64} respectively all 2^{32} Boolean functions of 6 or 5 inputs can be implemented! With the help of multiplexers, several of these memories can also be combined. Memories can also serve as ordinary RAM or as shift registers (SRLs). Each slice also includes four output registers and some special logic for fast additions (see fig. 3.34) [Xilinx, 2009].

Configuration data determines the setting of multiplexers in the slices, the clocking of registers and RAM, the content of RAM components and the con-

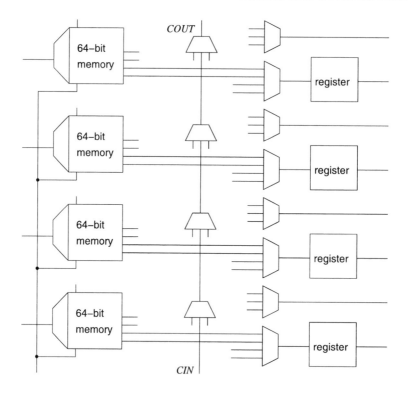

Figure 3.34.   Virtex-5 Slice (simplified)

nections between CLBs. Typically, this configuration data is generated from a high-level description of the functionality of the hardware, for example in VHDL. Ideally, the same description could also be used for generating ASICs automatically. In practice, some interaction is required.

Integration of reconfigurable computing with processors and software is simplified if processors are available in the FPGAs. There may be either hard cores or soft cores. For hard cores, the layout contains a special area implementing a core in a dense way. This area cannot be used for anything but the hard core. Soft cores are available as synthesizable models which are mapped to standard CLBs. Soft cores are more flexible, but less efficient than hard cores.

For example, the Virtex-5 FXT product line from Xilinx contains up to 2 Power-PC processors as hard cores.

Soft cores can be implemented on any FPGA chip. The MicroBlaze processor [Xilinx, 2008] is an example of such cores.

3.4   Memories

Data, programs, and FPGA configurations must be stored in some kind of memory. This must be done in an efficient way. Efficient means run-time, code-size and energy-efficient. Code-size efficiency requires a good compiler and can be improved with code compression (see page 138). Memory hierarchies can be exploited in order to achieve a good run-time and energy efficiency. The underlying reason is that large memories require more energy per access and are also slower than small memories.

Fig. 3.35 shows the cycle time and the power as a function of the size of memories used as register files [Rixner et al., 2000].

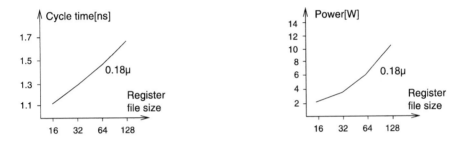

Figure 3.35.   Cycle time and power as a function of the register file size

Power and delay for caches can be computed with CACTI [Wilton and Jouppi, 1996]. Generated values include the power and the delay for the data RAM. These values can be used to predict power and delay for general RAM memories. Fig. 3.36 shows the results for a larger range of sizes [Banakar et al., 2002].

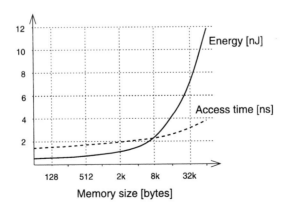

Figure 3.36.   Power and delay of RAM memory as predicted by CACTI

It has been observed that the difference in speeds between processors and memories is expected to increase (see fig. 3.37).

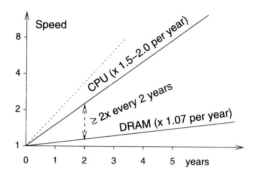

Figure 3.37. Increasing gap between processor and memory speeds

While the speed of memories is increasing by only a factor of about 1.07 per year, overall processor performance has increased by a factor of 1.5 to 2 per year [Machanik, 2002]. This means that the gap between processor performance and memory speeds is becoming larger. However, increasing processor performance further requires the use of multi-core systems.

Therefore, it is important to use smaller and faster memories that act as buffers between the main memory and the processor. In contrast to PC-like systems, the architecture of these small memories must guarantee a predictable real-time performance. A combination of small memories containing frequently used data and instructions and a larger memory containing the remaining data and instructions is generally also more energy efficient than a single, large memory. Memory partitioning has been considered, for example, by A. Macii [Macii et al., 2002].

Caches were initially introduced in order to provide good run-time efficiency. In the context of fig. 3.35 (right) however, it is obvious that caches potentially also improve the energy-efficiency of a memory system. Accesses to caches are accesses to small memories and therefore may require less energy per access than large memories. However, for caches it is required that the hardware checks whether or not the cache has a valid copy of the information associated with a certain address. This check involves comparing the tag fields of caches, containing a subset of the relevant address bits [Hennessy and Patterson, 2002]. Reading these tags requires additional energy. Also, the predictability of the real-time performance of caches is frequently low.

Alternatively, small memories can be mapped into the address space (see fig. 3.38).

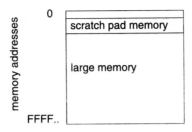

Figure 3.38. Memory map with scratch-pad included

Such memories are called **scratch pad memories** (SPMs). Frequently used variables and instructions should be allocated to that address space and no checking needs to be done in hardware. As a result, the energy per access is reduced. Fig. 3.39 shows a comparison between the energy required per access to the scratch-pad (SPM) and the energy required per access to the cache.

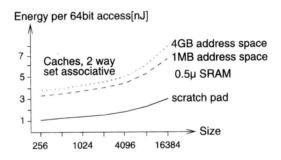

Figure 3.39. Energy consumption per scratch pad and cache access

For a two-way set associative cache, the two values differ by a factor of about three. The values in this example were computed using the energy consumption for RAM arrays as estimated by the CACTI cache estimation tool [Wilton and Jouppi, 1996].

SPMs can improve the memory access times very predictably, if the compiler is in charge of keeping frequently used variables in the SPM (see page 297).

3.5 Communication

Information must be available before it can be processed in an embedded system. Information can be communicated through various **channels**. Channels are abstract entities characterized by the essential properties of communication, like maximum information transfer capacity and noise parameters. The probability of communication errors can be computed using communication theory techniques. The physical entities enabling communication are called communication **media**. Important media classes include: wireless media (radio frequency media, infrared), optical media (fibers), and wires.

There is a huge variety of communication requirements between the various classes of embedded systems. In general, connecting the different embedded hardware components is far from trivial. Some common requirements can be identified.

3.5.1   Requirements

The following list contains some of the requirements that must be met:

- **Real-time behavior**: This requirement has far-reaching consequences on the design of the communication system. Several low-cost solutions such as standard Ethernet fail to meet this requirement.

- **Efficiency**: Connecting different hardware components can be quite expensive. For example, point to point connections in large buildings are almost impossible. Also, it has been found that separate wires between control units and external devices in cars significantly add to the cost and the weight of the car. With separate wires, it is also very difficult to add new components. The need of providing cost efficient designs also affects the way in which power is made available to external devices. There is frequently the need to use a central power supply in order to reduce the cost.

- **Appropriate bandwidth and communication delay**: Bandwidth requirements of embedded systems may vary. It is important to provide sufficient bandwidth without making the communication system too expensive.

- **Support for event-driven communication**: Polling-based systems provide a very predictable real-time behavior. However, their communication delay may be too large and there should be mechanisms for fast, event-oriented communication. For example, emergency situations should be communicated immediately and should not remain unnoticed until some central controller polls for messages.

- **Robustness**: Cyber-physical systems may be used at extreme temperatures, close to major sources of electromagnetic radiation etc. Car engines, for example, can be exposed to temperatures less than -20 and up to +180 degrees Celsius (-4 to 356 degrees Fahrenheit). Voltage levels and clock frequencies could be affected due to this large variation in temperatures. Still, reliable communication must be maintained.

- **Fault tolerance**: Despite all the efforts for robustness, faults may occur. Cyber-physical systems should be operational even after faults, if at all feasible. Restarts, like the ones found in personal computers, cannot be accepted. This means that retries may be required after attempts to communicate failed. A conflict exists with the first requirement: If we allow retries, then it is difficult to meet strict real-time requirements.

- **Maintainability, diagnosability**: Obviously, it should be possible to repair embedded systems within reasonable time frames.

- **Privacy:** Ensuring privacy of confidential information may require the use of encryption.

These communication requirements are a direct consequence of the general characteristics of embedded/cyber-physical systems mentioned in Chapter 1. Due to the conflicts between some of the requirements, compromises must be made. For example, there may be different communication modes: one high-bandwidth mode guaranteeing real-time behavior but no fault tolerance (this mode is appropriate for multimedia streams) and a second fault-tolerant, low-bandwidth mode for short messages that must not be dropped.

3.5.2   Electrical robustness

There are some basic techniques for electrical robustness. Digital communication within chips is normally using so-called single-ended signaling. For single-ended signaling, signals are propagated on a single wire (see fig. 3.40).

Figure 3.40.   Single-ended signaling

Such signals are represented by voltages with respect to a common ground (less frequently by currents). A single ground wire is sufficient for a number of single-ended signals. Single ended signaling is very much susceptible to external noise. If external noise (originating from, for example, motors being switched on) affects the voltage, messages can easily be corrupted. Also, it is difficult to establish high-quality common ground signals between a large number of communicating systems, due to the resistance (and inductance) on the ground wires. This is different for differential signaling. For differential signaling, each signal needs two wires (see fig. 3.41).

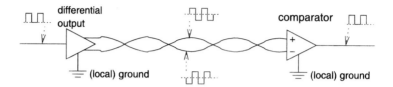

Figure 3.41. Differential signaling

Using differential signaling, binary values are encoded as follows: If the voltage on the first wire with respect to the second is positive, then this is decoded as '1', otherwise values are decoded as '0'. The two wires will typically be twisted to form so-called **twisted pairs**. There will be local ground signals, but a non-zero voltage between the local ground signals does not hurt. Advantages of differential signaling include:

- Noise is added to the two wires in essentially the same way. The comparator therefore removes almost all the noise.

- The logic value depends just on the polarity of the voltage between the two wires. The magnitude of the voltage can be affected by reflections or because of the resistance of the wires; this has no effect on the decoded value.

- Signals do not generate any currents on the ground wires. Hence, the quality of the ground wires becomes less important.

- No common ground wire is required. Hence, there is no need to establish a high quality ground wiring between a large number of communicating partners (this is one of the reasons for using differential signaling for Ethernet).

- As a consequence of the properties mentioned so far, differential signaling allows a larger throughput than single-ended signaling.

However, differential signaling requires two wires for every signal and it also requires negative voltages (unless it is based on complementary logic signals using voltages for single-ended signals).

Differential signaling is used, for example, in standard Ethernet-based networks.

3.5.3   Guaranteeing real-time behavior

For internal communication, computers may be using dedicated point-to-point communication or shared buses. Point-to-point communication can have a good real-time behavior, but requires many connections and there may be congestion at the receivers. Wiring is easier with common, shared buses. Typically, such buses use priority-based arbitration if several access requests to the communication media exists (see, for example, [Hennessy and Patterson, 2002]). Priority-based arbitration comes with poor timing predictability, since conflicts are difficult to anticipate at design time. Priority-based schemes can even lead to "starvation" (low-priority communication can be completely blocked by higher priority communication). In order to get around this problem, *time division multiple access* (TDMA) can be used. In a TDMA-scheme, each partner is assigned a fixed time slot. The partner is allowed to transmit during that particular time slot. Typically, communication time is divided into frames. Each frame starts with some time slot for frame synchronization, and possibly some gap to allow the sender to turn off (see fig. 3.42, [Koopman and Upender, 1995]).

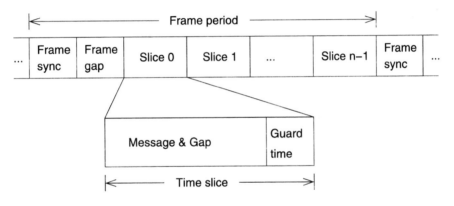

Figure 3.42. TDMA-based communication

This gap is followed by a number of slices, each of which serves for communicating messages. Each slice also contains some gap and guard time to take clock speed variations of the partners into account. Slices are assigned to communication partners. Variations of this scheme exist. For example, truncation of unused slices or the assignment of partners to several slides are feasible. TDMA reduces the maximum amount of data available per frame and partner, but guarantees a certain bandwidth for all partners. Starvation can be avoided. The ARM AMBA-bus [ARM Ltd., 2009a] includes TDMA-based bus allocation.

Communication between computers is frequently based on Ethernet standards. For 10 Mbit/s and 100 Mbit/s versions of Ethernet, there can be collisions between various communication partners. This means: several partners are trying to communicate at about the same time and the signals on the wires are corrupted. Whenever this occurs, the partners must stop communications, wait for some time, and then retry. The waiting time is chosen at random, so that it is not very likely that the next attempt to communicate results in another collision. This method is called **carrier-sense multiple access/collision detect** (CSMA/CD). For CSMA/CD, communication time can get huge, since conflicts can repeat a large number of times, even though this is not very likely. Hence, CSMA/CD cannot be used when real-time constraints must be met.

This problem can be solved with CSMA/CA (**carrier-sense multiple access/ collision avoidance**). As the name indicates, collisions are completely avoided, rather than just detected. For CSMA/CA, priorities are assigned to all partners. Communication media are allocated to communication partners during **arbitration phases**, which follow communication phases. During arbitration phases, partners wanting to communicate indicate this on the media. Partners finding such indications of higher priority must immediately remove their indication.

Provided that there is an upper bound on the time between arbitration phases, CSMA/CA guarantees a predictable real-time behavior for the partner having the highest priority. For other partners, real-time behavior can be guaranteed if the higher priority partners do not continuously request access to the media.

Note that high-speed versions of Ethernet (≥ 1 Gbit/s) also avoid collisions. TDMA-schemes are also used for wireless communication. For example, mobile phone standards like GSM use TDMA for accesses to the communication medium.

3.5.4   Examples

- **Sensor/actuator buses:** Sensor/actuator buses provide communication between simple devices such as switches or lamps and the processing equipment. There may be many such devices and the cost of the wiring needs special attention for such buses.

- **Field buses:** Field buses are similar to sensor/actuator buses. In general, they are supposed to support larger data rates than sensor/actuator buses. Examples of field buses include the following:

  - **Controller Area Network (CAN):** This bus was developed in 1981 by Bosch and Intel for connecting controllers and peripherals. It is popular in the automotive industry, since it allows the replacement of a large amount of wires by a single bus. Due to the size of the automotive market, CAN components are relatively cheap and are therefore also used in other areas such as smart homes and fabrication equipment. CAN has the following properties:
    * differential signaling with twisted pairs,

* arbitration using CSMA/CA,
* throughput between 10kbit/s and 1 Mbit/s,
* low and high-priority signals,
* maximum latency of 134 μs for high priority signals,
* coding of signals similar to that of serial (RS-232) lines of PCs, with modifications for differential signaling.

CSMA/CA-based arbitration does not prevent starvation. This is an inherent problem of the CAN protocol.

– The **Time-Triggered-Protocol (TTP)** [Kopetz and Grunsteidl, 1994] for fault-tolerant safety systems like airbags in cars.

– **FlexRay**^{TM} [FlexRay Consortium, 2002] is a TDMA protocol which has been developed by the FlexRay consortium (BMW, DaimlerChrysler, General Motors, Ford, Bosch, Motorola and Philips Semiconductors). FlexRay is a combination of a variant of the TTP and the byteflight [Byteflight Consortium, 2003] protocol.

FlexRay includes a static as well as a dynamic arbitration phase. The static phase uses a TDMA-like arbitration scheme. It can be used for real-time communication and starvation can be avoided. The dynamic phase provides a good bandwidth for non-real-time communication. Communicating partners can be connected to up to two buses for fault-tolerance reasons. **Bus guardians** may protect partners against partners flooding the bus with redundant messages, so-called **babbling idiots**. Partners may be using their own local clock periods. Periods common to all partners are defined as multiples of such local clock periods. Time slots allocated to partners for communication are based on these common periods.

The levi simulation allows simulating the protocol in a lab environment [Sirocic and Marwedel, 2007a].

– **LIN** (Local Interconnect Network) is a low-cost communication standard for connecting sensors and actuators in the automotive domain [LIN Administration, 2010].

– **MAP:** MAP is a bus designed for car factories.

– **EIB:** The European Installation Bus (EIB) is a bus designed for smart homes.

■ **Wired multimedia communication**: For wired multimedia communication, larger data rates are required. Example: **MOST** (Media Oriented Systems Transport) is a communication standard for multimedia and infotainment equipment in the automotive domain [MOST Cooperation, 2010]. Standards like IEEE 1394 (FireWire) may be used for the same purpose.

- **Wireless communication:** This kind of communication is becoming more popular. Currently (2010), 7 Mbit/s are widely available with HSPA (High Speed Packet Access). Even higher rates (based, for example, on the **long-term evolution (LTE)** technology) are on the horizon.

  Bluetooth is a standard for connecting devices such as mobile phones and their headsets.

  The wireless version of Ethernet is standardized as IEEE standard 802.11. It is being used in local area networks (LANs).

  DECT is a standard used for wireless phones in Europe.

3.6   Output

Output devices of embedded/cyber-physical systems include:

- **Displays:** Display technology is an area which is extremely important. Accordingly, a large amount of information [Society for Display Technology, 2003] exists on this technology. Major research and development efforts lead to new display technology such as organic displays [Gelsen, 2003]. Organic displays are emitting light and can be fabricated with very high densities. In contrast to LCD displays, they do not need back-light and polarizing filters. Major changes are therefore expected in these markets.

- **Electro-mechanical devices:** these influence the environment through motors and other electro-mechanical equipment.

Analog as well as digital output devices are used. In the case of analog output devices, the digital information must first be converted by digital-to-analog (D/A)-converters. These converters can be found on the path from analog inputs of embedded systems to their outputs. Fig. 3.43 shows the naming convention of signals along the path which we use. Purpose and function of the boxes will be explained in this section.

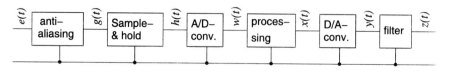

Figure 3.43.   Naming convention for signals between analog inputs and outputs

3.6.1   D/A-converters

D/A-converters are not very complex. Fig. 3.44 shows the schematic of a simple so-called weighted-resistor D/A converter.

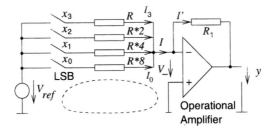

Figure 3.44. D/A-converter

The key idea of the converter is to first generate a current which is proportional to the value represented by a digital signal x. Such a current can hardly be used by a following system. Therefore, this current is converted into a proportional voltage y. This conversion is done with an operational amplifier (depicted by a triangle in fig. 3.44). Essential characteristics of operational amplifiers are described in Appendix B of this book.

How do we compute the output voltage y? Let us first consider the loop indicated by the dashed line in fig. 3.44. The current through any resistor is zero, if the corresponding element of digital signal x is '0'. If it is '1', the current corresponds to the weight of that bit, since resistor values are chosen accordingly. We can apply Kirchhoff's Loop Rule (see Appendix B) to the loop turned on by the least significant bit x_0 of x. We have

$$x_0 \cdot I_0 \cdot 8 \cdot R + V_- - V_{ref} \;=\; 0 \qquad (3.17)$$

V_- is approximately 0 (see Appendix B). Therefore, we have

$$I_0 \;=\; x_0 * \frac{V_{ref}}{8 * R} \qquad (3.18)$$

Corresponding equations hold for the currents I_1 to I_3 through the other resistors. We can now apply Kirchhoff's Node Rule (see Appendix B) to the circuit node connecting all resistors. At this node, the outgoing current must be equal to the sum of the incoming currents. Therefore, we have

$$I \;=\; I_3 + I_2 + I_1 + I_0 \qquad (3.19)$$

$$I \;=\; x_3 * \frac{V_{ref}}{R} + x_2 * \frac{V_{ref}}{2 * R} + x_1 * \frac{V_{ref}}{4 * R} + x_0 * \frac{V_{ref}}{8 * R}$$

$$= \frac{V_{ref}}{R} * \sum_{i=0}^{3} x_i * 2^{i-3} \tag{3.20}$$

Now, we can apply Kirchhoff's Loop Rule to the loop comprising R_1, y and V_-. Since V_- is approximately 0, we have:

$$y + R_1 * I' = 0. \tag{3.21}$$

Next, we can apply Kirchhoff's Node Rule to the node connecting I, I' and the inverting signal input of the operational amplifier. The current into this input is practically zero, and currents I and I' are equal: $I = I'$. Hence, we have:

$$y + R_1 * I = 0 \tag{3.22}$$

From equations 3.20 and 3.22 we obtain:

$$y = -V_{ref} * \frac{R_1}{R} * \sum_{i=0}^{3} x_i * 2^{i-3} = -V_{ref} * \frac{R_1}{8*R} * nat(x) \tag{3.23}$$

nat denotes the natural number represented by digital signal x. Obviously, y is proportional to the value represented by x. Positive output voltages and bit-vectors representing two's complement numbers require minor extensions.

From a DSP point of view, $y(t)$ is a function over a discrete time domain: it provides us with a **sequence** of voltage levels. In our running example, it is defined only over integer times. From a practical point of view, this is inconvenient, since we would typically observe the output of the circuit of fig. 3.44 continuously. Therefore, D/A-converters are frequently extended by a **"zero-order hold" functionality**. This means that the converter will keep the previous value until the next value is converted. Actually, the D/A-converter of fig. 3.44 will do exactly this if we do not change the settings of the switches until the next discrete time instant. Hence, the output of the converter is a step function $y'(t)$ corresponding to the sequence $y(t)$[9]. $y'(t)$ is a function over the continuous time domain.

As an example, let us consider the output resulting from the conversion of the signal of equation 3.3, assuming a resolution of 8 steps per polarity. For this case, fig. 3.45 shows $y'(t)$ instead of $y(t)$, since $y'(t)$ is a bit easier to visualize.

---

[9] In practice, due to rise and fall times being > 0, transitions from one step to the next will not be ideal, but take some time.

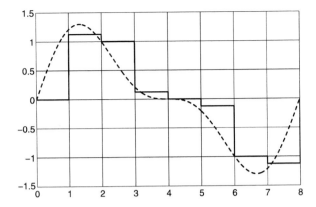

Figure 3.45. Step function $y'(t)$ generated from signal $e_3(t)$ (eq. 3.3) sampled at integer times

D/A-converters enable a conversion from time- and value-discrete signals to signals in the continuous time and value domain. However, neither $y(t)$ nor $y'(t)$ reflect the values of the input signal in-between the sampling instances.

3.6.2   Sampling theorem

Suppose that the processors used in the hardware loop forward values from A/D-converters unchanged to the D/A-converters. We could also think of storing values $x(t)$ on a CD and aiming at generating an excellent analog audio signal. Would it be possible to reconstruct the original analog voltage $e(t)$ (see fig. 3.8, fig. 3.20, and fig. 3.43) at the outputs of the D/A-converters?

It is obvious that reconstruction is not possible if we have aliasing of the type described in the section on sampling[10]. So, we assume that the sampling rate is larger than twice the highest frequency of the decomposition of the input signal into sine waves (sampling criterion, see equation 3.8). Does meeting this criterion allow us to reconstruct the original signal? Let us have a closer look!

Feeding D/A-converters with a discrete sequence of digital values will result in a sequence of analog values being generated. Values of the input signal in-between the sampling instances are not generated by D/A-converters. The simple zero-order hold functionality (if present) would generate only step functions. This seems to indicate that reconstruction of $e(t)$ would require an infinitely large sampling rate, such that all intermediate values can be generated.

---

[10]Reconstruction may be possible, if additional information about the signal is available, i.e. if we restrict ourselves to certain signal types.

However, there could be some kind of smart interpolation computing values in-between the sampling instances from the values at sampling instances. And indeed, sampling theory [Oppenheim et al., 2009] tells us that a corresponding time-continuous signal $z(t)$ can be constructed from the sequence $y(t)$ of analog values.

Let $\{t_s\}, s = ..., -1, 0, 1, 2, ...$ be the times at which we sample our input signal. Let us assume a constant sampling rate of $f_s = \frac{1}{p_s}$ ($\forall s : p_s = t_{s+1} - t_s$). Then, sampling theory tells us that we can approximate $e(t)$ from $y(t)$ as follows:

$$z(t) \quad = \quad \sum_{s=-\infty}^{\infty} \frac{y(t_s)\sin\frac{\pi}{p_s}(t-t_s)}{\frac{\pi}{p_s}(t-t_s)} \tag{3.24}$$

This equation is known as the **Shannon-Whittaker interpolation**. $y(t_s)$ is the contribution of signal y at sampling instance t_s. The influence of this contribution decreases the further t is away from t_s. The decrease follows a weighting factor, also known as the *sinc* function:

$$sinc(t - t_s) \quad = \quad \frac{sin(\frac{\pi}{p_s}(t-t_s))}{\frac{\pi}{p_s}(t-t_s)} \tag{3.25}$$

which decreases non-monotonically as a function of $|t - t_s|$. This weighting factor is used to compute values in-between the sampling instances. Fig. 3.46 shows the weighting factor for the case $p_s = 1$.

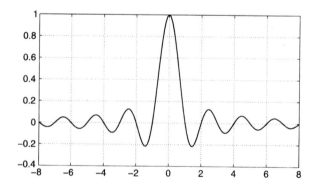

Figure 3.46.   Visualization of eq. 3.25 used for interpolation

Using the *sinc* function, we can compute the terms of the sum in eq. 3.24. Fig. 3.47 and fig. 3.48 show the resulting terms if $e(t) = e_3(t)$ and processing performs the identify function $(x(t) = w(t))$.

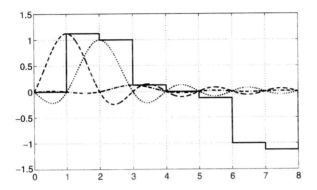

Figure 3.47. $y'(t)$ (solid line) and the first three terms of eq. 3.24

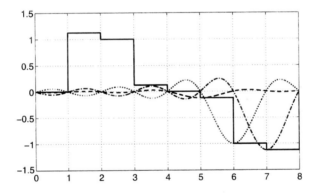

Figure 3.48. $y'(t)$ (solid line) and the last three non-zero terms of eq. 3.24

At each of the sampling instances t_s (integer times in our case), $z(t_s)$ is computed just from the corresponding value $y(t_s)$, since the *sinc* function is zero in this case for all other sampled values. In between the sampling instances, all of the adjacent discrete values contribute to the resulting value of $z(t)$. Fig. 3.49 shows the resulting $z(t)$ if $e(t) = e_3(t)$ and processing performs the identify function $(x(t) = w(t))$.

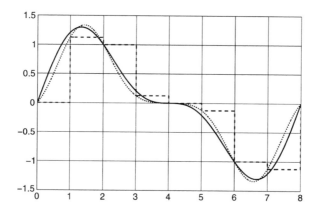

Figure 3.49. $e_3(t)$ (solid), $z(t)$ (dotted), $y'(t)$ (dashed)

The figure includes signals $e_3(t)$ (solid line), $z(t)$ (dotted line), and $y'(t)$ (dashed line). $z(t)$ is based on summing up the contributions of all sampling instances shown in the diagrams 3.47 and 3.48. $e_3(t)$ and $z(t)$ are very similar.

How close could we get to the original input signal by implementing equation 3.24? Sampling theory tells us (see, for example, [Oppenheim et al., 2009]), that **equation 3.24 computes an exact approximation**, if the sampling criterion (equation 3.8) is met. Therefore, let us see how we can implement equation 3.24.

How do we compute equation 3.24 in an electronic system? We cannot compute this equation in the discrete time domain using a digital signal processor for this, since this computation has to generate a time-continuous signal. Computing such a complex equation with analog circuits seems to be difficult at first sight.

Fortunately, the required computation is a so-called *folding operation* between signal $y(t)$ and the *sinc*-function. According to the classical theory of Fourier transforms, a folding operation in the time domain is equivalent to a multiplication with frequency-dependent filter function in the frequency domain. This filter function is the Fourier transform of the corresponding function in the time domain. Therefore, equation 3.24 can be computed with some appropriate filter. Fig. 3.50 shows the corresponding placement of the filter.

The remaining question is: which frequency-dependent filter function is the Fourier transform of the *sinc*-function? Computing the Fourier transform of the *sinc*-function yields a low-pass filter function [Oppenheim et al., 2009]. So, "all" we must do to compute equation 3.24 is to pass signal $y(t)$ through

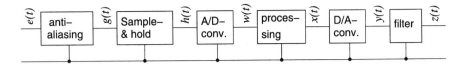

Figure 3.50. Converting signals $e(t)$ from the analog time and value domain to the digital domain and back

a low-pass filter, filtering frequencies as shown for the "ideal filter" in fig. 3.51. Note that the representation of function $y(t)$ as a sum of sine waves would require very high frequency components, making such a filtering non-redundant, even though we have already assumed an anti-aliasing filter to be present at the input.

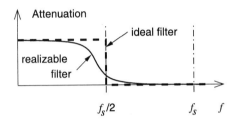

Figure 3.51. Low-pass filter: ideal (dashed) and realistic (solid)

There is still one problem, though: ideal low-pass filters do not exist. Therefore, we must live with compromises and design filters approximating the low pass filter characteristics. Actually, we must live with several imperfections preventing a precise reconstruction of the input signals:

■ Ideal low pass filters cannot be designed. Therefore, we must use approximations of such filters. Designing good compromises is an art (performed extensively, for example, for audio equipment).

■ For the same reason, we cannot completely remove input frequencies beyond the Nyquist frequency.

■ The impact of value quantization is visible in fig. 3.49. Due to value quantization, $e_3(t)$ is sometimes different from $z(t)$. Quantization noise, as introduced by A/D-converters, cannot be removed during output generation. Signal $w(t)$ from the output of the A/D-converter will remain distorted by the quantization noise. However, this effect does not affect the signal $h(t)$ from the output of sample-and-hold circuits.

■ Equation 3.24 is based on an infinite sum, involving also values at future instances in time. In practice, we can delay signals by some finite amount to know a finite number of "future" samples. Infinite delays are impossible. In fig. 3.49, we did not consider contributions of sampling instances outside the diagram.

The functionality provided by low-pass filters demonstrates the power of analog circuits: there would be no way of implementing the behavior of analog filters in the digital domain, due to the inherent restriction to discretized time and values.

Many authors have contributed to sampling theory. Therefore, many names can be associated with the sampling theorem. Contributors include Shannon, Whittaker, Kotelnikov, Nyquist, Küpfmüller, and others. Therefore, the fact that the original signal can be reconstructed should simply be called the sampling theorem, since there is no way of attaching all names of relevant contributors to the theorem.

3.6.3   Actuators

There is a huge amount of actuators [Elsevier B.V., 2010a]. Actuators range from huge ones that are able to move tons of weight to tiny ones with dimensions in the μm area, like the one shown in fig. 3.52.

Figure 3.52.   Microsystem technology based actuator motor (partial view; courtesy E. Obermeier, MAT, TU Berlin), ©TU Berlin

It is impossible to provide a complete overview. As an example, we mention only a special kind of actuators which will become more important in the future: microsystem technology enables the fabrication of tiny actuators, which can be put into the human body, for example.

Using such tiny actuators, the amount of drugs fed into the body can be adapted to the actual need. This allows a much better medication than needle-based injections. Fig. 3.52 shows a tiny motor manufactured with microsystem technology. The dimensions are in the μm range. The rotating center is controlled by electrostatic forces.

3.7 Secure hardware

The general requirements for embedded systems can often include security (see page 5). If security is a major concern, special secure hardware may need to be developed. Security may need to be guaranteed for communication and for storage [Krhovjak and Matyas, 2006]. Also, security might demand special equipment for the generation of cryptographic keys. Special hardware security modules have been designed. One of the goals for such modules is to resist side-channel attacks such as measurement of the supply current or electromagnetic radiation. Such modules include special mechanisms for physical protection (shielding, or sensors to detect tampering with the modules). Special processors may support encryption and decryption. In addition to the physical security, we need logical security, typically using cryptographic methods. Smart cards are a special case of secure hardware that must run using a very small amount of energy. In general, it is necessary to distinguish between different levels of security and levels of knowledge of "adversaries". A full presentation of the techniques for designing secure hardware is beyond the scope of this book. Interested readers are referred to Gebotys [Gebotys, 2010] and workshop proceedings [Clavier and Gaj, 2009].

3.8 Assignments

1 It is suggested that locally available small robots are used to demonstrate hardware in the loop, corresponding to fig. 3.2. The robots should includes sensors and actuators. Robots should run a program implementing a control loop. For example, an optical sensor could be used to let a robot follow a black line on the ground. The details of this assignment depend on the availability of robots.

2 Why is it so important to optimize embedded systems? Compare different technologies for processing information in an embedded system with respect to their efficiency!

3 Assume that we have an input signal x consisting of the sum of sine waves of 1.75 kHz and 2 kHz. We are sampling x at a rate of 3 kHz. Will we be able to reconstruct the original signal after discretization of time? Please explain your result!

4 Discretization of values is based on A/D-converters. Develop the schematic of a flash-based A/D-converter for positive and negative input voltages!

The output should be encoded as 3-bit two's complement numbers, allowing to distinguish between 8 different voltage intervals.

5 Compare the complexity of flashed-based and successive approximation-based A/D-converters. Assume that you would like to distinguish between n different voltage intervals. Enter the complexity into the table of fig. 3.53, using the O-notation.

	Flash-based converter	Successive approximation converter
Time complexity		
Space complexity		

Figure 3.53. Complexity of A/D-converters

6 Suppose that we are working with a successive approximation-based 4-bit A/D-converter. The input voltage range extends from V_{min} =1 V (="0000") to V_{max} =4.75 V (="1111"). Which steps are used to convert voltages of 2.25 V, 3.75 V, and 1.8 V? Draw a diagram similar to fig. 3.12 which depicts the successive approximation to these voltages!

7 Extend the flash-based A/D converter such that it can be used for negative voltages as well!

8 Suppose a sine wave is used as an input signal to the converter designed in assignment 4. Depict the quantization noise signal for this case!

9 Create a list of features of DSP-processors!

10 Which components do FPGA comprise? Which of these are used to implement Boolean function? How are FPGAs configured? Are FPGAs energy-efficient? Which kind of applications are FPGAs good for?

11 In the context of memories, we are sometimes saying "small is beautiful". What could be the reason for this?

12 Develop the following FlexRay^{TM} cluster: The cluster consists of the 5 nodes A, B, C, D and E. All nodes should be connected via two channels. The cluster uses a bus topology. The nodes A, B and C are executing a safety critical task and therefore their bus requests should be guaranteed at the time of 20 macroticks. The following is expected from you:

- Download the levi FlexRay simulator [Sirocic and Marwedel, 2007a]. Unpack the .zip file and install!

- Start the training module by executing the file leviFRP.jar.

- Design the described FlexRay cluster within the training module.

- Configure the communication cycle such that the nodes A, B and C have a guaranteed bus access within a maximal delay of 20 macroticks. The nodes D and E should use only the dynamic segment.

- Configure the node bus requests. The node A sends a message every cycle. The nodes B and C send a message every second cycle. The node D sends a message of the length of 2 minislots every cycle and the node E sends every second cycle a message of the length of 2 minislots.

- Start the visualization and check if the bus requests of the nodes A, B and C are guaranteed.

- Swap the positions of nodes D and E in the dynamic segment. What is the resulting behavior?

13 Develop the schematic of a 3-bit D/A-converter! The conversion should be done for a 3-bit vector x encoding positive numbers. Prove that the output voltage is proportional to the value represented by the input vector x. How would you modify the circuit if x represented two's complement numbers?

14 The circuit shown in fig. B.4 in Appendix B is an amplifier, amplifying input voltage V_1:

$$V_{out} = g_{closed} \cdot V_1$$

Compute the gain g_{closed} for the circuit of fig. B.4 as a function of R and R_1!

Chapter 4

SYSTEM SOFTWARE

Not all components of embedded systems need to be designed from scratch. Instead, there are standard components that can be reused. These components comprise knowledge from earlier design efforts and constitute **intellectual property** (IP). IP reuse is one key technique in coping with the increasing complexity of designs. The term "IP reuse" frequently denotes the reuse of hardware. However, reusing hardware is not enough. Sangiovanni-Vincentelli pointed out, that software components need to be reused as well. Therefore, the platform-based design methodology advocated by Sangiovanni-Vincentelli [Sangiovanni-Vincentelli, 2002] (see page 236) comprises the reuse of hardware and software IP.

Standard software components that can be reused include system software components such as embedded operating systems (OS) and **middleware**. The last term denotes software that provides an intermediate layer between the OS and application software. We include libraries for communication as a special case of middleware. Such libraries extend the basic communication facilities provided by operating systems. Also, we consider real-time databases (see Section 4.5) to be a second class of middleware. Calls to standard software components may already need to be included in the specification. Therefore, information about the application programming interface (API) of these standard components may already be needed for completing executable specifications of the SUD.

Consistent with the design information flow, we will describe embedded operating systems, and middleware in this chapter (see also fig. 4.1).

P. Marwedel, *Embedded System Design*, Embedded Systems,
DOI 10.1007/978-94-007-0257-8_4, © Springer Science+Business Media B.V. 2011

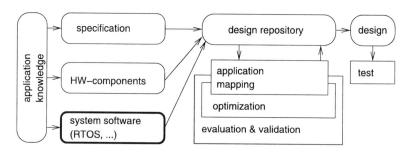

Figure 4.1. Simplified design information flow

4.1   Embedded Operating Systems

4.1.1   General requirements

Except for very simple systems, scheduling, task switching, and I/O require the support of an operating system suited for embedded applications. Task switch (or task "dispatch") algorithms multiplex processors such that each task seems to have its own processor.

For systems with virtual memory, we can distinguish between different address spaces, and between processes and threads. **Each process has its own address space, whereas several threads may share an address space**. Context switches which change the address space require more time than those which do not. Threads sharing an address space will typically communicate via shared memory. Operating systems must provide communication and synchronization methods for threads and processes. More information about the just touched standard topics in system software can be found in textbooks on operating systems, such as the book by Tanenbaum [Tanenbaum, 2001][1].

The following are essential features of embedded operating systems:

- Due to the large variety of embedded systems, there is also a large variety of requirements for the functionality of embedded OSs. Due to efficiency requirements, it is not possible to work with OSs which provide the union of all functionalities. For most applications, the OS must be small. Hence, we need operating systems which can be **flexibly tailored** towards the application at hand. **Configurability** is therefore one of the main characteristics of embedded OSs. There are various techniques of implementing configurability, including[2]:

---

[1]Students who have not attended a course on operating systems may have to browse through one of these textbooks before proceeding any further.

[2]This list is sorted by the position of the technique in the development process or tool chain.

- Object-orientation, used for a derivation of proper subclasses: for example, we could have a general scheduler class. From this class we could derive schedulers having particular features. However, object-oriented approaches typically come with an additional overhead. For example, dynamic binding of methods does create run-time overhead. Such overhead may be unacceptable for performance-critical system software.

- Aspect-oriented programming [Lohmann et al., 2009]: with this approach, orthogonal aspects of software can be described independently and then can be added automatically to all relevant parts of the program code. For example, some code for profiling can be described in a single module. It can then be automatically added to or dropped from all relevant parts of the source code. The CIAO family of operating systems has been designed in this way [Lohmann et al., 2006].

- Conditional compilation: In this case, we are using some macro preprocessor and we are taking advantage of #if and #ifdef preprocessor commands.

- Advanced compile-time evaluation: configurations could be performed by defining constant values of variables before compiling the OS. The compiler could then propagate the knowledge of these values as much as possible. Advanced compiler optimizations may also be useful in this context. For example, if a particular function parameter is always constant, this parameter can be dropped from the parameter list. Partial evaluation [Jones, 1996] provides a framework for such compiler optimizations. In a sophisticated form, dynamic data might be replaced by static data [Atienza et al., 2007]. A survey of operating system specialization was published by McNamee et al. [McNamee et al., 2001].

- Linker-based removal of unused functions: At link-time, there may be more information about used and unused functions than during earlier phases. For example, the linker can figure out, which library functions are used. Unused library functions can be accordingly dropped and specializations can take place [Chanet et al., 2007].

These techniques are frequently combined with a rule-based selection of files to be included in the operating system. Tailoring the OS can be made easy through a graphical user interface hiding the techniques employed for achieving this configurability. For example, VxWorks [Wind River, 2010a] from Wind River is configured via a graphical user interface.

Verification is a potential problem of systems with a large number of derived tailored OSs. Each and every derived OS must be tested thoroughly. Takada mentions this as a potential problem for eCos (an open source RTOS from Red Hat [Massa, 2002]), comprising 100 to 200 configuration points [Takada, 2001]. Software product line engineering [Pohl et al., 2005] can contribute towards solving this problem.

- There is a large variety of peripheral devices employed in embedded systems. Many embedded systems do not have a hard disk, a keyboard, a screen or a mouse. There is effectively **no device that needs to be supported by all variants of the OS**, except maybe the system timer. Frequently, applications are designed to handle particular devices. In such cases, devices are not shared between applications and hence there is no need to manage the devices by the OS. Due to the large variety of devices, it would also be difficult to provide all required device drivers together with the OS. Hence, it makes sense to decouple OS and drivers by using special tasks instead of integrating their drivers into the kernel of the OS. Due to the limited speed of many embedded devices, there is also no need for an integration into the OS in order to meet performance requirements. This may lead to a different stack of software layers. For PCs, some drivers, such as disk drivers, network drivers, or audio drivers are implicitly assumed to be present. They are implemented at a very low level of the stack. The application software and middleware are implemented on top of the application programming interface, which is standard for all applications. For an embedded OS, device drivers are implemented on top of the kernel. Applications and middleware may be implemented on top of appropriate drivers, not on top of a standardized API of the OS (see fig. 4.2).

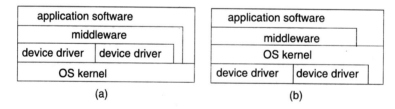

Figure 4.2. Device drivers implemented on top of (a) or below (b) OS kernel

VxWorks can again serve as an example here. Fig. 4.3 shows a fraction of the Wind River® Industrial Automation platform [Wind River, 2010a].

- **Protection mechanisms are not always necessary**, since embedded systems are frequently designed for a single purpose (they are not supposed to support so-called "multiprogramming"). Therefore, untested programs are hardly ever loaded. After the software has been tested, it could be assumed to be reliable. This also applies to input/output. In contrast to desktop applications, there is no desire to implement I/O instructions as privileged

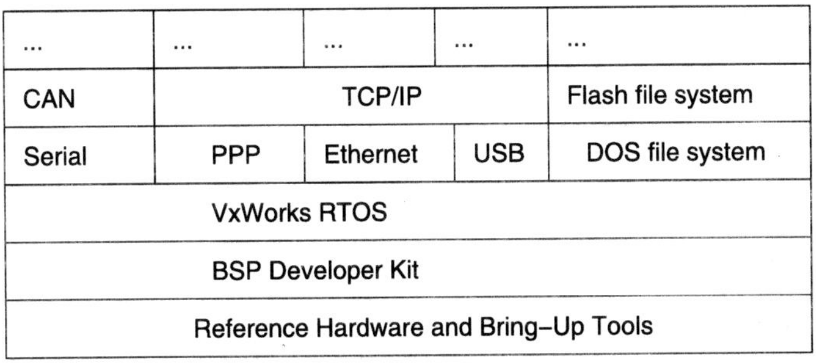

...	...	...	...	...	
CAN	TCP/IP			Flash file system	
Serial	PPP	Ethernet	USB	DOS file system	
VxWorks RTOS					
BSP Developer Kit					
Reference Hardware and Bring–Up Tools					

Figure 4.3. Software stack for Wind River® Industrial Automation Platform

instructions and tasks can be allowed to do their own I/O. This matches nicely with the previous item and reduces the overhead of I/O operations.

Example: Let switch correspond to the (memory-mapped) I/O address of some switch which needs to be checked by some program. We can simply use a

    load register,switch

instruction to query the switch. There is no need to go through an OS service call, which would create a lot of overhead for saving and restoring the task context (registers etc.).

However, there is a trend towards more dynamic embedded systems. Also, safety and security requirements might make protection necessary. Special memory protection units (MPUs) have been proposed for this (see Fiorin [Fiorin et al., 2007] for an example).

- **Interrupts can be connected to any process**. Using OS service calls, we can request the OS to start or stop tasks if certain interrupts happen. We could even store the start address of a task in the interrupt vector address table, but this technique is very dangerous, since the OS would be unaware of the task actually running. Also composability may suffer from this: if a specific task is directly connected to some interrupt, then it may be difficult to add another task which also needs to be started by some event. Application-specific device drivers (if used) might also establish links between interrupts and processes.

- Many embedded systems are real-time (RT) systems and, hence, the OS used in these systems **must be a real-time operating system** (RTOS).

Additional information about embedded operating systems can be found in a book chapter written by Bertolotti [Bertolotti, 2006]. This chapter comprises information about the architecture of embedded operating systems, the POSIX standard, open-source real-time operating systems and virtualization.

4.1.2   Real-time operating systems

Definition: (A) *"real-time operating system is an operating system that supports the construction of real-time systems"* [Takada, 2001].

What does it take to make an OS an RTOS? There are four key requirements[3]:

- **The timing behavior of the OS must be predictable.** For each service of the OS, an upper bound on the execution time must be guaranteed. In practice, there are various levels of predictability. For example, there may be sets of OS service calls for which an upper bound is known and for which there is not a significant variation of the execution time. Calls like "get me the time of the day" may fall into this class. For other calls, there may be a huge variation. Calls like "get me 4MB of free memory" may fall into this second class. In particular, the scheduling policy of any RTOS must be deterministic.

  There may also be times during which interrupts must be disabled to avoid interferences between components of the OS. Less importantly, they can also be disabled to avoid interferences between tasks. The periods during which interrupts are disabled must be quite short in order to avoid unpredictable delays in the processing of critical events.

  For RTOSs implementing file-systems on hard disks, it may be necessary to implement contiguous files (files stored in contiguous disk areas) to avoid unpredictable disk head movements.

- **The OS must manage the scheduling of tasks**. Scheduling can be defined as mapping from sets of tasks to intervals of execution time, including the mapping to start times as a special case. Also, the OS possibly has to be aware of task deadlines so that the OS can apply appropriate scheduling techniques (there are, however, cases in which scheduling is done completely off-line and the OS only needs to provide services to start tasks at specific times or priority levels). Scheduling algorithms will be discussed in detail in Chapter 6.

- **Some systems require the OS to manage time**. This management is mandatory if internal processing is linked to an absolute time in the physical environment. Physical time is described by real numbers. In computers,

---

[3]This section includes information from Hiroaki Takada's tutorial [Takada, 2001].

discrete time standards are typically used instead. The precise requirements may vary:

1 In some systems, synchronization with global time standards is necessary. In this case, **global clock synchronization** is performed. Two
   standards are available for this:

   – Universal Time Coordinated (UTC): UTC is defined by astronom-
     ical standards. Due to variations regarding the movement of the
     Earth, this standard has to be adjusted from time to time. Several
     seconds have been added during the transition from one year to the
     next. The adjustments can be problematic, since incorrectly imple-
     mented software could get the impression that the next year starts
     twice during the same night.

   – International atomic time (in French: *temps atomic internationale*,
     or TAI). This standard is free of any artificial artifacts.

   Some connection to the environment is used to obtain accurate time
   information. External synchronization is typically based on wireless
   communication standards such as the global positioning system (GPS)
   [National Space-Based Positioning, Navigation, and Timing Coordination Office, 2010] or mobile networks.

2 If embedded systems are used in a network, it is frequently sufficient
   to synchronize time information within the network. Local clock synchronization can be used for this. In this case, connected embedded
   systems try to agree on a consistent view of the current time.

3 There may be cases in which provision for precise local delays is all
   that is needed.

For several applications, precise time services with a high resolution must
be provided. They are required for example in order to distinguish between
original and subsequent errors. For example, they can help to identify the
power plant(s) that are responsible for blackouts (see [Novosel, 2009]). The
precision of time services depends on how they are supported by a particular execution platform. They are very imprecise (with precisions in the
millisecond range) if they are implemented through tasks at the application level and very precise (with precisions in the microsecond range) if
they are supported by communication hardware. More information about
time services and clock synchronization is contained in the book by Kopetz
[Kopetz, 1997].

■ **The OS must be fast**. An operating system meeting all the requirements
  mentioned so far would be useless, if it were very slow. Therefore, the OS
  must obviously be fast.

Each RTOS includes a so-called real-time OS **kernel**. This kernel manages the resources which are found in every real-time system, including the processor, the memory and the system timer. Major functions in the kernel include the task management, inter-task synchronization and communication, time management and memory management.

While some RTOSs are designed for general embedded applications, others focus on a specific area. For example, OSEK/VDX-compatible operating systems focus on automotive control. Operating systems for a selected area can provide a dedicated service for that particular area and can be more compact than operating systems for several application areas.

Similarly, while some RTOSs provide a standard API, others come with their own, proprietary API. For example, some RTOSs are compliant with the standardized POSIX RT-extension [Harbour, 1993] for UNIX, with the OSEK/VDX standard, or with the ITRON specification developed in Japan. Many RT-kernel type of OSs have their own API. ITRON, mentioned in this context, is a mature RTOS which employs link-time configuration.

Available RTOSs can further be classified into the following three categories [Gupta, 2002]:

- **Fast proprietary kernels:** According to Gupta, *"for complex systems, these kernels are inadequate, because they are designed to be fast, rather than to be predictable in every respect"*. Examples include QNX, PDOS, VCOS, VTRX32, VxWorks.

- **Real-time extensions to standard OSs:** In order to take advantage of comfortable main stream operating systems, hybrid systems have been developed. For such systems, there is an RT-kernel running all RT-tasks. The standard operating system is then executed as one of these tasks (see fig. 4.4).

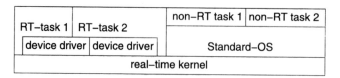

Figure 4.4.   Hybrid OSs

This approach has some advantages: the system can be equipped with a standard OS API, can have graphical user interfaces (GUIs), file-systems etc. and enhancements to standard OSs become quickly available in the embedded world as well. Also, problems with the standard OS and its non-RT tasks do not negatively affect the RT-tasks. The standard OS can even crash and this would not affect the RT-tasks. On the down side, and this is already visible from fig. 4.4, there may be problems with device drivers, since the standard OS will have its own device drivers. In order to avoid interference between the drivers for RT-tasks and those for the other tasks, it may be necessary to partition devices into those handled by RT-tasks and those handled by the standard OS. Also, RT-tasks cannot use the services of the standard OS. So all the nice features like file-system access and GUIs are normally not available to those tasks, even though some attempts may be made to bridge the gap between the two types of tasks without losing the RT-capability. RT-Linux is an example of such hybrid OSs.

According to Gupta [Gupta, 2002], trying to use a version of a standard OS is *"not the correct approach because too many basic and inappropriate underlying assumptions still exist such as optimizing for the average case (rather than the worst case), ... ignoring most if not all semantic information, and independent CPU scheduling and resource allocation"*. Indeed, dependences between tasks are not very frequent for most applications of standard operating systems and are therefore frequently ignored by such systems. This situation is different for embedded systems, since dependences between tasks are quite common and they should be taken into account. Unfortunately, this is not always done if extensions to standard operating systems are used. Furthermore, resource allocation and scheduling are rarely combined for standard operating systems. However, integrated resource allocation and scheduling algorithms are required in order to guarantee meeting timing constraints.

- There is a number of **research systems** which aim at avoiding the above limitations. These include Melody [Wedde and Lind, 1998], and (according to Gupta [Gupta, 2002]) MARS, Spring, MARUTI, Arts, Hartos, and DARK.

Takada [Takada, 2001] mentions low overhead memory protection, temporal protection of computing resources (targeting at preventing tasks from computing for longer periods of time than initially planned), RTOSs for on-chip multiprocessors (especially for heterogeneous multiprocessors and multi-threaded processors) and support for continuous media and quality of service control as research issues.

Due to the potential growth in the embedded system market, vendors of standard OSs are actively trying to sell variations of their products (like Windows Embedded [Microsoft Inc., 2003]) and obtain market shares from traditional vendors such as Wind River Systems [Wind River, 2010b].

4.1.3   Virtual machines

In certain environments, it may be useful to emulate several processors on a
single real processor. This is possible with **virtual machines** executed on the
bare hardware. On top of such a virtual machine, several operating systems
can be executed. Obviously, this allows several operating systems to be run on
a single processor. For embedded systems, this approach has to be used with
care since the temporal behavior of such an approach may be more problematic and timing predictability may be lost. Nevertheless, there may be cases in
which this approach is useful. For example, there may be the need to integrate
several legacy applications using different operating systems on a single hardware processor. A full coverage of virtual machines is beyond the scope of this
book. Interested readers should refer to books by Smith et al. [Smith and Nair,
2005] and Craig [Craig, 2006]. PikeOS is an example of a virtualization concept dedicated toward embedded systems [SYSGO AG, 2010]. PikeOS allows
the system's resources (e.g. memory, I/O devices, CPU-time) to be divided
into separate subsets. PikeOS comes with a small micro-kernel. Several operating systems, application programming interfaces (APIs) and run-time environments (RTEs) can be implemented on top of this kernel (see fig. 4.5).

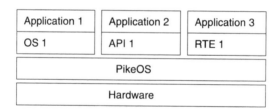

Figure 4.5.   PikeOS virtualization (©SYSGO)

4.1.4   Resource access protocols

4.1.4.1   Priority inversion

There are cases in which tasks must be granted exclusive access to resources
such as global shared variables or devices in order to avoid non-deterministic or
otherwise unwanted program behavior. Such exclusive access is very important
for embedded systems, e.g. for implementing shared memory-based communication or exclusive access to some special hardware device. Program sections
during which such exclusive access is required are called **critical sections**.
Critical sections should be short. Operating systems typically provide primitives for requesting and releasing exclusive access to resources, also called
mutex primitives. Tasks not being granted exclusive access must wait until

the resource is released. Accordingly, the release operation has to check for waiting tasks and resume the task of highest priority.

In this book, we will call the request operation P(S) and the release operation V(S), where S corresponds to the particular resource requested. P(S) and V(S) are so-called **semaphore** operations. Semaphores allow up to n (with n being a parameter) threads or processes to use a particular resource protected by S concurrently. S is a data structure maintaining a count on how many resources are still available. P(S) checks the count and blocks the caller if all resources are in use. Otherwise, the count is modified and the caller is allowed to continue. V(S) increments the number of available resources and makes sure that a blocked caller (if it exists) is unblocked. The names P(S) and V(S) are derived from the Dutch language. We will use these operations only in the form of binary semaphores with $n = 1$, i.e. we will allow only a single caller to use the resource.

For embedded systems, dependencies between tasks is a rule, rather than an exception. Also, the effective task priority of real-time applications is more important than for non-real applications. Mutually exclusive access can lead to priority inversion, an effect which changes the effective priority of tasks. Priority inversion exists on non-embedded systems as well. However, due to the reasons just listed, the priority inversion problem can be considered a more serious problem in embedded systems.

A first case of the consequences resulting from the combination of "mutual exclusion" with "no pre-emption" can be seen in fig. 4.6.

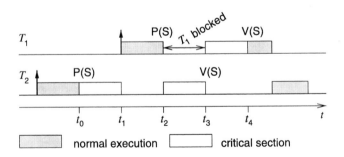

Figure 4.6. Blocking of a task by a lower priority task

Bold upward pointing arrows indicate the times at which tasks become executable, or "ready". At time t_0, task T_2 enters a critical section after requesting exclusive access to some resource via an operation P. At time t_1, task T_1 becomes ready and preempts T_2. At time t_2, T_1 fails getting exclusive access to the resource in use by T_2 and becomes blocked. Task T_2 resumes and after some time releases the resource. The release operation checks for pending tasks of higher priority and preempts T_2. During the time T_1 has been blocked, a lower priority task has effectively blocked a higher priority task. The necessity of providing exclusive access to some resources is the main reason for this effect. Fortunately, in the particular case of figure 4.6, the duration of the blocking cannot exceed the length of the critical section of T_2. This situation is problematic, but difficult to avoid.

In more general cases, the situation can be even worse. This can be seen, for example, from fig. 4.7.

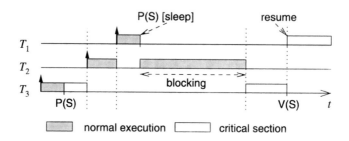

Figure 4.7.   Priority inversion with potentially large delay

We assume that tasks T_1, T_2 and T_3 are given. T_1 has the highest priority, T_2 has a medium priority and T_3 has the lowest priority. Furthermore, we assume that T_1 and T_3 require exclusive use of some resource via operation P(S). Now, let T_3 be in its critical section when it its preempted by T_2. When T_1 preempts T_2 and tries to use the same resource that T_3 is having exclusive access of, it blocks and lets T_2 continue. As long as T_2 is continuing, T_3 cannot release the resource. Hence, T_2 is effectively blocking T_1 even though the priority of T_1 is higher than that of T_2. In this example, the blocking of T_1 continues as long as T_2 executes. T_1 is blocked by a task of lower priority, which is not in its critical section. This effect is called **priority inversion**[4]. In fact, priority inversion happens even though T_2 is unrelated to T_1 and T_3. The duration of the priority inversion situation is not bounded by the length of any critical section. This example and other examples can be simulated with the levi simulation software [Sirocic and Marwedel, 2007c].

One of the most prominent cases of priority inversion happened in the Mars Pathfinder, where an exclusive use of a shared memory area led to priority inversion on Mars [Jones, 1997].

---

[4]Some authors do already consider the case of fig. 4.6 as a case of priority inversion. This was also done in earlier versions of this book.

4.1.4.2 Priority inheritance

One way of dealing with priority inversion is to use the priority inheritance protocol. This protocol is a standard protocol available in many real-time operating systems. It works as follows:

- Tasks are scheduled according to their active priorities. Tasks with the same priorities are scheduled on a first-come, first-served basis.

- When a task T_1 executes P(S) and exclusive access is already granted to some other task T_2, then T_1 will become blocked. If the priority of T_2 is lower than that of T_1, T_2 inherits the priority of T_1. Hence, T_2 resumes execution. In general, every task inherits the highest priority of tasks blocked by it.

- When a task T_2 executes V(S), its priority is decreased to the highest priority of the tasks blocked by it. If no other task is blocked by T_2, its priority is reset to the original value. Furthermore, the highest priority task so far blocked on S is resumed.

- Priority inheritance is transitive: if T_x blocks T_y and T_y blocks T_z, then T_x inherits the priority of T_z.

This way, high priority tasks being blocked by low priority tasks propagate their priority to the low priority tasks such that the low priority tasks can release semaphores as soon as possible.

In the example of fig. 4.7, T_3 would inherit the priority of T_1 when T_1 executes P(S). This would avoid the problem mentioned since T_2 could not preempt T_3 (see fig. 4.8).

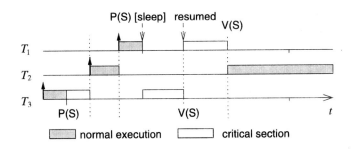

Figure 4.8. Priority inheritance for the example of fig. 4.7

Fig. 4.9 shows an example of nested critical sections [Buttazzo, 2002].

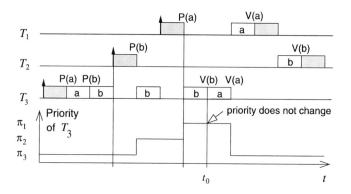

Figure 4.9.   Nested critical sections

Note that the priority of task T_3 is not reset to its original value at time t_0. Instead, its priority is decreased to the lowest priority of the tasks blocked by it, in this case the priority π_1 of T_1.

Transitiveness of priority inheritance is shown in fig. 4.10 [Buttazzo, 2002].

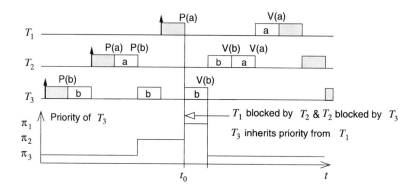

Figure 4.10.   Transitiveness of priority inheritance

At time t_0, T_1 is blocked by T_2 which in turn is blocked by T_3. Therefore, T_3 inherits the priority π_1 of T_1.

Priority inheritance is also used by ADA: during a rendez-vous, the priority of both tasks is set to their maximum.

Priority inheritance also solved the Mars Pathfinder problem: the VxWorks operating system used in the pathfinder implements a flag for the calls to mutex primitives. This flag allows priority inheritance to be set to "on". When the software was shipped, it was set to "off". The problem on Mars was corrected by using the debugging facilities of VxWorks to change the flag to "on", while the Pathfinder was already on Mars [Jones, 1997]. Priority inheritance can be simulated with the levi simulation software [Sirocic and Marwedel, 2007c].

While priority inheritance solves some problems, it does not solve others. There may be a large number of tasks having a high priority and there may even be deadlocks. The **priority ceiling protocol** [Sha et al., 1990] can be used instead, but requires processes to be known at design time.

4.2 ERIKA

Several embedded systems (such as automotive systems and home appliances) require the entire application to be hosted on small micro-controllers[5]. For that reason, the operating system services provided by the firmware on such systems must be limited to a minimal set of features allowing multi-threaded execution of periodic and aperiodic tasks, with support for shared resources to avoid the priority inversion phenomenon.

Such requirements have been formalized in the 1990s by the OSEK/VDX Consortium [OSEK Group, 2010], which defined the minimal services of a multi-threaded real-time operating system allowing implementations of 1-10 kilobytes code footprint on 8-bit micro-controllers. The OSEK/VDX API has been recently extended by the AUTOSAR Consortium [AUTOSAR, 2010] which provided enhancements to support time protection, scheduling tables for time triggered systems, and memory protection to protect the execution of different applications hosted on the same micro-controller. This section briefly describes the main features and requirements of such systems, considering as a reference implementation the open-source ERIKA Enterprise real-time kernel [Evidence, 2010].

The first feature that distinguishes an OSEK kernel from other operating systems is that all kernel objects are *statically* defined at compile time. In particular, most of these systems do not support dynamic memory allocation, and dynamic creation of tasks. To help the user in configuring the system, the OSEK/VDX standard provides a configuration language, named OIL, to specify the objects that must be instantiated in the application. When the application is compiled, the OIL Compiler generates the operating system data structures, allocating the exact amount of memory needed. This approach allows allocating only the data really needed by the application, to be put in flash memory (which is less expensive than RAM memory on most micro-controllers).

---

[5]This section was contributed by G. Buttazzo and P. Gai (Pisa).

The second feature distinguishing an OSEK/VDX system is the support for *Stack Sharing*. The reason for providing stack sharing is that RAM memory is very expensive on small micro-controllers. The possibility of implementing a stack sharing system is related to how the task code is written.

In traditional real-time systems, the typical implementation of a periodic task is structured according to the following scheme:

```
task(x) {
  int local;
  initialization();
  for (;;) {
   do_instance();
   end_instance();
}}
```

Such a scheme is characterized by a forever loop containing an instance of the periodic task that terminates with a blocking primitive (end_instance()), which has the effect of blocking the task until the next activation. When following such a programming scheme (called *extended task* in OSEK/VDX), the task is always present in the stack, even during waiting times. In this case, the stack cannot be shared and a separate stack space must be allocated for each task.

The OSEK/VDX standard also provides support for *basic tasks*, which are special tasks that are implemented in a way more similar to functions, according to the following scheme:

```
int local;
Task x() {
  do_instance();
}
System_initialization() {
  initialization();
  ...}
```

With respect to extended tasks, in basic tasks, the persistent state that must be maintained between different instances is not stored in the stack, but in global variables. Also, the initialization part is moved to system initialization, because tasks are not dynamically created, but they exist since the beginning. Finally, no synchronization primitive is needed to block the task until its next period, because the task is activated every time a new instance starts. Also, the task cannot call any blocking primitive, therefore it can either be preempted by higher priority tasks or execute until completion. In this way, the task behaves like a function, which allocates a frame on the stack, runs, and then cleans the frame. For this reason, the task does not occupy stack space between two executions, allowing the stack to be shared among all tasks in the system. ERIKA Enterprise supports stack sharing, allowing all basic tasks in the system to share a single stack, so reducing the overall RAM memory used for this purpose.

Concerning task management, OSEK/VDX kernels provide support for Fixed Priority Scheduling with Immediate Priority Ceiling to avoid the Priority Inversion problem. The usage of Immediate Priority Ceiling is supported through the specification of the resource usage of each task in the OIL configuration file. The OIL Compiler computes the resource ceiling of each task based on the resource usage declared by each task in the OIL file.

OSEK/VDX systems also support Non Preemptive Scheduling and Preemption Thresholds to limit the overall stack usage. The main idea is that limiting the preemption between tasks reduces the number of tasks allocated on the system stack at the same time, further reducing the overall amount of required RAM. Note that reducing preemptions may degrade the schedulability of the tasks set, hence the degree of preemption must be a traded off with the system schedulability and the overall RAM memory used in the system.

Another requirement for operating systems designed for small micro-controllers is *scalability*, which means supporting reduced versions of the API for smaller footprint implementations. In mass production systems, in fact, the footprint significantly impacts on the overall cost. In this context, scalability is provided through the concept of *Conformance Classes*, which define specific subsets of the operating system API. Conformance Classes are also accompanied by an upgrade path between them, with the final objective of supporting partial implementation of the standard with reduced footprint. The conformance classes supported by the OSEK/VDX standard (and by ERIKA Enterprise) are:

- BCC1: This is the smallest Conformance class, supporting a minimum of 8 tasks with different priority and 1 shared resource.

- BCC2: Compared to BCC1, this conformance class adds the possibility to have more than one task at the same priority. Each task can have pending activations, that is, the operating system records the number of instances that have been activated but not yet executed.

- ECC1: Compared to BCC1, this conformance class adds the possibility to have Extended tasks that can wait for an event to appear.

- ECC2: This Conformance class adds both multiple activations and Extended tasks.

ERIKA Enterprise further extends these conformance classes by providing the following two conformance classes:

- EDF: This conformance class does not use a fixed priority scheduler but an Earliest Deadline First (EDF) Scheduler (see section 6.2.2.3) optimized for the implementation on small micro-controllers.

- FRSH: This conformance class extends the EDF scheduler class by providing a resource reservation scheduler based on the IRIS scheduling algorithm [Marzario et al., 2004].

Anther interesting feature of OSEK/VDX systems is that the system provides an API for controlling interrupts. This is a major difference when compared to POSIX-like systems, where the interrupts are exclusive domain of the operating system and are not exported to the operating system API. The rationale for this is that on small micro-controllers users often want to directly control interrupt priorities, hence it is important to provide a standard way to deal with interrupt disabling/enabling. Moreover, the OSEK/VDX standard specifies two types of Interrupt Service Routines (ISR):

- Category 1: simpler and faster, does not implement a call to the scheduler at the end of the ISR;

- Category 2: this ISR can call some primitives that change the scheduling behavior. The end of the ISR is a rescheduling point. ISR1 has always a higher priority of ISR2.

An important feature of OSEK/VDX kernels is the possibility to fine tune the footprint by removing error checking code from the production versions, as well as to define hooks that will be called by the system when specific events occur. These features allow for a fine tuning of the application footprint that will be larger (and safer) when debugging and smaller in production when most bugs will be found and removed from the code.

To support a better debugging experience, the OSEK/VDX standard defines a textual language, named ORTI, which describes where the various objects of the operating system are allocated. The ORTI file is typically generated by the OIL compiler and is used by debuggers to print detailed information about operating system objects defined in the system (for example, the debugger could print the list of the tasks in an application with their current status).

All the features defined by the OSEK/VDX standard have been implemented in the open-source ERIKA Enterprise kernel [Evidence, 2010], for a set of embedded micro-controllers, with a final footprint ranging between 1 and 5 kilobytes of object code. ERIKA Enterprise also implements additional features, like the EDF scheduler, providing an open and free of charge operating system that can be used to learn, test and implement real applications for industrial and educational purposes.

4.3   Hardware abstraction layers

Hardware abstraction layers (HALs) provide a means for accessing hardware through a hardware-independent application programming interface (API). For example, we could come up with a hardware-independent technique for accessing timers, irrespective of the addresses to which timers are mapped. Hardware abstraction layers are used mostly between the hardware and operating system layers. They provide software intellectual property (IP), but they are neither part of operating systems nor can they be classified as middleware. A survey over work in this area is provided by Ecker, Müller and Dömer [Ecker et al., 2009].

4.4   Middleware

Communication libraries provide a means for adding communication functionality to languages lacking this feature. They add communication functionality on top of the basic functionality provided by operating systems. Due to being added on top of the OS, they can be independent of the OS (and obviously also of the underlying processor hardware). As a result, we will obtain **networked embedded systems**. There is a trend towards supporting communication within some local system as well as communication over longer distances. The use of Internet protocols is becoming more popular.

4.4.1   OSEK/VDX COM

OSEK/VDX® COM is a special communication standard for the OSEK automotive operating systems [OSEK Group, 2004][6]. OSEK COM provides an "Interaction Layer" as an application programming interface (API) through which internal communication (communication within one ECU) and external communication (communication with other ECUs) can be performed. OSEK COM specifies just the functionality of the Interaction layer. Conforming implementations must be developed separately.

---

[6]OSEK is a trademark of Continental Automotive GmbH.

The Interaction layer communicates with other ECUs via a "Network Layer" and a "Data Link" layer. Some requirements for these layers are specified by OSEK COM, but these layers themselves are not part of OSEK COM. This way, communication can be implemented on top of different network protocols.

OSEK COM is an example of communication middleware dedicated toward embedded systems. In addition to middleware dedicated toward embedded systems, many communication standards developed for non-embedded applications can be adopted for embedded systems as well.

4.4.2   CORBA

CORBA® (Common Object Request Broker Architecture) [Object Management Group (OMG), 2003] is one example of such adopted standards. CORBA facilitates the access to remote services. With CORBA, remote objects can be accessed through standardized interfaces. Clients are communicating with local stubs, imitating the access to the remote objects. These clients send information about the object to be accessed as well as parameters (if any) to the Object Request Broker (ORB, see fig. 4.11). The ORB then determines the location of the object to be accessed and sends information via a standardized protocol, e.g. the IIOP protocol, to where the object is located. This information is then forwarded to the object via a skeleton and the information requested from the object (if any) is returned using the ORB again.

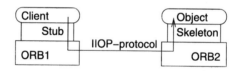

Figure 4.11.   Access to remote objects using CORBA

Standard CORBA does not provide the predictability required for real-time applications. Therefore, a separate real-time CORBA (RT-CORBA) standard has been defined [Object Management Group (OMG), 2005a]. A very essential feature of RT-CORBA is to provide *end-to-end predictability of timeliness in a fixed priority system*. This involves *respecting thread priorities between client and server for resolving resource contention*, and bounding the latencies of operation invocations. One particular problem of real-time systems is that thread priorities might not be respected when threads obtain mutually exclusive access to resources. The priority inversion problem (see page 186) has to be addressed in RT-CORBA. RT-CORBA includes provisions for bounding the time during which such priority inversion can happen. RT-CORBA also includes facilities for thread priority management. This priority is independent of the priorities of the underlying operating system, even though it is compatible with the real-time extensions of the POSIX standard for operating systems [Harbour, 1993]. The thread priority of clients can be propagated to the server side. Priority management is also available for primitives providing mutually exclusive access to resources. The priority inheritance protocol just described must be available in implementations of RT-CORBA. Pools of pre-existing threads avoid the overhead of thread creation and thread-construction.

4.4.3   MPI

As an alternative to CORBA, the message passing interface (MPI) can be used for communication between different processors. MPI is a very frequently used library, initially designed for high-performance computing. It is based on message passing and allows a choice between synchronous and asynchronous message passing. For example, synchronous message passing is possible with the MPI_Send library function [MHPCC, 2010]:

MPI_Send(buffer,count,type,dest,tag,comm) where:

- buffer: is the address of data to be sent,
- count: is the number of data elements to be sent,
- type: is the data type of data to be sent (e.g. MPI_CHAR, MPI_SHORT, MPI_INT),
- dest: is the process id of the target process,
- tag: is a message id (for sorting incoming messages),
- comm: is the communication context (set of processes for which destination field is valid) and
- function result: indicates success.

The following is an asynchronous library function:

MPI_Isend(buffer,count,type,dest,tag,comm,request) where

- buffer, count, type, dest, tag, comm: are same as above, and
- the system issues a unique "request number". The programmer uses this system assigned "handle" later (in a WAIT type routine) to determine completion of the non-blocking operation.

For MPI, the partitioning of computations among various processors must be done explicitly and the same is true for the communication and the distribution of data. Synchronization is implied by communication, but explicit synchronization is also possible. As a result, much of the management code is explicit and causes a major amount of work for the programmer. Also, it does not scale well when the number of processors is significantly changed [Verachtert, 2008].

In order to apply the MPI-style of communication to real-time systems, a real-time version of MPI, called MPI/RT has been defined [MPI/RT forum, 2001]. MPI-RT does not cover some of the issues covered in RT-CORBA, such as thread creation and termination. MPI/RT is conceived as a potential layer between the operating system and standard (non real-time) MPI.

MPI is available on a variety of platforms and also considered for multiple processors on a chip. However, it is based on the assumption that memory accesses are faster than communication operations. Also, MPI is mainly targeting at homogeneous multi-processors. These assumptions are not true for multiple processors on a chip.

4.4.4   POSIX Threads (Pthreads)

The POSIX thread (Pthread) library is an application programming interface (API) to threads at the operating system level [Barney, 2010]. Pthreads are consistent with the IEEE POSIX 1003.1c operating system standard. A set of threads can be run in the same address space. Therefore, communication can be based on shared memory communication. This avoids the memory copy operations typically required for MPI. The library is therefore appropriate for programming multi-core processors sharing the same address space. The library includes a standard API with mechanisms for mutual exclusion. Pthreads use completely explicit synchronization [Verachtert, 2008]. The exact semantics depends on the memory consistency model used. Synchronization is hard to program correctly. The library can be employed as a back-end for other programming models.

4.4.5   OpenMP

For OpenMP, parallelism is mostly explicit, whereas computation partitioning, communication, synchronization etc. are implicit. Parallelism is expressed with pragmas: for example, loops can be preceded by pragmas indicating that they should be parallelized. The following program demonstrates a small parallel loop [OpenMP Architecture Review Board, 2008]:

```
void a1(int n, float *a, float *b)
    {int i;
```

```
#pragma omp parallel for
    for (i=1; i<n; i++) /* i is private by default */
        b[i] = (a[i] + a[i-1]) / 2.0;
}
```

This means that (among the approaches just introduced) OpenMP requires the least amount of effort for parallelization for the user. However, this also means that the user cannot control partitioning [Verachtert, 2008]. OpenMP is targeted towards shared memory hardware. There are first applications for MP-SoCs (see, for example [Marongiu and Benini, 2009]).

4.4.6   UPnP, DPWS and JXTA

Universal Plug-and-Play (UPnP) is an extension of the plug-and-play concept of PCs towards devices connected within a network. Connecting network printers, storage space and switches in homes and offices easily can be seen as the key target [UPnP Forum, 2010]. Due to security concerns, only data is exchanged. Code cannot be transfered.

Devices Profile for Web Services (DPWS) aims at being more general than UPnP. *"The Devices Profile for Web Services (DPWS) defines a minimal set of implementation constraints to enable secure Web Service messaging, discovery, description, and eventing on resource-constrained devices"* [ws4d, 2010]. DPWS specifies services for discovering devices connected to a network, for exchanging information about available services, and for publishing and subscribing to events.

In addition to libraries designed for high-performance computing (HPC), several comprehensive network communication libraries can be used. These are typically designed for a loose coupling over Internet-based communication protocols. JXTA^{TM} [JXTA Community, 2010] is an open source peer-to-peer protocol specification. It defines a protocol by a set of XML messages that allow any device connected to a network peer to exchange messages and collaborate independently of the network topology. JXTA creates a virtual overlay network, allowing a peer to interact with other peers even when some of the peers and resources are behind firewalls. The name is derived from the word "juxtapose".

CORBA, MPI, Pthreads, OpenMP, UPnP, DPWS and JXTA are special cases of communication middleware (software to be used at a layer between the operating system and applications). Initially, they were essentially designed for communication between desktop computers. However, there are attempts to leverage the knowledge and techniques also for embedded systems. For mobile devices like smart phones, this approach may be appropriate. For "hard real-time systems", their overhead, their real-time capabilities and their services may be inappropriate.

4.5   Real-time databases

Data bases provide a convenient and structured way of storing and accessing information. Accordingly, databases provide an API for writing and reading information. A sequence of read and write operations is called a **transaction**. Transactions may have to be aborted for a variety of reasons: there could be hardware problems, deadlocks, problems with concurrency control etc. A frequent requirement is that transactions do not affect the state of the database unless they have been executed to their very end. Hence, changes caused by transactions are normally not considered to be final until they have been **committed**. Most transactions are required to be **atomic**. This means that the end result (the new state of the database) generated by some transaction must be the same as if the transaction has been fully completed or not at all. Also, the database state resulting from a transaction must be **consistent**. Consistency requirements include, for example, that the values from read requests belonging to the same transaction are consistent (do not describe a state which never existed in the environment modeled by the database). Furthermore, to some other user of the database, no intermediate state resulting from a partial execution of a transaction must be visible (the transactions must be performed as if they were executed in **isolation**). Finally, the results of transactions should be persistent. This property is also called **durability** of the transactions. Together, the four properties printed in bold are known as ACID properties (see the book by Krishna and Shin [Krishna and Shin, 1997], Chapter 5).

For some databases, there are soft real-time constraints. For example, time-constraints for airline reservation systems are soft. In contrast, there may also be hard constraints. For example, automatic recognition of pedestrians in automobile applications and target recognition in military applications must meet hard real-time constraints. The above requirements make it very difficult to guarantee hard real-time constraints. For example, transactions may be aborted various times before they are finally committed. For all databases relying on demand paging and on hard disks, the access times to disks are hardly predictable. Possible solutions include main memory databases and predictable use of flash memory. Embedded databases are sometimes small enough to make this approach feasible. In other cases, it may be possible to relax the ACID requirements. For further information, see the book by Krishna and Shin.

4.6   Assignments

1 Which requirements must be met for a real-time operating system? How do they differ from the requirements of a standard OS?

2 How many seconds have been added at New Year's Eve to compensate for the differences between UTC and TAI since 1958? You may search the Internet for an answer to this question.

3 Which features of a standard OS like Windows or Linux could be missing in an RTOS?

4 Find processors for which memory protection units are available! How are memory protection units different from the more frequently used memory management units (MMUs)? You may search the Internet for an answer to this question.

5 Describe classes of embedded systems for which protection should definitely be provided! Describe classes of systems, for which we would possibly not need protection!

6 Provide an example demonstrating priority inversion for a system comprising three tasks!

7 Download the levi learning module leviRTS from the levi web site [Sirocic and Marwedel, 2007c]. Model a task set as described in figure 4.12.

Task	Priority	Arrival	Run time	Printer		Comm line	
				$t_{P,P}$	$t_{V,P}$	$t_{P,C}$	$t_{V,C}$
T_1	1 (high)	3	4	1	4	-	-
T_2	2	10	3	-	-	1	2
T_3	3	5	6	-	-	4	6
T_4	4 (low)	0	7	2	5	-	-

Figure 4.12. Task set requesting exclusive use of resources

$t_{P,P}$ and $t_{P,C}$ are the times relative to the start times, at which a task requests exclusive use of the printer or the communication line, respectively (called ΔtP in levi). $t_{V,P}$ and $t_{V,C}$ are the times relative to the start times at which these resources are released. Use priority-based, preemptive scheduling! Which problem occurs? How can it be solved?

8 Which impact does the priority inversion problem have on the design of network middleware?

9 How could flash memory have an influence on the design of real-time databases?

Chapter 5

EVALUATION AND VALIDATION

5.1 Introduction

5.1.1 Scope

Specification, hardware platforms and system software provide us with the basic ingredients which we need for designing embedded systems. During the design process, we must validate and evaluate designs rather frequently. Therefore, we will describe validation and evaluation before we talk about design steps. Validation and evaluation, even though different from each other, are very much linked.

Definition: **Validation** is the process of checking whether or not a certain (possibly partial) design is appropriate for its purpose, meets all constraints and will perform as expected.

Definition: Validation with mathematical rigor is called **(formal) verification**.

Validation is important for any design procedure, and hardly any system would work as expected, had it not been validated during the design process. Validation is extremely important for safety-critical embedded systems. In theory, we could try to design verified tools which always generate correct implementations from the specification. In practice, this verification of tools does not work, except in very simple cases. As a consequence, each and every design has to be validated. In order to minimize the number of times that we must validate a design, we could try to validate it at the very end of the design process. Unfortunately, this approach normally does not work either, due to the large differences between the level of abstraction used for the specification and that used for the implementation. Therefore, validation is required at various

P. Marwedel, *Embedded System Design*, Embedded Systems,
DOI 10.1007/978-94-007-0257-8_5, © Springer Science+Business Media B.V. 2011

phases during the design procedure (see fig. 5.1). Validation and design should be intertwined and not be considered as two completely independent activities.

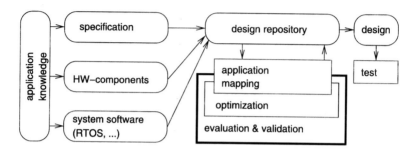

Figure 5.1.   Context of the current Chapter

It would be nice to have a single validation technique applicable to all validation problems. In practice, none of the available techniques solves all the problems, and a mix of techniques has to be applied. In this Chapter, starting in Section 5.6, we will provide a brief overview of key techniques which are available. This material will be preceded by an overview of evaluation techniques.

Definition: **Evaluation** is the process of computing quantitative information of some key characteristics (or "objectives") of a certain (possibly partial) design.

5.1.2   Multi-objective optimization

Design evaluations will, in general, lead to a characterization of the design by several criteria, such as average and worst case execution time, energy consumption, code size, dependability and safety. Merging all these criteria into a single objective function (e.g. by using a weighted average) is usually not advisable, as this would hide some of the essential characteristics of designs. Rather, it is advisable to return to the designer a set of designs among which the designer can then select an appropriate design. Such a set should, however, only contain "reasonable" designs. Finding such sets of designs is the purpose of **multi-objective optimization techniques**.

In order to perform multi-objective optimization, we do consider an m-dimensional space X of possible solutions of the optimization problem. These dimensions could, for example, reflect the number of processors, the sizes of memory, types and number of buses. For this space X, we define an n-dimensional function

$$f(x) = (f_1(x), \ldots, f_n(x)) \text{ where } x \in X$$

which evaluates designs with respect to several criteria or objectives (e.g. cost and performance). Let F be the n-dimensional space of values of these objectives (the so-called **objective space**). Suppose that, for each of the objectives, some total order $<$ and the corresponding $\leq$-order are defined. In the following, we assume that the goal is to **minimize** our objectives.

Definition: Vector $u = (u_1, ..., u_n) \in F$ **dominates** vector $v = (v_1, ..., v_n) \in F$ iff u is "better" than v with respect to at least one objective and not worse than v with respect to all other objectives:

$$\forall i \in \{1, ...n\} \quad : \quad u_i \leq v_i \ \wedge \tag{5.1}$$
$$\exists i \in \{1, .., n\} \quad : \quad u_i < v_i \tag{5.2}$$

Definition: Vector $u \in F$ is called **indifferent** with respect to vector $v \in F$ iff neither u dominates v nor v dominates v.

Definition: A design $x \in X$ is called **Pareto-optimal** with respect to X iff there is no design $y \in X$ such that $u = f(x)$ is dominated by $v = f(y)$.

The previous definition defines Pareto-optimality in the solution space. The next definition serves the same purpose in the objective space.

Definition: Let $S \subseteq F$ be a subset of vectors in the objective space. $v \in F$ is called a **non-dominated solution** with respect to S iff v is not dominated by any element $\in S$. v is called Pareto-optimal iff v is non-dominated with respect to all solutions F.

Fig. 5.2 highlights the different areas in the objective space, relative to design point (1).

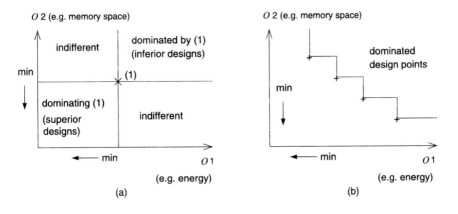

Figure 5.2.   (a) Pareto point   (b) Pareto front

The upper right area corresponds to designs that would be dominated by design (1), since they would be "worse" with respect to both objectives. Designs in the lower left rectangle (if they would exist) would dominate design (1), since the would be "better" with respect to both objectives. Designs in the upper left and the lower right area are indifferent: they are "better" with respect to one objective and "worse" with respect to the other. Fig. 5.2 (right) shows a set of Pareto points, i.e., the so-called **Pareto front**.

Design space exploration (DSE) based on Pareto points is the process of finding and returning a set of Pareto-optimal solutions to the designer, enabling the designer to select the most appropriate implementation.

5.1.3   Relevant objectives

For PC-like systems, the expected average performance plays a dominating role during the design of new systems. For embedded and cyber-physical systems, multiple objectives need to be considered. The following list explains if and where this objective is discussed in this book:

1 **Average performance:** An analysis of this objective is frequently based on simulations. Section 5.6 briefly presents some issues in simulations. An abundant amount of additional information on the simulation of systems (in particular of heterogeneous, cyber-physical systems) is available. Due to the large number of physical effects, it is impossible to provide a complete list of references.

2 **Worst case performance/real-time behavior:** Details of the aiT timing analysis tool will be presented in section 5.2.2.

3 **Energy/power consumption:** A brief overview of techniques for evaluating this objective will be presented in section 5.3.

4 **Temperatures/thermal behavior:** A brief introduction to this topic will be presented in section 5.4.

5 **Reliability:** An introduction to the theory of reliability can be found in section 5.5.

6 **Electromagnetic compatibility:** This objective will not be considered in this book.

7 **Numeric precision:** A minor loss in numerical precision can be tolerated in several applications. Accepting such a loss can improve the design in terms of other objectives. As an example, we will discuss the transformation from floating point to fixed point arithmetic in section 7.2.1. Several similar other cases exist.

8 **Testability:** Costs for testing systems can be very large, sometimes larger even than production costs. Hence, testability should be considered as well, preferably already during the design. Testability will be discussed in Chapter 8.

9 **Cost:** Cost in terms of silicon area or real money will not be considered in this book.

10 **Weight, robustness, usability, extendibility, security, safety, environmental friendliness:** These objectives will also not be considered.

There may be even more objectives than the ones listed above. The next section presents several approaches for performance evaluation, with a focus on the worst case performance.

5.2 Performance evaluation

Performance evaluation aims at predicting the performance of systems. This is a major challenge (especially for cyber-physical systems) since we might need worst case information, rather than just average case information. Such information is necessary in order to guarantee real-time constraints.

5.2.1 Early phases

Two different classes of techniques have been proposed for obtaining performance information already during early design phases:

- **Estimated cost and performance values:** Quite a number of estimators have been developed for this purpose. Examples include the work by Jha and Dutt [Jha and Dutt, 1993] for hardware, Jain et al. [Jain et al., 2001], and Franke [Franke, 2008] for software. Generating sufficiently precise estimates requires considerable efforts.

- **Accurate cost and performance values:** We can also use the real software code (in the form of some binary) on a close-to-real hardware platform. This is only possible if interfaces to "software synthesis tools" (compilers) and hardware synthesis tools exist. This method can be more precise than the previous one, but may be significantly (and sometimes prohibitively) more time consuming.

In order to obtain sufficiently precise information, communication needs to be considered as well. Unfortunately, it is typically difficult to compute communication cost already during early design phases.

5.2.2   WCET estimation

Formal performance evaluation techniques have been proposed by many researchers. For embedded systems, the work of Thiele et al., Henia and Ernst et al., and Wilhelm et al. is particularly relevant (see, for example, [Thiele, 2006b], [Henia et al., 2005], and [Wilhelm, 2006]). These techniques require some knowledge of architectures. They are less appropriate for very early design phases, but some of them can still be used without knowing all the details about target architectures. These approaches model real, physical time.

Scheduling of tasks requires some knowledge about the duration of task executions, especially if meeting time constraints has to be guaranteed, as is in real-time (RT) systems. The **worst case execution time** (WCET) is the basis for most scheduling algorithms. Some definitions related to the WCET are shown in fig. 5.3.

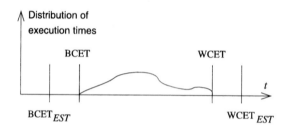

Figure 5.3.   WCET-related terms

The worst case execution time is the largest execution time of a program for any input and any initial execution state. Unfortunately, the WCET is extremely difficult to compute. In general, it is undecidable whether or not the WCET is finite. This is obvious from the fact that it is undecidable whether or not a program terminates. Hence, the WCET can only be computed for certain programs/tasks. For example, for programs without recursion, without while loops and with loops having statically known iteration counts, the WCET can be computed. But even with such restrictions, it is usually practically impossible to compute the WCET. The effect of modern processor architectures' pipelines with their different kinds of hazards and memory hierarchies with limited predictability of hit rates is difficult to precisely predict at design time. Computing the WCET for systems containing caches, pipelines, interrupts and virtual memory is an even greater challenge. As a result, we must be happy if we are able to compute good **upper bounds** on the WCET.

Such upper bounds are usually called **estimated worst case execution time**s, or WCET$_{EST}$. Such bounds should have at least two properties:

1 The bounds should be safe ($\text{WCET}_{EST} \geq \text{WCET}$).

2 The bounds should be tight ($\text{WCET}_{EST}\text{-WCET} \ll \text{WCET}$)

Note that the term "estimated" does not mean that the resulting times are unsafe.

Sometimes, architectural features which reduce the average execution time but cannot guarantee to reduce the WCET are completely omitted from the real-time designs (see page 145). Computing tight upper bounds on the execution time may still be difficult. The architectural features described above also present problems for the computation of WCET_{EST}.

Accordingly, the **best-case execution time** (BCET) and the corresponding estimate BCET_{EST} are defined in an analogous manner. The BCET_{EST} is a safe and tight lower bound on the execution time.

Computing tight bounds from a program written in a high-level language such as C without any knowledge of the generated assembly code and the underlying architectural platform is impossible. Therefore, a safe analysis must start from real machine code. Any other approach would lead to unsafe results.

In the following, we will study WCET estimation more closely. The presentation is based on the description of the tool aiT by R. Wilhelm [Wilhelm, 2006]. The architecture of aiT is shown in fig. 5.4.

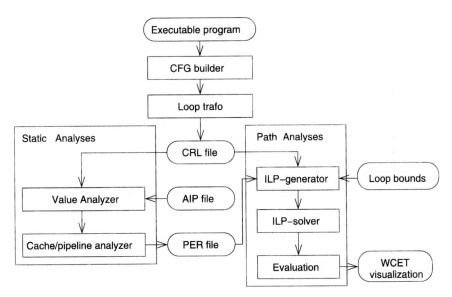

Figure 5.4. Architecture of the aiT timing analysis tool

Consistent with our remark about the problems with high-level code, aiT starts from an executable object file comprising the code to be analyzed. From this code, a control-flow graph (CFG) is extracted. Next, loop transformations are applied. These include transformations between loops and recursive function calls as well as virtual loop unrolling. This unrolling is called "virtual" since it is performed internally, without actually modifying the code to be executed. Results are represented in the CRL (control flow representation language) format. The next phase employs different static analyses. Static analyses read the AIP-file comprising designer's annotations. These annotations contain information which is difficult or impossible to extract automatically from the program (for example, bounds of complex loops). Static analyses include value-, cache-, and pipeline analyses.

A **value analysis** computes enclosing intervals for possible values in registers and local variables. The resulting information can be used for control-flow analysis and for data-cache analysis. Frequently, values such as addresses are precisely known (especially for "clean" code) and this helps in predicting accesses to memories.

The next step is **cache** and **pipeline analysis**. In the following, we will present a few details about the cache analysis.

Suppose that we are using an n-way set associative cache (see fig. 5.5)[1].

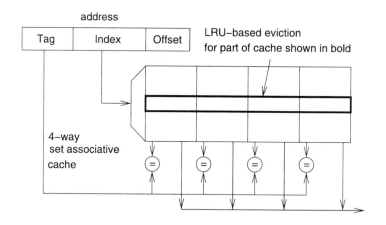

Figure 5.5. Set associative cache (for $n=4$)

We consider that part (row) of the cache corresponding to a certain index (shown in bold in fig. 5.5). We assume that eviction from that part of the cache

---

[1]We assume that students are familiar with concepts of caches.

is controlled by the least recently used (LRU) strategy. This means that among all references for a particular index, the last n referenced memory blocks are stored in that part of the cache. We assume that the necessary LRU management hardware is available for each index and that each index is handled independently of other indexes. Under this assumption, all evictions for a particular index are completely independent of decisions for other indexes. This independence is extremely important, since it allows us to consider each of the indexes independently.

Let us now consider a partial cache and a particular index. Suppose that we have information about potential entries for each of the cache ways (columns). Furthermore, consider control flow joins. What do we know about the content of the partial cache after the join? We must distinguish between *may-* and *must*-information and the corresponding analysis. Must-analysis reveals the entries which **must** be in the cache. This information is useful for computing the WCET. May-analysis identifies the entries which **may** be in the cache. This information is typically used to conclude that certain information will definitely not be in the cache. This knowledge is then exploited during the computation of the BCET. As an example of must- and may-analysis, we consider must information at control flow joins. Fig. 5.6 shows the corresponding situation. Entries on the left are assumed to be younger than the ones on the right.

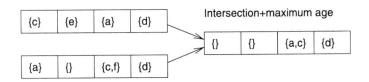

Figure 5.6. Must-analysis at program joins for LRU-caches

In fig. 5.6, memory object c is assumed to be the youngest object for one path to the join and a assumed to be the youngest object for the other path to the join. The age of the other entries is defined accordingly. What do we know about the "worst" case after the join? A certain entry is guaranteed to be in the cache only if it is guaranteed to be in the cache for both paths. This means that the **intersection** of the memory objects defines the result of the must-analysis after the join. As a worst case, we must assume the **maximum of the ages** along the two paths. Fig. 5.6 shows the result. Obviously, this analysis has to use sets of entries for each of the cache ways.

Let us now consider may-analysis for control flow joins. Fig. 5.7 depicts the situation.

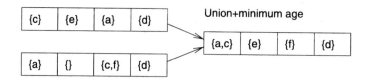

Figure 5.7. May-analysis at program joins for LRU-caches

Some object being in the cache on either of the two paths to the join *may* be in the cache after the join. This means that the set of objects which may be in the cache after the join consists of the **union** of the objects that were in the cache before the join. As a best case, we use the **minimum of the ages** before the join. Fig. 5.7 shows the result.

For any reference to a memory block b, the accessed memory block moves to the youngest position, and all the other memory blocks age by 1.

Static analyses also comprise pipeline analysis. Pipeline analysis has to compute safe bounds on the number of cycles required in order to execute machine code in the machine pipeline. Details of pipeline analysis are explained by R. Wilhelm [Wilhelm, 2006] and S. Thesing [Thesing, 2004].

The overall result of static analyses consists of bounds on the execution times for each of the basic blocks of a program. Results are written to the PER-file shown in fig. 5.4.

aiT's next phase uses these bounds in order to derive worst case execution times for the entire program. This step is based on an ILP model (see page 335). In this model, the overall execution time is used as the objective function. The overall execution time is calculated as the sum over the execution-time estimates of basic blocks multiplied by their execution frequencies. The execution time of basic blocks is defined as the WCET of a single execution of the block (as computed during static analysis) multiplied by the the worst case execution count of that block. Only some of the execution counts of blocks can be determined automatically. Therefore, building the ILP model relies on additional designer-provided information, e.g. about loop bounds. This information is read in from the external AIP-file. Constraints model relations between blocks. This technique for modeling execution time is called **implicit path enumeration**, since the problem of enumerating the potentially large number of execution paths is avoided. The ILP problem defined in this way can be solved with some standard ILP solver maximizing the objective function. The generated maximum yields a safe upper bound on the overall execution time. aiT also provides a visualization of the results in the form of annotated control flow graphs. These graphs can be analyzed by the designer in order to optimize the system under design.

5.2.3   Real-time calculus

Thiele's **real-time calculus (RTC)** is based on the description of the rate of incoming events[2]. This description also includes fluctuations of this rate. Towards this end, the timing characteristics of a sequence (or stream) of events are represented by a tuple of *arrival curves*:

$$\overline{\alpha}^{\,u}(\Delta), \overline{\alpha}^{\,l}(\Delta) \in \mathbb{R} \geq 0, \Delta \in \mathbb{R} \geq 0$$

These curves represent the maximal resp. the minimal number of events arriving within a time interval of length Δ. There are at most $\overline{\alpha}^{\,u}(\Delta)$ and at least $\overline{\alpha}^{\,l}(\Delta)$ events arriving within the time interval $(t, t + \Delta)$ for all $t \geq 0$. Fig. 5.8 shows the number of possibly arriving events for some possible models of arriving events.

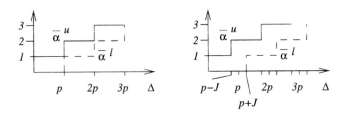

Figure 5.8.   Arrival curves: periodic stream (left), periodic stream with jitter J (right)

For example, in the case of periodic event streams with period p, there is a maximum of a single event happening in time interval $(0, p)$[3]. Similarly, there is an upper bound of two events within time interval $(p, 2p)$. Now, let us consider the lower bound for time interval $(0, p)$. There is possibly not a single event in this interval. Hence, the bound is zero. For time interval $(p, 2p)$, there has to be at least one event. Therefore, the bound is one. So, for $\Delta = 0.5p$, there will be at least zero and at most one incoming event (see fig. 5.8 (left)). In the case of periodic event streams with jitter J, these curves are shifted by this amount (see fig. 5.8 (right)). The upper bound is shifted to the left, the lower bound is shifted to the right. The jitter is assumed not to be accumulating. We

---

[2]Our presentation of the real-time calculus is based on Thiele's presentation in the book edited by Zurawski [Thiele, 2006b]. Resulting considerations at the system level have been called *modular performance analysis* (MPA).

[3]We leave out the subtle discussion of dis-continuities at $\Delta = n * p$.

are using bars on top of symbols (like $\bar{\alpha}$) for all entities referring to incoming events.

Available computational and communication service capacity can be described by *service functions*:

$$\beta^u(\Delta), \beta^l(\Delta) \in \mathbb{R} \geq 0, \Delta \in \mathbb{R} \geq 0$$

These functions allow us to model situations in which the available service capacity is fluctuating. Fig. 5.9 shows the communication capacity of some *time division multiple access* (TDMA) bus (see page 161). Allocation is done periodically with a period of p. Bus arbitration allocates this bus during a time window s time units long. During this window, the bus achieves a band width of b.

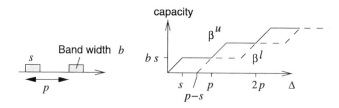

Figure 5.9.   Service functions for a TDMA bus

The upper bound is obtained if the bus is allocated exactly at the time we are starting our observation. The transfered amount is then increasing linearly. The lower bound is obtained if the bus was just deallocated when we started our observation of length Δ. Then we must wait $p-s$ time units until the bus gets allocated again.

Separate methods are required to determine $\bar{\alpha}$ and β for streams of ("external") events arriving at the system to be modeled. Their computation is not part of RTC. In contrast, bounds for events generated within the system are derived by the calculus (see below).

Up till now, there is no information about the **workload** required by each of the incoming events. This workload is represented by additional functions $\gamma^u(e), \gamma^l(e) \in \mathbb{R} \geq 0$ for each sequence e of incoming events. This information can be derived from bounds on the execution time of code required for each of the events. Fig. 5.10 shows an example of such functions. This example is based on the assumption that between three and four time units are required for processing a single event.

Accordingly, the workload for a single event varies between three and four time units, the work load for two events varies between six and eight time units, etc.

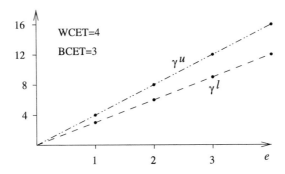

Figure 5.10. Work load characterization

The dashed lines are not part of the function, since it is defined only for an integer number of events. The work load resulting from an incoming stream of events can now be easily computed. Upper and lower bounds are characterized by the functions

$$\alpha^{u}(\Delta) = \gamma^{u}(\overline{\alpha}^{u}(\Delta)) \text{ and} \tag{5.3}$$
$$\alpha^{l}(\Delta) = \gamma^{l}(\overline{\alpha}^{l}(\Delta)) \tag{5.4}$$

There should be enough computational or communication capacity to handle this work load. The number of events which can be processed with the available computational capacity can be computed as

$$\overline{\beta}^{u}(\Delta) = (\gamma^{l})^{-1}(\beta^{u}(\Delta)) \text{ and} \tag{5.5}$$
$$\overline{\beta}^{l}(\Delta) = (\gamma^{u})^{-1}(\beta^{l}(\Delta)) \tag{5.6}$$

Equations 5.5 and 5.6 use the inverse of functions γ^{u} and γ^{l} to convert bounds on the available capacity (measured in real time units) into bounds measured in terms of the number of events that can be processed.

Based on this information, it is possible to derive the properties of outgoing streams of events from incoming streams of events. Suppose the incoming stream is characterized by bounds $[\overline{\alpha}^{l}, \overline{\alpha}^{u}]$. We can then compute characteristics of the outgoing streams such as the corresponding bounds $[\overline{\alpha}^{l'}, \overline{\alpha}^{u'}]$ of the outgoing stream of events and the remaining service capacity, available for other tasks. This remaining capacity is derived by transforming *service curves* $[\overline{\beta}^{l}, \overline{\beta}^{u}]$ into *service curves* $[\overline{\beta}^{l'}, \overline{\beta}^{u'}]$ (see fig. 5.11). This remaining service capacity can be employed for lower priority tasks to be executed on the same processor.

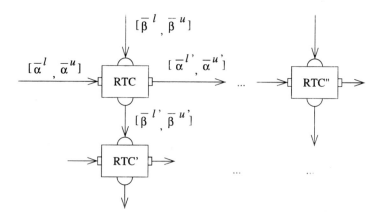

Figure 5.11.   Transformation of event stream and service capacities by real-time components

According to Thiele et al., outgoing streams and remaining service capacities are bounded by the following functions [Thiele, 2006b]:

$$\overline{\alpha}^{u\prime} = [(\overline{\alpha}^{u}\otimes\overline{\beta}^{u})\oslash\overline{\beta}^{l}]\wedge\overline{\beta}^{u} \qquad (5.7)$$

$$\overline{\alpha}^{l\prime} = [(\overline{\alpha}^{l}\oslash\overline{\beta}^{u})\otimes\overline{\beta}^{l}]\wedge\overline{\beta}^{l} \qquad (5.8)$$

$$\overline{\beta}^{u\prime} = (\overline{\beta}^{u}-\overline{\alpha}^{l})\oslash0 \qquad (5.9)$$

$$\overline{\beta}^{l\prime} = (\overline{\beta}^{l}-\overline{\alpha}^{u})\overline{\otimes}0 \qquad (5.10)$$

Operators used in these equations are defined as follows:

$$(f\otimes g)(t) = inf_{0\leq u\leq t}\{f(t-u)+g(u)\} \qquad (5.11)$$

$$(f\overline{\otimes} g)(t) = sup_{0\leq u\leq t}\{f(t-u)+g(u)\} \qquad (5.12)$$

$$(f\overline{\oslash} g)(t) = sup_{u\geq 0}\{f(t+u)-g(u)\} \qquad (5.13)$$

$$(f\underline{\oslash} g)(t) = inf_{u\geq 0}\{f(t+u)-g(u)\} \qquad (5.14)$$

$\wedge$ denotes the minimum operator.

In essence, these equations characterize outgoing streams and capacities. These equations have been adopted from communications theory. Proofs regarding these equations are provided by Network Calculus [Le Boudec and Thiran, 2001]. The easiest way of using these equations is to download a Matlab toolbox [Wandeler and Thiele, 2006].

The same theory also allows to compute the delay caused by the real-time components as well as the size of the buffer required to temporarily store incoming/outgoing events. This way, performance and other characteristics of the system can be computed from information about the components.

A second performance analysis method has been proposed by Henia, Ernst et al. In this so-called SymTA/S approach [Henia et al., 2005], the different curves in Thiele's approach are replaced by standard models of event streams such as periodic event streams, periodic event streams with jitter and periodic streams with bursts. SymTA/S explicitly supports the combination and integration of different kinds of analysis techniques known from real-time research.

5.3 Energy and power models

Energy models and **power models** are essential for evaluating the corresponding objectives. The two models are closely related, as can be seen from equation 3.13. Such models are needed for optimizations aiming at a reduction of power and energy consumptions. They are also required for optimizations trying to reduce operating temperatures.

- One of the first power models was proposed by Tiwari [Tiwari et al., 1994]. It is based on measurements on a real system. Measured values are then associated with executed instructions. The model includes so-called base costs and inter-instruction costs. Base costs of an instruction correspond to the energy consumed per instruction execution if an infinite sequence of instances of that instruction is executed. Inter-instruction costs model the additional energy consumed by the processor if instructions change. This additional energy is required, for example, due to switching functional units on and off. This power model focuses on the consumption in the processor and does not consider the power consumed in the memory or in other parts of the system.

- Another power model was proposed by Simunic et al. [Simunic et al., 1999]. That model is based on data sheets. The advantage of this approach is that the contribution of all components of an embedded system to the energy consumption can be computed. However, the information in data sheets about average values may be less precise than the information about maximal or minimal values.

- A third model has been proposed by Rusell and Jacome [Rusell and Jacome, 1998]. This model is based on precise measurements of two fixed configurations.

- Still another model was proposed by Lee [Lee et al., 2001]. This model includes a detailed analysis of the effects of the pipeline. It does not include multicycle operations and pipeline stalls.

- The energy model by Steinke et al. [Steinke et al., 2001] is based on precise measurements using real hardware. The consumption of the processor as well as that of the memory are included. This model has been integrated into the energy-aware compiler encc from TU Dortmund.

- The energy consumption of caches can be computed with CACTI [Wilton and Jouppi, 1996].

- The Wattch power estimation tool [Brooks et al., 2000] estimates the power consumption of microprocessor systems at the architectural level, without requiring detailed information at the circuit or layout level.

- Several commercial tools provide power estimation.

Power estimation is used in **power management** algorithms (see page 313).

These examples lead to the following general conclusion: for some real, existing hardware, precise power models can be generated with a limited effort. However, during design space exploration, such hardware is typically not available and the resulting power models may be imprecise[4].

5.4   Thermal models

The quest for higher performances of embedded systems has increased the chances of components becoming hot during their operation. Temperatures of the various components of embedded systems can have a serious impact on their usability. In the worst case, overheated components can cause damages to other systems. For example, they may cause fire hazards. Overheated components can also cause the embedded systems themselves to fail. However, hot components might also have other consequences, even in the absence of immediate failures. For example, the useful system life might be shortened, sometimes by rather large factors.

The thermal behavior of embedded systems is closely linked to the transformation of electrical energy into heat. Therefore, thermal models are usually linked to energy models. Thermal models are based on the laws of physics. **Thermal conductance** is the key quantity considered in thermal modeling. The thermal conductance of a certain material reflects the amount of heat transfered through a plate (made from that material) of area A and thickness L when the temperatures at the opposite sides differ by one Kelvin. The reciprocal of thermal conductance is called **thermal resistance**. For stacked plates in close contact, the effective overall thermal resistance is the sum of the individual resistances.

---

[4]Deviations of about 50% are sometimes mentioned in discussions.

This means, thermal resistances add up like electrical resistances in an electrical network. This correspondence also extends to masses storing heat: such masses correspond to capacitors of electrical networks. As a result, thermal modeling typically uses equivalent electrical models and employs well-known techniques for solving electrical network equations (see, for example, Chen et al. [Chen et al., 2010]).

Tools for thermal modeling include HotSpot [Skadron et al., 2009], a tool which can be integrated with power simulators such as Wattch (see page 218). Both tools can be interfaced to the SimpleScalar functional simulator [Simple Scalar LLC, 2004]. Validation of thermal models requires precise temperature measurements [Mesa-Martinez et al., 2010].

5.5   Risk- and dependability analysis

Embedded and cyber-physical systems (like other products) can cause damages to properties and lives. It is not possible to reduce the risk of damages to zero. The best that we can do is to make the probability of damages small, hopefully orders of magnitude smaller than other risks. This task is expected to become more difficult in the future, since **decreasing feature sizes of semiconductors will be resulting in a reduced reliability of semiconductor devices** [ITRS Organization, 2009]. Transient as well as permanent faults are expected to become more frequent. Shrinking feature sizes will also cause an increased variability among device parameters. Therefore, dependability analysis and fault tolerant designs are becoming extremely important [Mukherjee, 2008], [Garg and Khatri, 2009] . Faults within semiconductors might lead to failures of the system. The terms **faults**, **failures** and the related terms **error** and **service** were defined by Laprie et al. [Laprie, 1992], [Avižienis et al., 2004].

Definitions:

- *"The **service** delivered by a system (in its role as a **provider**) is its behavior as it is perceived by its user(s); ... The delivered service is a sequence of the provider's external states. ... **Correct service** is delivered when the service implements the system function"*.

- *"A **service failure**, often abbreviated here to **failure**, is an event that occurs when the delivered service of a system deviates from the correct service. ... A service failure is a transition from correct service to incorrect service"*.

- An **error** exists if one of the system's states is incorrect and may lead to its subsequent service failure.

- *"The adjudged or hypothesized cause of an error is called a **fault**. Faults can be internal or external of a system."*

Some faults will not cause a system failure.

As an example, we might consider a transient *fault* flipping a bit in memory. After this bit flip, the memory cell will be in *error*. A *failure* will occur if the system service is affected by this error.

In line with these definitions, we will talk about *failure* rates when we consider systems that do not provide the expected system function. We will talk about *faults* whenever we consider the underlying **reasons** that might cause failures. There is a large number of possible reasons for faults, some of them resulting from reduced feature sizes of semiconductors. *Errors* will not be considered in the remaining part of this book.

For many applications, a rate of a catastrophe has to be less than 10^{-9} per hour [Kopetz, 1997], corresponding to one case per 100,000 systems operating for 10,000 hours. Reaching this level of dependability is only feasible if design evaluation also comprises the analysis of the reliability, the expected life-time and related objectives. Such an analysis is usually based on the probability of failures.

More precisely, we consider the probability densities of failures. Let x be the time until the first failure. x is a random variable. Let $f(x)$ be the probability density of this random variable.

As an example, we are frequently using the exponential probability density $f(x) = \lambda e^{-\lambda x}$. For this density function, failures are becoming less and less likely over time (after some time, it is likely that the system is not working any more and a system which is not working cannot fail). This density function is frequently used since it has nice mathematical properties and since the actual time dependency of the failure rate is often unknown. In the absence of knowledge about the latter, a constant rate is assumed, leading to the exponential density function. The exponential distribution will possibly be inexact, but it is assumed that it does typically provide at least a first rough approximation of the real system. Fig. 5.12 (left) shows this density function.

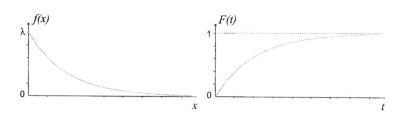

Figure 5.12. Density function and probability distribution for exponential distributions

The probability distribution is frequently more interesting than the density. This distribution represents the probability of a system not working at time t. It can be obtained by integrating the density function until time t.

$$F(t) = Pr(x \le t) \tag{5.15}$$

$$F(t) = \int_0^t f(x)dx \tag{5.16}$$

For example, for the exponential distribution we obtain:

$$F(t) = \int_0^t \lambda e^{-\lambda x}dx = -[e^{-\lambda x}]_0^t = 1 - e^{-\lambda t} \tag{5.17}$$

Fig. 5.12 (right) contains the corresponding function. As time advances, this probability approaches 1. This means that, as time progresses, it becomes likely that the system will have failed.

Definition: The **reliability** $R(t)$ of a system is the probability of the time until the first failure being larger than t:

$$R(t) = Pr(x > t), t \ge 0 \tag{5.18}$$

$$R(t) = \int_t^\infty f(x)dx \tag{5.19}$$

$$F(t) + R(t) = \int_0^t f(x)dx + \int_t^\infty f(x)dx = 1 \tag{5.20}$$

$$R(t) = 1 - F(t) \tag{5.21}$$

$$f(x) = -\frac{dR(t)}{dt} \tag{5.22}$$

For the exponential distribution, we have $R(t) = e^{-\lambda t}$ (see fig. 5.13).

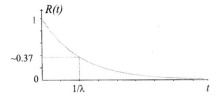

Figure 5.13. Reliability for exponential distributions

The probability for the system to be functional after time $t = 1/\lambda$ is about 37%.

Definition: The **failure rate** $\lambda(t)$ is the probability of a system failing between time t and time $t + \Delta t$.

$$\lambda(t) \quad = \quad \lim_{\Delta t \to 0} \frac{Pr(t < x \le t + \Delta t | x > t)}{\Delta t} \qquad (5.23)$$

$Pr(t < x \le t + \Delta t | x > t)$ is the conditional probability for the system failing within this time interval provided that it was working at time t. For conditional probabilities, there is the general equation $Pr(A|B) = Pr(AB)/Pr(B)$, where $Pr(AB)$ is the probability of A *and* B happening. $Pr(AB)$ is equal to $F(t + \Delta t) - F(t)$ in our case. $Pr(B)$ is the probability of the system working at time t, which is $R(t)$ in our notation. Therefore, equation 5.23 leads to:

$$\lambda(t) \quad = \quad \lim_{\Delta t \to 0} \frac{F(t + \Delta t) - F(t)}{\Delta t R(t)} = \frac{f(t)}{R(t)} \qquad (5.24)$$

For example, for the exponential distribution we obtain[5]:

$$\lambda(t) \quad = \quad \frac{f(t)}{R(t)} = \frac{\lambda e^{-\lambda t}}{e^{-\lambda t}} = \lambda \qquad (5.25)$$

Failure rates are frequently measured as multiples (or fractions) of 1 FIT, where "FIT" stands for *Failure unIT* and is also known as *Failures In Time*. 1 FIT corresponds to 1 failure per 10^9 hours.

However, failure rates of real systems are usually not constant. For many systems, we have a "bath tub"-like behavior (see fig. 5.14).

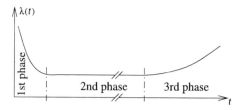

Figure 5.14.   Bath tub-like failure rates

---

[5]This result motivates denoting the failure rate and the constant of the exponential distribution with the same symbol.

For this behavior, we are starting with an initially larger failure rate. This higher rate is a result of an imperfect production process or "infant mortality". The rate during the normal operating life is then essentially constant. At the end of the useful product life, the rate is then increasing again, due to wear-out.

Definition: The **Mean Time To Failure** (**MTTF**) is the average time until the next failure, provided that the system was initially working. This average can be computed as the expected value of random variable x:

$$\text{MTTF} \quad = \quad E\{x\} = \int_0^\infty xf(x)dx \qquad (5.26)$$

For example, for the exponential distribution we obtain:

$$\text{MTTF} \quad = \quad \int_0^\infty x\lambda e^{-\lambda x}dx \qquad (5.27)$$

This integral can be computed using the product rule ($\int uv' = uv - \int u'v$ where in our case we have $u = x$ and $v' = \lambda e^{-\lambda x}$). Therefore, equation 5.27 leads to the following equation:

$$\text{MTTF} \quad = \quad -[xe^{-\lambda x}]_0^\infty + \int_0^\infty e^{-\lambda x}dx \qquad (5.28)$$

$$= \quad -\frac{1}{\lambda}[e^{-\lambda x}]_0^\infty = -\frac{1}{\lambda}[0-1] = \frac{1}{\lambda} \qquad (5.29)$$

This means that, for the exponential distribution, the expected time until the next failure is the reciprocal value of the failure rate.

Definition: The **Mean Time To Repair** (**MTTR**) is the average time to repair a system, provided that the system is initially not working. This time is the expected value of the random variable denoting the time to repair.

Definition: The **Mean Time Between Failures** (**MTBF**) is the average time between two failures.

MTBF is the sum of MTTF and MTTR:

$$\text{MTBF} = \text{MTTF} + \text{MTTR} \qquad (5.30)$$

Figure 5.15 shows a simplistic view of this equation: it is not reflecting the fact that we are dealing with probabilistic events and actual MTBF, MTTF, and MTTR values may vary randomly.

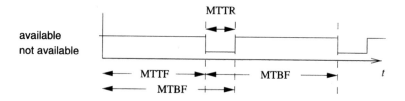

Figure 5.15.   Illustration of MTTF, MTTR and MTBF

For many systems, repairs are not considered. Also, if they are considered, the MTTR should be much smaller than the MTTF. Therefore, the terms MTBF and MTTF are frequently mixed up. For example, the life-time of a hard disk may be quoted as a certain MTBF, even though it will never be repaired. Quoting this number as the MTTF would be more correct. Still, the MTTF provides only very rough information about dependability, especially if there are large variations in the failure rates over time.

Definition: The **availability** is the probability of a system being in an operational state.

The availability varies over time (just consider the bath tub curve!). Therefore, we can model availability by a time-dependent function $A(t)$. However, we are frequently only considering the availability A for large time intervals. Hence, we define

$$ A \ = \ \lim_{t \to \infty} A(t) = \frac{\text{MTTF}}{\text{MTBF}} \qquad (5.31) $$

For example, assume that we have a system which is repeatedly available for 999 days and then needs one day for repair. Such a system would have an availability of $A = 0.999$.

Allowed failure rates can be in the order of 1 FIT. This may be several orders of magnitude less than the failure rates of chips. This means that systems must be more reliable than their components! Obviously, the required level of reliability makes fault tolerance techniques a must!

Obtaining actual failure rates is difficult. Fig. 5.16 shows one of the few published results [TriQuint Semiconductor Inc., 2010].

This figure contains failure rates for different Gallium-Arsenide (GaAs) devices with the hottest transistor operating at a temperature of 150 C. This example is used here to demonstrate that there exist devices for which the assumptions of constant failure rates or a bath tub-like behavior are oversim-

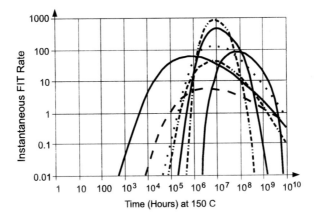

Figure 5.16. Failure rates of TriQuint Gallium-Arsenide devices (courtesy of TriQuint, Inc., Hillsboro), ©TriQuint

plifying. As a result, citing a single MTTF number may be misleading. The actual distribution of failures over time should be used instead. In the particular case of this example, failure rates are less than 100 FIT for the first 20 years (175,300 hrs) of product life time, despite the high temperature. FIT numbers are actually very much temperature dependent and temperatures up to 275 C and known temperature dependences have been used at Triquint to compute failure rates for periods larger than the time available for testing. Triquint claims that their GaAs devices are more reliable than average silicon devices. Reports on FIT testing are also available for Xilinx FPGAs (see, for example, [Xilinx, 2009]).

It is frequently not possible to experimentally verify failure rates of complete systems. Requested failure rates are too small and failures may be unacceptable. We cannot fly 10^5 airplanes 10^4 hours each in an attempt to check if we reach a failure rate of less than 10^{-9}! The only way out of this dilemma is to use a combination of checking failure rates of components and formally deriving from this guarantees for a reliable operation of the system. Design- and user-generated failures also must be taken into account. It is state of the art to use decision diagrams to compute the reliability of a system from that of its components [Israr and Huss, 2008].

Damages are resulting from **hazards** (chances for a failure). For each possible damage caused by a failure, there is a severity (the cost) and a probability. Risk can be defined as the product of the two. Information concerning the damages resulting from component failures can be derived with at least two techniques [Dunn, 2002], [Press, 2003]:

- **Fault tree Analysis (FTA):** FTA is a top-down method of analyzing risks. The analysis starts with a possible damage and then tries to come up with possible scenarios that lead to that damage. FTA is based on modeling a Boolean function reflecting the operational state of the system (operational or not operational). FTA typically includes symbols for AND- and OR-gates, representing conditions for possible damages. OR-gates are used if a single event could result in a hazard. AND-gates are used when several events or conditions are required for that hazard to exist. Fig. 5.17 shows an example[6]. FTA is based on a **structural** model of the system, i.e. it reflects the partitioning of the system into components.

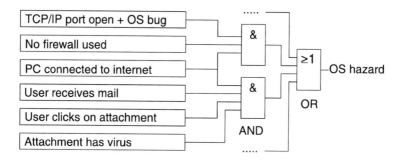

Figure 5.17. Fault tree

The simple AND- and OR-gates cannot model all situations. For example, their modeling power is exceeded if shared resources of some limited amount (like energy or storage locations) exist. Markov models [Bremaud, 1999] may have to be used to cover such cases. Markov models are based the notion of **states**, rather than on the structure of the system.

- **Failure mode and effect analysis (FMEA):** FMEA starts at the components and tries to estimate their reliability. Using this information, the reliability of the system is computed from the reliability of its parts (corresponding to a bottom-up analysis). The first step is to create a table containing components, possible failures, probability of failures and consequences on the system behavior. Risks for the system as a whole are then computed from the table. Figure 5.18 shows an example.

Tools supporting both approaches are available. Both approaches may be used in "safety cases". In such cases, an independent authority has to be convinced

---

[6]Consistent with the ANSI/IEEE standard 91, we use the symbols &, =1 and ≥1 to denote and-, xor-, and or-gates, respectively.

Component	Failure	Consequences	Probability	Critical?
...	...	...	...	...
Processor	metal migration	no service	10^{-7} /h	yes
...	...	...	...	...

Figure 5.18. FMEA table

that certain technical equipment is indeed safe. One of the commonly requested properties of technical systems is that no single failing component should potentially cause a catastrophe.

Safety requirements cannot come in as an afterthought, but must be considered right from the beginning. The design of safe and dependable systems is a topic by its own. This book can only provide a few hints into this direction.

According to Kopetz [Kopetz, 2003], the following must be taken into account: For safety-critical systems, the system as a whole must be more dependable than any of its parts. Allowed failures may be in the order of 1 failure per 10^9 hours of operation. This may be in the order of 1000 times less than the failure rates of chips. Obviously, fault-tolerance mechanisms must be used. Due to the low acceptable failure rate, systems are not 100% testable. Instead, safety must be shown by a combination of testing and reasoning. Abstraction must be used to make the system explainable using a hierarchical set of behavioral models. Design faults and human faults must be taken into account. In order to address these challenges, Kopetz proposed the following twelve design principles:

1 Safety considerations may have to be used as **the** important part of the specification, driving the entire design process.

2 Precise specifications of design hypotheses must be made right at the beginning. These include expected failures and their probability.

3 Fault containment regions (FCRs) must be considered. Faults in one FCR should not affect other FCRs.

4 A consistent notion of time and state must be established. Otherwise, it will be impossible to differentiate between original and follow-up errors.

5 Well-defined interfaces must hide the internals of components.

6 It must be ensured that components fail independently.

7 Components should consider themselves to be correct unless two or more other components pretend the contrary to be true (principle of self-confidence).

8 Fault tolerance mechanisms must be designed such that they do not create any additional difficulty in explaining the behavior of the system. Fault tolerance mechanisms should be decoupled from the regular function.

9 The system must be designed for diagnosis. For example, it has to be possible to identify existing (but masked) errors.

10 The man-machine interface must be intuitive and forgiving. Safety should be maintained despite mistakes made by humans.

11 Every anomaly should be recorded. These anomalies may be unobservable at the regular interface level. This recording should involve internal effects, since otherwise they may be masked by fault-tolerance mechanisms.

12 Provide a never-give up strategy. Embedded systems may have to provide uninterrupted service. The generation of pop-up windows or going off line is unacceptable.

For further information about dependability and safety issues, consult books [Laprie, 1992], [Neumann, 1995], [Leveson, 1995], [Storey, 1996], [Geffroy and Motet, 2002] on those areas.

There is an abundant amount of recent publications on the impact of reliability issues on system design. Examples include publications by Huang [Huang and Xu, 2010], Zhuo [Zhuo et al., 2010], and Pan [Pan et al., 2010].

5.6   Simulation

Simulation is a very common technique for **evaluating and validating** designs. Simulation consists of executing a design model on appropriate computing hardware, typically on general purpose digital computers. Obviously, this requires models to be executable. All the executable models and languages introduced in Chapter 2 can be used in simulations, and they can be used at various levels as described starting at page 107. The level at which designs are simulated is always a compromise between simulation speed and accuracy. The faster the simulation, the less accuracy is available.

So far, we have used the term behavior in the sense of the functional behavior of systems (their input/output behavior). There are also simulations of some non-functional behaviors of designs, including the thermal behavior and the electro-magnetic compatibility (EMC) with other electronic equipment. Due to the integration with physics, there is a large range of physical effects which may have to be included in the simulation model. As a result, it is impossible to cover all relevant approaches for simulating cyber-physical systems in this book. Law [Law, 2006] provides an overview of approaches and topics in simulations on digital systems.

For cyber-physical systems, simulations have serious limitations:

- Simulations are typically a lot slower than the actual design. Hence, if we interface the simulator with the actual environment, we can have quite a number of **violations of timing constraints**.

- Simulations in the physical environment may even be **dangerous** (who would want to drive a car with unstable control software?).

- For many applications, there may be huge amounts of data and it may be impossible to simulate enough data in the available time. Multimedia applications are notoriously known for this. For example, simulating the compression of some video stream takes an enormous amount of time.

- Most actual systems are too complex to allow simulating all possible cases (inputs). Hence, simulations can help us to find errors in our designs. They cannot guarantee absence of errors, since simulations cannot exhaustively be done for all possible combinations of inputs and internal states.

Due to these limitations, there is an increased emphasis on validation by formal verification (see page 231). Nevertheless, sophisticated simulation techniques continue to play a key role for validation (see, for example, Braun et al. [Braun et al., 2010]).

5.7 Rapid prototyping and emulation

Simulations are based on models, which are approximations of real systems. In general, there will be some difference between the real system and the model. We can reduce the gap by implementing some parts of our SUD more precisely than in a simulator (for example, in a real, physical component).

Definition: Adopting a definition phrased by McGregor [McGregor, 2002], we define **emulation** as the process of executing a model of the SUD where at least one component is **not** represented by simulation on some kind of host computer.

According to McGregor, "*Bridging the credibility gap is not the only reason for a growing interest in emulation—the above definition of an emulation model remains valid when turned around— an emulation model is one where part of the real system is replaced by a model. Using emulation models to test control systems under realistic conditions, by replacing the ... (real system) ... with a model, is proving to be of considerable interest to those responsible for commissioning, or the installation and start-up of automated systems of many kinds.*"

In order to improve credibility further, we can continue replacing simulated components by real components. These components do not have to be the final components. They can be approximations of the real system itself, but should exceed the precision of simulations.

Note that it is now common to discuss the "emulation" of one computer on another computer by means of software. There is a lack of a precise definition of the use of the term in this context. However, it can be considered consistent with our definition, since the emulated computer is not just simulated. Rather, a speed faster than simulation speed is expected.

Definition: Fast prototyping is the process of executing a model of the SUD where **no** component is represented by simulation on some kind of host computer. Rather, all components are represented by realistic components. Some of these components should not yet be the finally used components (otherwise, this would be the real system).

There are many cases in which the designs should be tried out in realistic environments before final versions are manufactured. Control systems in cars are an excellent example for this. Such systems should be used by drivers in different environments before mass production is started. Accordingly, the automotive industry designs prototypes. These prototypes should essentially behave like the final systems, but they may be larger, more power consuming and have other properties which test drivers can accept. The term "prototype" can be associated with the entire system, comprising electrical and mechanical components. However, the distinction between rapid prototyping and emulation is also blurring. Rapid prototyping is by itself a wide area which cannot be comprehensively covered in this book.

Prototypes and emulators can be built, for example, using FPGAs. Racks containing FPGAs can be stored in the trunk while test drivers exercise the car. This approach is not limited to the automotive industry. There are several other cases in which prototypes are built from FPGAs. Commercially available **emulators** consist of a large number of FPGAs. They come with the required mapping tools which map specifications to these emulators. Using these emulators, experiments with systems which behave "almost" like the final systems can be run. However, catching errors by prototyping and emulation is already a problem for non-distributed systems. For distributed systems, the situation is even more difficult (see, for example, Tsai [Tsai and Yang, 1995]).

5.8 Formal Verification

Formal verification[7] is concerned with formally proving a system correct, using the language of mathematics. First of all, a formal model is required to make formal verification applicable. This step can hardly be automated and may require some effort. Once the model is available, we can try to prove certain properties.

Formal verification techniques can be classified by the type of logic employed:

- **Propositional logic:** In this case, models consist of Boolean functions. Tools are called **Boolean checkers**, **tautology checkers** or **equivalence checkers**. They can be used to verify that two representations of Boolean functions (or sets of Boolean functions) are equivalent. Since propositional logic is decidable, it is also decidable whether or not the two representations are equivalent (there will be no cases of doubt). For example, one representation might correspond to gates of an actual circuit and the other to its specification. Proving the equivalence then proves the effect of all design transformations (for example, optimizations for power or delay) to be correct. Boolean checkers can cope with designs which are too large to allow simulation-based exhaustive validation. The key reason for the power of Boolean checkers is the use of Binary Decision Diagrams (BDDs) [Wegener, 2000]. The complexity of equivalence checks of Boolean functions represented with BDDs is linear in the number of BDD-nodes. In contrast, the equivalence check for functions represented by sums of products is NP-hard. BDD-based equivalence checkers have therefore replaced simulators for this application and handle circuits with millions of transistors.

- **First order logic (FOL):** FOL includes ∃ and ∀ operators. Typically, integers are also allowed. Some automation for verifying FOL models is feasible. However, since FOL is undecidable, there may be cases of doubt. Popular techniques include the **Hoare calculus**.

- **Higher order logic (HOL):** Higher order is based on lambda-calculus and allows functions to be manipulated like other objects [University of Cambridge, 2010]. For higher order logic, proofs can hardly ever be automated and typically must be done manually with some proof-support.

Propositional logic can be used to verify stateless logic networks, but cannot directly model finite state machines. For short input sequences, it may be sufficient to cut the feed-back loop in FSMs and to effectively deal with several

---

[7]This text on formal verification is based on a guest lecture given by Tiziana Margaria-Steffen at TU Dortmund.

copies of these FSMs, each copy representing the effect of one input pattern. However, this method does not work for longer input sequences. Such sequences can be handled with **model checking**.

For model checking, we have two inputs to the verification tool:

1 the model to be verified, and

2 properties to be verified.

States can be quantified with $\exists$ and $\forall$; numbers cannot. Verification tools can prove or disprove the properties. In the latter case, they can provide a counterexample. Model checking is easier to automate than FOL. It has been implemented for the first time in 1987, using BDDs. It was possible to locate several errors in the specification of the *future bus* protocol [Clarke et al., 2005].

As a next step, there have been attempts to integrate model checking and higher order logic. In this integrated model, HOL is used only where it is absolutely necessary.

Clarke's EMC-system [Clarke and et al., 2003] (see fig. 5.19) is an example of this approach.

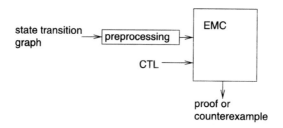

Figure 5.19.   Clarke's EMC system

This system accepts properties to be described as CTL formulas. CTL-formulas include two parts:

- a **path quantifier** (this part specifies paths in the state transition diagram), and

- a **state quantifier** (this part specifies states).

Example: $M, s \models AGg$ means: In the transition graph M, property g holds for all paths (denoted by A) starting at state s and all states (denoted by G). Extensions are needed in order to also cover real-time behavior and numbers.

This technique could be used, for example, to prove properties of the railway model of fig. 2.52 (see page 75). It should be possible to convert the Petri net into a state chart and then confirm that the number of trains commuting between Cologne and Paris is indeed constant, confirming our discussion of Petri net place invariants on page 73.

5.9   Assignments

1 Let us consider an example demonstrating the concept of Pareto-optimality. In this example, we study the results generated by task concurrency management (TCM) tools designed at the IMEC research center (*Interuniversitair Micro-Electronica Centrum*). TCM tools aim at establishing efficient mappings from applications to processors. Different multi-processor systems are evaluated and represented as sets of Pareto-optimal designs. Wong et al. [Wong et al., 2001] describe different options for the design of an MPEG-4-player. The authors assume that a combination of StrongARM-Processors and specialized accelerators should be used. Four designs meet the timing constraint of 30 ms (see figure 5.20).

Processor combination	1	2	3	4
Number of high speed processors	6	5	4	3
Number of low speed processors	0	3	5	7
Total number of processors	6	8	9	10

Figure 5.20.   Processor configurations

These different designs are shown in fig. 5.21.

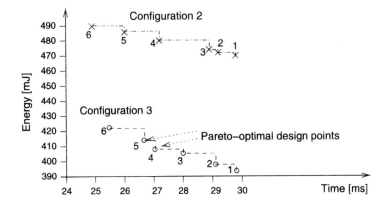

Figure 5.21.   Pareto points for multi processor systems 2 und 3

For combinations 1 and 4, the authors report that only one mapping of tasks to processors meets the timing constraints. For combinations 2 and 3, different time budgets lead to different task to processor mappings and different energy consumptions.

Which area in the objective space is dominated by at least one design of configuration 3? Is there any design belonging to configuration 2 which is not dominated by at least one design of configuration 3? Which area in the objective space dominates at least one design of configuration 3?

2 Which conditions must be met by computations of $WCET_{EST}$?

3 Let's consider cache states at a control flow join! Fig. 5.22 shows abstract cache states before the join.

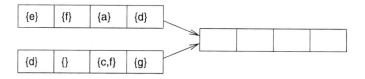

Figure 5.22.   Abstract cache states

Now let us look at abstract cache states after the join. Which state would a *must*-analysis derive? Which state would a *may*-analysis derive?

4 Consider an incoming "bursty" event stream. The stream is periodic with a period of p. At the beginning of each period, two events arrive with a separation of d time units. Develop *arrival curves* for this stream! Resulting graphs should display times from 0 up to $3*p$.

5 Suppose that you are working with a processor having a maximum performance of b.

   (a) How do the *service curves* look like if the performance can deteriorate to b', due to cache conflicts?

   (b) How do the *service curves* change if some timer is interrupting the executed program every 100 ms and if servicing the interrupt takes 10 ms? Assume that there are no cache conflicts.

   (c) How do the *service curves* look like if you consider cache conflicts like in (a) **and** interrupts like in (b)?

   Resulting graphs should display times from 0 up to 300 ms.

Chapter 6

APPLICATION MAPPING

6.1 Problem definition

Once the specification has been completed, design activities can start. This is consistent with the simplified design information flow (see fig. 6.1). Mapping applications to execution platforms is really a key activity. Therefore we underline the importance of this book chapter.

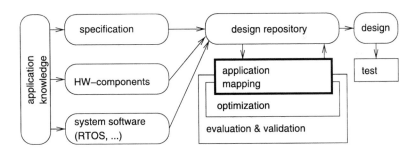

Figure 6.1. Simplified design flow

For embedded systems, we are frequently expecting that the system works with a certain combination of applications. For example, for a mobile phone, we expect being able to make a phone call while the Bluetooth stack is transmitting the audio signals to a head set and while we are looking up information in our "personal information manager" (PIM). At the same time, there may be a concurrent file transfer or even a video connection. We must make sure that these applications can be used together and that we are keeping the deadlines (no lost audio samples!). This is feasible through an analysis of the use cases.

P. Marwedel, *Embedded System Design*, Embedded Systems,
DOI 10.1007/978-94-007-0257-8_6, © Springer Science+Business Media B.V. 2011

It is a characteristic of embedded and cyber-physical systems that both hardware and software must be considered during their design. Therefore, this type of design is also called **hardware/software codesign**. The overall goal is to find the right combination of hardware and software resulting in the most efficient product meeting the specification. Therefore, embedded systems cannot be designed by a synthesis process taking only the behavioral specification into account. Rather, available components must be accounted for. There are also other reasons for this constraint: in order to cope with the increasing complexity of embedded systems and their stringent time-to-market requirements, reuse is essentially unavoidable. This led to the term **platform-based design:**

"A platform is a family of architectures satisfying a set of constraints imposed to allow the reuse of hardware and software components. However, a hardware platform is not enough. Quick, reliable, derivative design requires using a platform application programming interface (API) to extend the platform toward application software. In general, a platform is an abstraction layer that covers many possible refinements to a lower level. Platform-based design is a meet-in-the-middle approach: In the top-down design flow, designers map an instance of the upper platform to an instance of the lower, and propagate design constraints" [Sangiovanni-Vincentelli, 2002]. The mapping is an iterative process in which performance evaluation tools guide the next assignment.

In this book, we focus on embedded system design based on available execution platforms. This reflects the fact that many modern systems are being built on top of some existing platform. Techniques other that the ones described in this book must be used in cases where the execution platform needs to be designed as well. Due to our focus, the **mapping of applications to execution platforms** can be seen as the **main design problem**.

In the general case, mapping will be performed onto multiprocessor systems. We can distinguish between two different classes of multiprocessor systems:

- **Homogeneous multiprocessor systems**: In this case, all the processors in the system provide the same functionality. This is the case for the multi- or many-core architectures considered in PC-like systems. Code compatibility between the different processors is the key advantage here: it can be exploited during run-time scheduling of tasks (including load balancing) and is also an advantage for fault-tolerant designs. We can just reallocate processors at run-time, if a processor fails. Also, the design of the processor platform and development tools is easier, if all processors are of the same type.

- **Heterogeneous multiprocessor systems**: In this case, processors are of different types. The improved efficiency of this approach is the key reason for accepting not to have all the advantages of homogeneous multiprocessor systems. Heterogeneous processors are the most efficient programmable platforms.

Even for platform-based design, there may be a number of design options. We might be able to select between different variants of a platform, where each variant might have a different number of processors, different speeds of processors or a different communication architecture. Moreover, there may be different applicable scheduling policies. Appropriate options must be selected.

This leads us to the following definition of our mapping problem [Thiele, 2006a]:

Given:

- a set of applications,

- use cases describing how the applications will be used,

- a set of possible candidate architectures:

    - (possibly heterogeneous) processors,

    - (possibly heterogeneous) communication architectures, and

    - possible scheduling policies.

Find:

- a mapping of applications to processors,

- appropriate scheduling techniques (if not fixed), and

- a target architecture (if not fixed).

Objectives:

- Keeping deadlines and/or maximizing performance, as well as

- minimizing cost, energy consumption, and possibly other objectives.

The exploration of possible architectural options is called **design space exploration** (DSE). The case of a completely fixed platform architecture can be considered as a special case.

Designing an AUTOSAR-based automotive system can be seen as an example: In AUTOSAR [AUTOSAR, 2010], we have a number of homogeneous execution units (called ECUs) and a number of software components. The question is: how do we map these software components to the ECUs such all real-time constraints are met? We would like to use the minimum number of ECUs.

The application mapping problem is a very difficult one and currently only approximations for an automated mapping are available. In the following, we will present building blocks for such a mapping:

- standard scheduling techniques,

- hardware/software partitioning, and

- advanced techniques for mapping sets of applications onto multi-processor systems.

We will start with standard scheduling techniques which can be used in various contexts.

6.2   Scheduling in real-time systems

As indicated above, scheduling is one of the key issues in implementing embedded systems. Scheduling algorithms may be required a number of times during the design of such systems. Very rough calculations may already be required while fixing the specification. Later, more detailed predictions of execution times may be required. After compilation, even more detailed knowledge exists about the execution times and accordingly, more precise schedules can be made. Finally, it may be necessary to decide at run-time which task is to be executed next. In contrast, in time-triggered systems, RTOS scheduling may be limited to simple table look-ups for tasks to be executed. Scheduling is similar to performance evaluation in that it cannot be constrained to a single design step.

Scheduling defines start times for each task and therefore defines a mapping τ from nodes of a task graph $G = (V, E)$ to time domain D_t:

$$\tau : V \to D_t \tag{6.1}$$

6.2.1   Classification of scheduling algorithms

Scheduling algorithms can be classified according to various criteria. Fig. 6.2 shows a possible classification of algorithms (similar schemes are described in books on the topic [Balarin et al., 1998], [Kwok and Ahmad, 1999], [Stankovic et al., 1998], [Liu, 2000], [Buttazzo, 2002]).

The following is a list of criteria, the first four of which are linked to fig. 6.2.

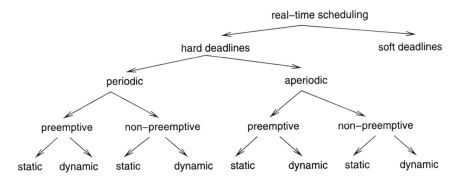

Figure 6.2. Classes of scheduling algorithms

- **Soft and hard deadlines:** Scheduling for soft deadlines is frequently based on extensions to standard operating systems. For example, providing task and operating system call priorities may be sufficient for systems with soft deadlines. We will not discuss these systems further in this book. More work and a detailed analysis are required for hard deadline systems. For these, we can consider periodic and aperiodic systems.

- **Scheduling for periodic and aperiodic tasks:** In the following, we will distinguish between periodic and aperiodic tasks.

  Definition: Tasks which must be executed once every p units of time are called **periodic tasks**, and p is called their **period**. Each execution of a periodic task is called a **job**.

  Definition: Tasks which are not periodic are called **aperiodic**.

  Definition: Aperiodic tasks requesting the processor at unpredictable times are called **sporadic**, if there is a minimum separation between the times at which they request the processor.

  This minimum separation is important, since tasks sets without such a separation can possibly not be scheduled. There may be not enough time to execute tasks if tasks become executable at arbitrarily short time intervals.

- **Preemptive and non-preemptive scheduling:** Non-preemptive schedulers are based on the assumption that tasks are executed until they are done. As a result the response time for external events[1] may be quite long if some tasks have a large execution time. Preemptive schedulers must be used if some tasks have long execution times or if the response time for external events

---

[1] This is the time from the occurrence of an external event until the completion of the reaction required for the event.

is required to be short. However, preemption can result in unpredictable execution times of the preempted tasks. Therefore, restricting preemptions may be required in order to guarantee meeting the deadline of hard real-time tasks.

- **Static and dynamic scheduling:** Dynamic schedulers take decisions at run-time. They are quite flexible, but generate overhead at run-time. Also, they are usually not aware of global contexts such as resource requirements or dependences between tasks. For embedded systems, such global contexts are typically available at design time and they should be exploited.

Static schedulers take their decisions at design time. They are based on planning the start times of tasks and generate tables of start times forwarded to a simple dispatcher. The dispatcher does not take any decisions, but is just in charge of starting tasks at the times indicated in the table. The dispatcher can be controlled by a timer, causing the dispatcher to analyze the table. Systems which are totally controlled by a timer are said to be **entirely time triggered** (TT systems). Such systems are explained in detail in the book by Kopetz [Kopetz, 1997]:

*"In an entirely time-triggered system, the temporal control structure of all tasks is established **a priori** by off-line support-tools. This temporal control structure is encoded in a **Task-Descriptor List (TDL)** that contains the cyclic schedule for all activities of the node[2] (Figure 6.3). This schedule considers the required precedence and mutual exclusion relationships among the tasks such that an explicit coordination of the tasks by the operating system at run time is not necessary."* Figure 6.3 includes scheduled task start, task stop and send message (send) activities.

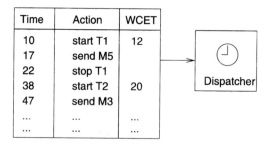

Time	Action	WCET
10	start T1	12
17	send M5	
22	stop T1	
38	start T2	20
47	send M3	
...	...	...
...	...	...

Figure 6.3.   Task descriptor list in a time-triggered system

---

[2]This term refers to a processor in this case.

"The dispatcher is activated by the synchronized clock tick. It looks at the TDL, and then performs the action that has been planned for this instant ...."

The main advantage of static scheduling is that it can be easily checked if timing constraints are met:

"For satisfying timing constraints in hard real-time systems, predictability of the system behavior is the most important concern; pre-run-time scheduling is often the only practical means of providing predictability in a complex system" [Xu and Parnas, 1993].

The main disadvantage is that the response to sporadic events may be quite poor.

- **Independent and dependent tasks:**

  It is possible to distinguish between tasks without any inter-task communication and other tasks. For embedded systems, dependencies between tasks are the rule rather than an exception.

- **Mono- and multi-processor scheduling:** Simple scheduling algorithms handle the case of single processors, whereas more complex algorithms also handle systems comprising multiple processors. For the latter, we can distinguish between algorithms for homogeneous multi-processor systems and algorithms for heterogeneous multi-processor systems. The latter are able to handle target-specific execution times and can also be applied to mixed hardware/software systems, in which some tasks are mapped to hardware.

- **Centralized and distributed scheduling:** Multiprocessor scheduling algorithms can either be executed locally on one processor or can be distributed among a set of processors.

- **Type and complexity of schedulability test:** In practice, it is very important to know whether or not a schedule exists for a given set of tasks and constraints.

  A set of tasks is said to be **schedulable** under a given set of constraints, if a schedule exists for that set of tasks and constraints. For many applications, **schedulability tests** are important. Tests which always return precise results (called exact tests) are NP-hard in many situations [Garey and Johnson, 1979]. Therefore, sufficient and necessary tests are used instead. For sufficient tests, sufficient conditions for guaranteeing a schedule are checked. There is a (hopefully small) probability of indicating that scheduling cannot be guaranteed even if a schedule exists. Necessary tests are based on checking necessary conditions. They can be used to show that no schedule exists. However, there may be cases in which necessary tests are passed and the schedule still does not exist.

- **Cost functions:** Different algorithms aim at minimizing different functions. Maximum lateness is a frequently used cost function.

  Definition: Maximum lateness is defined as the difference between the completion time and the deadline, maximized over all tasks. Maximum lateness is negative if all tasks complete before their deadline.

6.2.2   Aperiodic scheduling without precedence constraints

6.2.2.1   Definitions

Let $\{T_i\}$ be a set of tasks. Let (see fig. 6.4)

- c_i be the execution time of T_i,

- d_i be the **deadline interval**, that is, the time between T_i becoming available and the time until which T_i has to finish execution.

- l_i be the **laxity** or **slack**, defined as

$$l_i \;\; = \;\; d_i - c_i \tag{6.2}$$

Again, upward pointing arrows denote the time at which tasks becomes available. Downward pointing arrows represent deadlines.

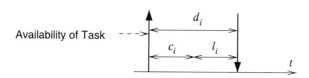

Figure 6.4.   Definition of the laxity of a task

If $l_i = 0$, then T_i has to be started immediately after it becomes executable.

Let us first consider[3] the case of uni-processor systems for which all tasks arrive at the same time. If all tasks arrive at the same time, preemption is obviously useless.

---

[3]We are using some of the material from the book by Buttazzo [Buttazzo, 2002] for this section. Refer to this book for additional references.

6.2.2.2   Earliest Due Date (EDD)-Algorithm

A very simple scheduling algorithm for this case was found by Jackson in 1955 [Jackson, 1955]. The algorithm is based on Jackson's rule:

Given a set of n independent tasks, any algorithm that executes the tasks in order of nondecreasing deadlines is optimal with respect to minimizing the maximum lateness.

The algorithm following this rule is called **Earliest Due Date** (EDD). If the deadlines are known in advance, EDD can be implemented as a static scheduling algorithm. EDD requires all tasks to be sorted by their deadlines. Hence, its complexity is $O(n \log(n))$.

Proof of the optimality of EDD:

Let τ be a schedule generated by any algorithm A. Suppose A does not lead to the same result as EDD. Then, there are tasks T_a and T_b such that the execution of T_b precedes the execution of T_a in τ, even though the deadline of T_a is earlier than that of T_b ($d_a < d_b$). Now, let us consider a schedule τ'. τ' is generated from τ by swapping the execution orders of T_a and T_b (see fig. 6.5).

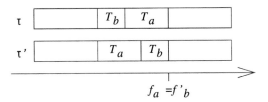

Figure 6.5.   Schedules τ and τ'

$L_{max}(a,b) = f_a - d_a$ is the maximum lateness of T_a and T_b in schedule τ. For schedule τ', $L'_{max}(a,b) = max(L'_a, L'_b)$ is the maximum lateness among tasks T_a and T_b. L'_a is the maximum lateness of task T_a in schedule τ'. L'_b is defined accordingly. There are two possible cases:

1 $L'_a > L'_b$: In this case, we have

$$L'_{max}(a,b) = f'_a - d_a$$

T_a terminates earlier in the new schedule. Therefore, we have

$$L'_{max}(a,b) = f'_a - d_a < f_a - d_a.$$

The right side of this inequality is the maximum lateness in schedule τ. Hence, the following holds:

$$L'_{max}(a,b) < L_{max}(a,b)$$

2 $L'_a \leq L'_b$:

In this case, we have:

$L'_{max}(a,b) = f'_b - d_b = f_a - d_b$ (see fig. 6.5).

The deadline of T_a is earlier than the one of T_b. This leads to

$L'_{max}(a,b) < f_a - d_a$

Again, we have

$L'_{max}(a,b) < L_{max}(a,b)$

As a result, any schedule (which is not an EDD-schedule) can be turned into an EDD-schedule by a finite number of swaps. Maximum lateness can only decrease during these swaps. Therefore, EDD is optimal among all scheduling algorithms.

6.2.2.3   Earliest Deadline First (EDF)-Algorithm

Let us consider the case of different arrival times for uni-processor systems next. Under this scenario, preemption can potentially reduce maximum lateness.

The Earliest Deadline First (EDF) algorithm is optimal with respect to minimizing the maximum lateness. It is based on the following theorem [Horn, 1974]:

Given a set of n independent tasks with arbitrary arrival times, any algorithm that at any instant executes the task with the earliest absolute deadline among all the ready tasks is optimal with respect to minimizing the maximum lateness.

EDF requires that, each time a new ready task arrives, it is inserted into a queue of ready tasks, sorted by their deadlines. Hence, EDF is a dynamic scheduling algorithm. If a newly arrived task is inserted at the head of the queue, the currently executing task is **preempted**. If sorted lists are used for the queue, the complexity of EDF is $O(n^2)$. Bucket arrays could be used for reducing the execution time.

Fig. 6.6 shows a schedule derived with the EDF algorithm. Vertical arrows indicate the arrival of tasks.

At time 4, task T_2 has an earlier deadline. Therefore it preempts T_1. At time 5, task T_3 arrives. Due to its later deadline it does not preempt T_2.

Proof of the optimality of EDF:

Let τ be a schedule generated by some algorithm A, where A is different from EDF. Let τ_{EDF} be a schedule generated by EDF. Now, we partition time into disjoint intervals of length 1. Each interval comprises times within the range $[t, t+1)$. Let $\tau(t)$ be the task which -according to schedule τ- is executed during the interval $[t, t+1)$. Let $E(t)$ be the task which at time t has the earliest deadline

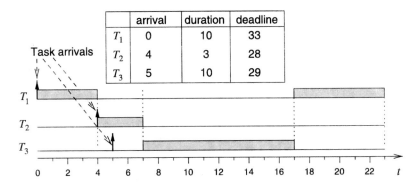

Figure 6.6. EDF schedule

among all tasks. Let $t_E(t)$ be the time $(\geq t)$ at which task $E(t)$ is starting its execution in schedule τ.

τ is not an EDF-schedule. Therefore, there must be a time t at which we are not executing the task having the earliest deadline. For t, we have $\tau(t) \neq E(t)$ (see fig. 6.7). Deadlines are represented by downward pointing arrows.

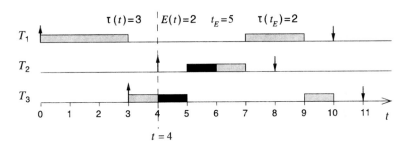

Figure 6.7. Schedule τ

The basic idea of the proof is to show that swapping $\tau(t)$ and $E(t)$ (see fig. 6.8) cannot increase maximum lateness.

Let D be the latest deadline. Then, we can generate τ_{EDF} from τ by at most D swaps of the following algorithm:

```
for (t=0 to D-1) {
    if (τ(t) ≠ E(t)) {
        τ(tE) = τ(t);
        τ(t) = E(t); }}
```

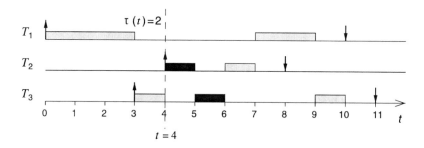

Figure 6.8. Schedule after swapping tasks $\tau(t)$ and $E(t)$

Using the same arguments as for Jackson's Rule we can show that swapping does not increase maximum lateness. Therefore, any non-EDF schedule can be turned into an EDF-schedule with increasing maximum lateness. This proves that EDF is optimal among all possible scheduling algorithms. We can show that swapping will keep all deadlines, provided they were kept in schedule τ. First of all, we consider task $E(t)$. It will be executed earlier than in the old schedule and, hence, it will meet the deadline in the new schedule if it did so in the old schedule. Next, we consider task $\tau(t)$. $\tau(t)$ has a deadline larger than $E(t)$. Hence, $\tau(t)$ will meet the deadline in the new schedule if $E(t)$ met the deadline in the old schedule.

6.2.2.4 Least Laxity (LL) algorithm

Least Laxity (LL), Least Slack Time First (LST), and Minimum Laxity First (MLF) are three names for another scheduling strategy [Liu, 2000]. According to LL scheduling, task priorities are a monotonically decreasing function of the laxity (see equation 6.2; the less laxity, the higher the priority). The laxity is dynamically changing and needs to be dynamically recomputed. Negative laxities provide an early warning for deadlines to be missed. LL scheduling is also preemptive. Preemptions are not restricted to times at which new tasks become available.

Fig. 6.9 shows an example of an LL schedule, together with the computations of the laxity.

At time 4, task T_1 is preempted, as before. At time 5, T_2 is now also preempted, due to the lower laxity of task T_3.

It can be shown (this is left as an exercise in [Liu, 2000]) that LL is also an optimal scheduling policy for mono-processor systems in this sense that it will find a schedule if one exists. Due to its dynamic priorities, it cannot be used with a standard OS providing only fixed priorities. Furthermore, LL scheduling -in contrast to EDF scheduling- requires the knowledge of the execution

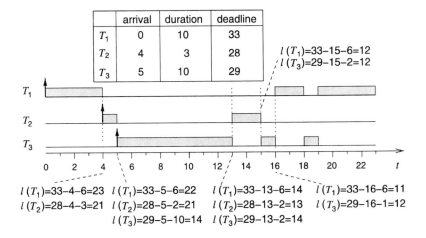

Figure 6.9. Least laxity schedule

time. Its use is therefore restricted to special situations where its properties are attractive.

6.2.2.5 Scheduling without preemption

If preemption is not allowed, optimal schedules may must leave the processor idle at certain times in order to finish tasks with early deadlines arriving late.

Proof: Let us assume that an optimal non-preemptive scheduler (not having knowledge about the future) never leaves the processor idle. This scheduler must schedule the example of fig. 6.10 optimally (it must find a schedule if one exists).

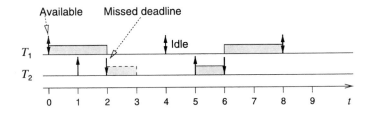

Figure 6.10. Scheduler needs to leave processor idle

For the example of fig. 6.10 we assume we are given two tasks. Let T_1 be a periodic process with an execution time of 2, a period of 4 and a deadline interval of 4. Let T_2 be a task occasionally becoming available at times $4*n+1$ and having an execution time and a deadline interval of 1. Let us assume that the concurrent execution of T_1 and T_2 is not possible (for example, since we are using a single processor). Under the above assumptions our scheduler has to start the execution of task T_1 at time 0, since it is supposed not to leave any idle time. Since the scheduler is non-preemptive, it cannot start T_2 when it becomes available at time 1. Hence, T_2 misses its deadline. If the scheduler had left the processor idle (as shown in fig. 6.10 at time 4), a legal schedule would have been found. Hence, the scheduler is not optimal. This is a contradiction to the assumptions that optimal schedulers not leaving the processor idle at certain times exist. q.e.d.

We conclude: In order to avoid missed deadlines the scheduler needs knowledge about the future. If no knowledge about the arrival times is available a priori, then no online algorithm can decide whether or not to keep the processor idle. It has been shown that EDF is still optimal among all scheduling algorithms not keeping the processor idle at certain times. If arrival times are known a priori, the scheduling problem becomes NP-hard in general and branch and bound techniques are typically used for generating schedules.

6.2.3   Aperiodic scheduling with precedence constraints

6.2.3.1   Latest Deadline First (LDF) algorithm

We start with a task graph reflecting tasks dependences (see fig. 6.11). Task T_3 can be executed only after tasks T_1 and T_2 have completed and sent messages to T_3.

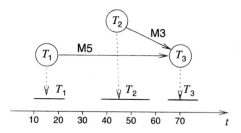

Figure 6.11.   Precedence graph and schedule

This figure also shows a legal schedule. For static scheduling, this schedule can be stored in a table, indicating to the dispatcher the times at which tasks must be started and at which messages must be exchanged.

An optimal algorithm for minimizing the maximum lateness for the case of simultaneous arrival times was presented by Lawler [Lawler, 1973]. The algorithm is called **Latest Deadline First** (LDF). LDF reads the task graph and inserts tasks with no successors into a queue. It then repeats this process, putting tasks whose successors have all been selected into the queue. At runtime, the tasks are executed in an order **opposite** to the order in which tasks have been entered into the queue. LDF is non-preemptive and is optimal for mono-processors.

The case of asynchronous arrival times can be handled with a modified EDF algorithm. The key idea is to transform the problem from a given set of dependent tasks into a set of independent tasks with different timing parameters [Chetto et al., 1990]. This algorithm is again optimal for uni-processor systems.

If preemption is not allowed, the heuristic algorithm developed by Stankovic and Ramamritham [Stankovic and Ramamritham, 1991] can be used.

6.2.3.2   As-soon-as-possible (ASAP) scheduling

A number of scheduling algorithms have been developed in other communities. For example, as-soon-as-possible (ASAP), as-late-as-possible (ALAP), list (LS), and force-directed scheduling (FDS) are very popular in the high-level synthesis (HLS) community (see [Coussy and Morawiec, 2008] for recent HLS results). ASAP and ALAP scheduling do not consider any resource or time constraints. LS considers resource constraints while FDS considers a global time constraint.

We will demonstrate the first three of these using a simple expression as an example. Consider a 3×3 matrix (see fig. 6.12).

$$A = \begin{vmatrix} a & b & c \\ d & e & f \\ g & h & i \end{vmatrix}$$

Figure 6.12. 3×3 matrix

The determinant $det(A)$ of this matrix can be computed as

$$det(A) = a*(e*i - f*h) + b*(f*g - d*i) + c*(d*h - e*g)$$

The computation can be represented as a data flow graph (see fig. 6.13). We assume that each arithmetic computation represents a simple "task".

We assume that all matrix values are available immediately (for example, they might be stored in registers).

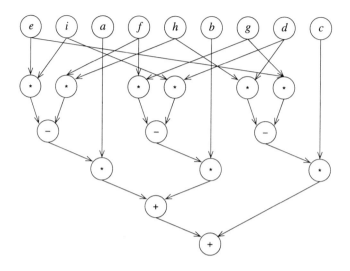

Figure 6.13.   Computation of the determinant of *A*

ASAP, as used in HLS, considers a mapping of tasks to integer start times[4] > 0. Therefore, scheduling provides a mapping:

$$\tau : V \rightarrow I\!N \tag{6.3}$$

where $G = (V, E)$ is the data flow graph.

For ASAP scheduling, all tasks are started as early as possible. The algorithm works as follows:

> **for** (t=1; all tasks are scheduled; t++) {
>
> s={all tasks for which all inputs are available};
>
> set start time of all tasks in s to t;
>
> }

For the sake of simplicity, we assume that all additions and subtractions of our example have an execution time of 1, whereas multiplications have an execution time of 2. Fig. 6.14 shows the resulting scheduled data flow graph for our example of fig. 6.13.

During the first iteration of the ASAP algorithm, all tasks not depending on other computations are set to start at time 1. During the second round, inputs

---

[4]Each integer is assumed to correspond to one clock cycle of some synchronous automaton.

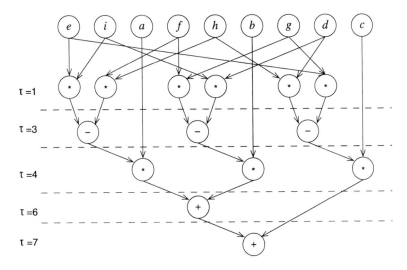

Figure 6.14. ASAP schedule for the example of fig. 6.13

from multiplications are not yet available. During the third round, subtractions are scheduled to start at time 3. This process continues until the final addition is scheduled to start at time 7.

ASAP scheduling can also be applied to real life: it means that all tasks are started as early as possible, without any consideration of resource constraints.

6.2.3.3   As-late-as-possible (ALAP) scheduling

As-late-as-possible is the second simple scheduling algorithm. For ALAP scheduling, all tasks are started as late as possible. The algorithm works as follows:

> **for** (t=0; all tasks are scheduled; t- -) {
>
> s={all tasks on which no unscheduled task depends};
>
>   set start time of all tasks in s to t - their execution time + 1;
>
> }
> Add the total number of time steps needed to all start times.

The algorithm starts with tasks on which no other task depends. These tasks are assumed to finish at time 0. Their start time is then computed from their execution time. The loop then iterates backwards over time steps. Whenever we reach a time step, at which a task should finish the latest, its start time is computed and the task is scheduled. After finishing the loop, all times are shifted towards positive times such that the first task starts at time 1. We could also consider ALAP scheduling as a case of ASAP scheduling starting at the "other" end of the graph.

Fig. 6.15 shows the resulting scheduled data flow graph for our example of fig. 6.13.

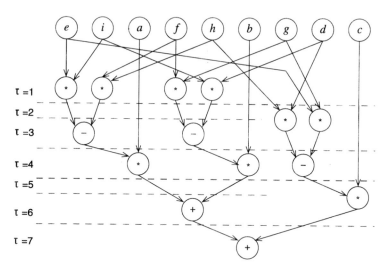

Figure 6.15. ALAP schedule for the example of fig. 6.13

For the ALAP schedule, the four "tasks" at the right start one time unit later.

6.2.3.4   List scheduling (LS) scheduling

List scheduling is a resource-constrained scheduling technique. We assume that we have a set M of resource types. List scheduling assumes that each task can be executed only on a particular resource type. List scheduling respects upper bounds B_m on the number of resources for each type $m \in M$.

List scheduling requires the availability of some priority function reflecting the urgency of scheduling a particular "task" $v \in V, G = (V, E)$. The following urgency metrics are in use [Teich, 1997]:

■ Number of successor nodes: this is the number of nodes below the current node v in the tree.

■ Path length: the path length for a node $v \in V$ is defined as the length of the path from starting v to finishing the entire graph G. In fig. 6.16, this information has been added. Path length is typically weighted by the execution time associated with the nodes, assuming that this information is known.

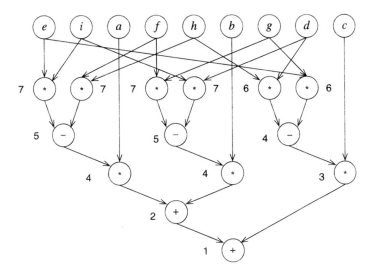

Figure 6.16. Path lengths for the example of fig. 6.13

- Mobility: mobility is defined as the difference between the start times for the ASAP and ALAP schedule. Fig. 6.17 shows the mobility for our example. Obviously, scheduling is *urgent* for all but four nodes. This means that all other nodes will have the same priority and that mobility provides only rough information about the order in which we should schedule tasks.

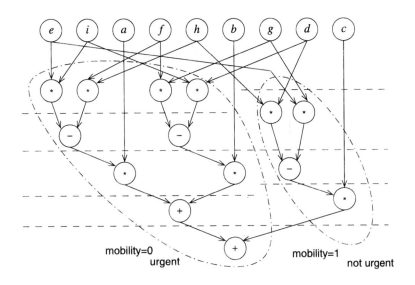

Figure 6.17. Mobility for the example of fig. 6.13

List scheduling requires the knowledge of the graph $G = (V, E)$ to be scheduled, a mapping from each node of the graph to the corresponding resource type $m \in M$, an upper bound B_m for each m, a priority function u reflecting the urgency of the nodes $v \in V$, and the execution time of each node $v \in V$. List scheduling then tries to fit nodes of maximum priority into each of the time steps such that the resource constraints are not violated [Teich, 1997]:

> **for** (t=0; all tasks are scheduled; t++) { //loop over time steps
>
> > **for** ($m \in M$) {   //loop over resource types
> >
> > $C_{t,m}$ = set of tasks of type m still executing at time t;
> >
> > $A_{t,m}$ = set of tasks of type m ready to start execution at time t;
> >
> > Compute set $S_t \subseteq A_{i,m}$ of maximum priority such that
> >
> > $|S_t| + |C_{t,m}| \leq B_m$.
> >
> > Set start times of all $v \in S_t$ to t: $\tau(v) = t$;
> >
> > } }

Fig. 6.18 shows the result of list scheduling as applied to our example in fig. 6.13.

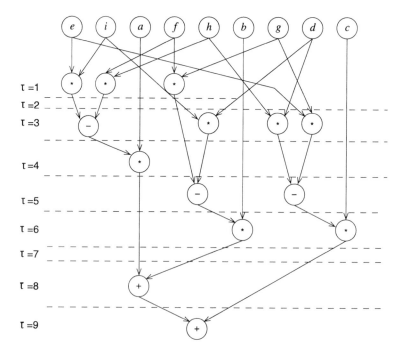

Figure 6.18.   Result of list scheduling for the example of fig. 6.13

In fig. 6.13 we assume a resource constraint of $B_* = 3$ for multiplications and multipliers and of $B_{+,-} = 2$ for all other "tasks". Due to the resource constraint, three multiplications are starting at time 3 instead of at time 1. The resource constraint for other operations does not have any impact. Remember that multiplications need two time steps.

6.2.3.5 Force-directed scheduling

For force-directed scheduling (FDS) [Paulin and Knight, 1987], we assume that a time constraint is given and that we would like to find a schedule keeping that resource constraint while minimizing the demand for resources. FDS considers each resource type separately.

FDS starts with a "probability" $P(v,t)$ reflecting the likelihood that a certain operation v is scheduled at a certain time step t. This "probability" is equal to 1 divided by the size of $R(v)$, where "range" $R(v)$ is the set of time steps at which this operation could be started:

$$P(v,t) = \begin{cases} \frac{1}{|R(v)|} & \text{if } t \in R(v) \\ 0 & \text{otherwise} \end{cases}$$

$R(v)$ is the interval between the time step allocated by ASAP scheduling and the time step allocated by ALAP scheduling. From this "probability", we compute a so-called "distribution" reflecting the total resource pressure for a certain resource m at control step t. This "distribution" is simply the sum of the probabilities over all operations requiring resource type m:

$$D(t) = \sum_{v \in V} P(v,t)$$

Fig. 6.19 shows distributions for our running example.

For example, for a time constraint of 8, the three multiplications on the right (which are not on the critical path) have a probability of 0.5 for the two time steps for which they are feasible. The distribution $D(1)$ is 5, due to the four multiplications on the critical path and the two multiplications having a probability of 0.5.

Next, FDS defines "forces" such that operations (or tasks) are moved away from time steps of high resource pressure to time steps with a lower resource pressure. In our example, multiplications which are not on the critical path are shifted towards later start times. However, for a total time constraint of 8, this does not lower the number of multipliers needed, since multiplies are assumed to last 2 time steps. For a time constraint of 9, we would reduce the number of

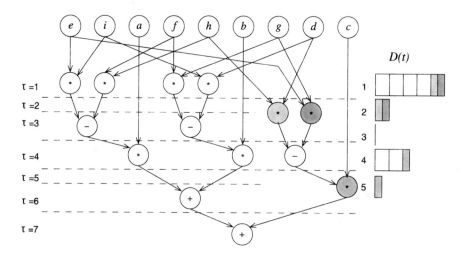

Figure 6.19.   Distributions for the example of fig. 6.13

multipliers if compared to the multipliers needed for ASAP scheduling. Details about FDS can be found in [Paulin and Knight, 1987] and follow-up papers.

FDS has a number of restrictions. For example, FDS is still based on the simple resource model where each task can be mapped only to a single resource type.

At this time, some general remarks regarding the applicability of scheduling techniques from high-level synthesis (HLS) to task scheduling are appropriate. HLS techniques

- are designed to take dependencies between "tasks" into account,

- are designed for "multi-processor" scheduling,

- are usually based on simplified resource (processor) models (i.e. require a one-to-one mapping between "tasks" and "processors"),

- typically use heuristics not guaranteeing optimality,

- are typically fast,

- almost never exploit global information about periodicity etc. and

- techniques more advanced than ASAP, ALAP and LS include techniques for handling control (loops etc).

6.2.4 Periodic scheduling without precedence constraints

6.2.4.1 Notation

Next, we will consider the case of periodic tasks. For periodic scheduling, objectives relevant for aperiodic scheduling are less useful. For example, minimization of the total length of the schedule is not an issue if we are talking about an infinite repetition of jobs. The best that we can do is to design an algorithm which will always find a schedule if one exists. This motivates the definition of optimality for periodic schedules.

Definition: For periodic scheduling, a scheduler is defined to be **optimal** iff it will find a schedule if one exists.

Let $\{T_i\}$ be a set of tasks. Each execution of some task T_i is called a **job**. The execution time for each job corresponding to one task is assumed to be the same. Let (see fig. 6.20)

- p_i be the period of task T_i,

- c_i be the execution time of T_i,

- d_i be the **deadline interval**, that is, the time between a job of T_i becoming available and the time after which the same job T_i has to finish execution.

- l_i be the **laxity** or **slack**, defined as

$$l_i \;=\; d_i - c_i \tag{6.4}$$

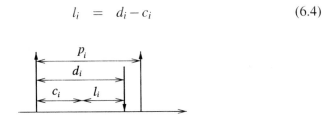

Figure 6.20. Notation used for time intervals

If $l_i = 0$, then T_i has to be started immediately after it becomes executable.

Let μ denote the **utilization** for a set of n processes, that is, the accumulated execution times of these processes divided by their period:

$$\mu \;=\; \sum_{i=1}^{n} \frac{c_i}{p_i} \tag{6.5}$$

Let us assume that the execution times are equal for a number of m processors. Obviously, equation 6.6 represents a necessary condition for a schedule to exist:

$$\mu \ \leq \ m \tag{6.6}$$

Initially, we will restrict ourselves to a description of the case in which tasks are independent.

6.2.4.2   Rate monotonic scheduling

Rate monotonic (RM) scheduling [Liu and Layland, 1973] is probably the most well-known scheduling algorithm for independent periodic processes. Rate monotonic scheduling is based on the following assumptions ("**RM assumptions**"):

1 All tasks that have hard deadlines are periodic.

2 All tasks are independent.

3 $d_i = p_i$, for all tasks.

4 c_i is constant and is known for all tasks.

5 The time required for context switching is negligible.

6 For a single processor and for n tasks, the following equation holds for the accumulated utilization μ:

$$\mu \ = \ \sum_{i=1}^{n} \frac{c_i}{p_i} \leq n(2^{1/n} - 1) \tag{6.7}$$

Fig. 6.21 shows the right hand side of equation 6.7.

The right hand side is about 0.7 for large n:

$$\lim_{n \to \infty} n * (2^{1/n} - 1) \ = \ log_e(2) = ln(2) \ (=\sim 0.7) \tag{6.8}$$

Then, according to the policy for rate monotonic scheduling, **the priority of tasks is a monotonically decreasing function of their period**. In other words, tasks with a short period will get a high priority and tasks with a long period

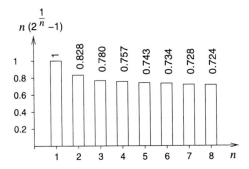

Figure 6.21. Right hand side of equation 6.7

will be assigned a low priority. RM scheduling is a **preemptive scheduling policy** with **fixed priorities**.

Fig. 6.22 shows an example of a schedule generated with RM scheduling. Task T_2 is preempted several times.

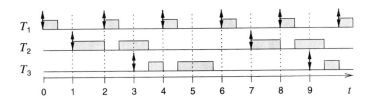

Figure 6.22. Example of a schedule generated with RM scheduling

Double-headed arrows indicate the arrival time of a job as well as the deadline of the previous job. Tasks 1 to 3 have a period of 2, 6 and 6, respectively. Execution times are, 0.5, 2, and 1.75. Task 1 has the shortest period and, hence, the highest rate and priority. Each time task 1 becomes available, its jobs preempt the currently active task. Task 2 has the same period as task 3, and neither of them preempts the other.

Equation 6.7 requires that some of the computing power of the processor is not used in order to make sure that all requests are honored in time. What is the reason for this bound on the utilization? The key reason is that RM scheduling, due to its static priorities, will possibly preempt a task which is close to its deadline in favor of some higher priority task with a much later deadline. The task having a lower priority can then miss its deadline.

Fig. 6.23 shows a case for which not enough idle time is available to guarantee schedulability for RM scheduling. One task has a period of 5, and an execution time of 3, whereas the second task has a period of 8, and an execution time of 3.

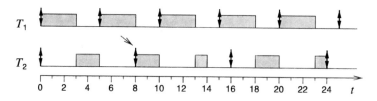

Figure 6.23.   RM schedule does not meet deadline at time 8

For this particular case we have $\mu = \frac{3}{5} + \frac{3}{8} = \frac{39}{40}$, which is 0.975. $2 * (2^{\frac{1}{2}} - 1)$ is about 0.828. Hence, schedulability is not guaranteed for RM scheduling and, in fact, the deadline is missed at time 8. We assume that the missing computations are not scheduled in the next period.

Such missed deadlines cannot happen if the utilization of the processor is very low and, obviously, they can happen when the utilization is high, as in fig. 6.23. If the condition of equation 6.7 is met, the utilization is guaranteed to be low enough to prevent problems like that of fig. 6.23. Equation 6.7 is a **sufficient** condition. This means: we might still find a schedule if the condition is not met. Other sufficient conditions exist [Bini et al., 2001].

RM scheduling has the following important advantages:

- It is possible to prove that rate monotonic scheduling is optimal for mono-processor systems.

- RM scheduling is based on **static** priorities. This opens opportunities for using RM scheduling in an operating system providing fixed priorities, such as Windows NT (see Ramamritham [Ramamritham et al., 1998], [Ramamritham, 2002]).

- If the above six RM-assumptions (see page 258) are met, all deadlines will be met (see Buttazzo [Buttazzo, 2002]).

RM scheduling is also the basis for a number of formal proofs of schedulability.

The idle time or *spare capacity* of the processor is not always required. It is possible to show that RM scheduling is also optimal, iff instead of equation (6.7) we have

$$\mu \leq 1 \tag{6.9}$$

provided that the period of all tasks is a multiple of the period of the task having the next higher priority. This requirement is met, for example, if tasks in a TV set must be executed at rates of 25, 50 and 100 Hertz.

Equations 6.7 or 6.9 provide easy means to check conditions for schedulability.

Designing examples and proofs is facilitated if the most problematic situations for RM scheduling are known.

Definition: Time t is called **critical time instant** for task T_j if the response time of this task is maximized if the task becomes available at this time.

Lemma: For each task T_j, the response time is maximized if T_j becomes available at the same time as all tasks having a higher priority.

Proof: Let $T = \{T_1, ..., T_n\}$ be a set of periodical tasks for which we have: $\forall i : p_i \leq p_{i+1}$. The response time of T_n will be increased by tasks of a higher priority . Consider task T_n and some task T_i of a higher priority (see fig. 6.24).

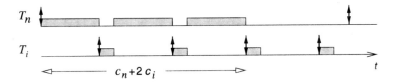

Figure 6.24. Delaying task T_n by some T_i of higher priority

The number of preemptions is potentially increasing if the time interval between the availability of T_n and T_i is reduced (see fig. 6.25). For example, the delay is $2c_i$ for fig. 6.24 and $3c_i$ for fig. 6.25.

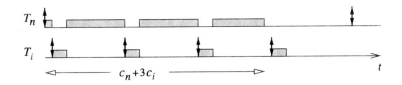

Figure 6.25. Increasing delay of task T_n

The number of preemptions and hence also the response time will be maximized if both tasks become available at the same time.

Arguments concerning T_n and T_i can be repeated for all pairs of tasks. As a result, T_n becomes available at its critical instant in time if it is released concurrently with all other tasks of higher priority. q.e.d.

Therefore, the proof of optimality of RM scheduling needs to consider only the case in which tasks are released concurrently with all other tasks of higher priority.

6.2.4.3   Earliest deadline first scheduling

EDF can also be applied to periodic task sets. Toward this end, we may consider a hyper period.

Definition: Hyper periods are defined as the least common multiple (lcm) of the periods of the individual tasks.

For example, the hyper period for the example of fig. 6.23 is 40. Obviously, it is sufficient to solve the scheduling problem for a single hyper period. This schedule can then be repeated for the other hyper periods. It follows from the optimality of EDF for non-periodic schedules that EDF is also optimal for a single hyper period and therefore also for the entire scheduling problem. No additional constraints must be met to guarantee optimality. This implies that EDF is optimal also for the case of $\mu = 1$. Accordingly, no deadline is missed if the example of fig. 6.23 is scheduled with EDF (see fig. 6.26). At time 5, the behavior is different from that of RM scheduling: due to the earlier deadline of T_2, it is not preempted.

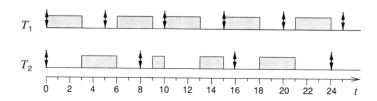

Figure 6.26.   EDF generated schedule for the example of 6.23

Since EDF uses dynamic priorities, it cannot be used with an operating system providing only fixed priorities. However, it has been shown that operating systems can be extended to simulate an EDF policy at the application level [Diederichs et al., 2008].

EDF can be easily extended to handle the case when deadlines are different from the periods.

6.2.5   Periodic scheduling with precedence constraints

Scheduling dependent tasks is more difficult than scheduling independent tasks. The problem of deciding whether or not a schedule exists for a given set of dependent tasks and a given deadline is NP-complete [Garey and Johnson, 1979]. In order to reduce the scheduling effort, different strategies are used:

- adding additional resources such that scheduling becomes easier, and

- partitioning of scheduling into static and dynamic parts. With this approach, as many decisions as possible are taken at design time and only a minimum of decisions is left for run-time.

Obviously, we can also try to exploit HLS-based techniques for periodic processes as well.

6.2.6   Sporadic events

We could connect sporadic events to interrupts and execute them immediately if their interrupt priority is the highest in the system. However, quite unpredictable timing behavior would result for all the other tasks. Therefore, special **sporadic task servers** are used which execute at regular intervals and check for ready sporadic tasks. This way, sporadic tasks are essentially turned into periodic tasks, thereby improving the predictability of the whole system.

6.3   Hardware/software partitioning

6.3.1   Introduction

According to the general problem description on page 237, application mapping techniques must support the mapping to heterogeneous processors. Standard scheduling techniques do not support such a mapping very well. It is supported, however, by hardware/software partitioning techniques. Therefore, we will present an example of such a technique in this section.

By hardware/software partitioning we mean the mapping of task graph nodes to either hardware or software. Applying hardware/software partitioning, we will be able to decide which parts must be implemented in hardware and which in software. A standard procedure for embedding hardware/software partitioning into the overall design flow is shown in fig. 6.27. We start from a common representation of the specification, e.g. in the form of task graphs and information about the platform.

For each of the nodes of the task graphs, we need information concerning the effort required and the benefits received from choosing a certain implementation of these nodes. For example, execution times must be predicted (see page 207). It is very hard to predict times required for communication. Nevertheless, two tasks requiring a very high communication bandwidth should preferably be mapped to the same components. Iterative approaches are used

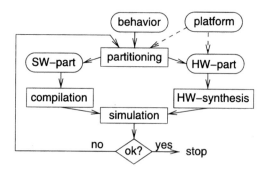

Figure 6.27.   General view of hardware/software partitioning

in many cases. An initial solution to the partitioning problem is generated, analyzed and then improved.

Some approaches for partitioning are restricted to mapping task graph nodes either to special purpose hardware or to software running on a single processor. Such partitioning can be performed with bipartitioning algorithms for graphs [Kuchcinski, 2002].

More elaborate partitioning algorithms are capable of mapping graph nodes to multi-processor systems and hardware. In the following, we will describe how this can be done using a standard optimization technique from operations research, **integer linear programming** (see Appendix A). Our presentation is based on a simplified version of the optimization proposed for the codesign tool COOL [Niemann, 1998].

6.3.2   COOL

For COOL, the input consists of three parts:

- **Target technology:** This part of the input to COOL comprises information about the available hardware platform components. COOL supports multiprocessor systems, but requires that all processors are of the same type, since it does not include automatic or manual processor selection. The type of the processors used (as well as information about the corresponding compiler) must be included in this part of the input to COOL. As far as the application-specific hardware is concerned, the information must be sufficient for starting automatic hardware synthesis with all required parameters. In particular, information about the technology library must be given.

- **Design constraints:** The second part of the input comprises design constraints such as the required throughput, latency, maximum memory size, or maximum area for application-specific hardware.

- **Behavior:** The third part of the input describes the required overall behavior. Hierarchical task graphs are used for this. We can think of, e.g. using the hierarchical task graph of fig. 2.6.

  COOL uses two kinds of edges: communication edges and timing edges. Communication edges may contain information about the amount of information to be exchanged. Timing edges provide timing constraints. COOL requires the behavior of each of the leaf nodes[5] of the graph hierarchy to be known. COOL expects this behavior to be specified in VHDL[6].

For partitioning, COOL uses the following steps:

1 **Translation** of the behavior **into an internal graph model.**

2 **Translation** of the behavior of each node **from VHDL into C.**

3 **Compilation of all C programs** for the selected target processor type, computation of the resulting program size, estimation of the resulting execution time. If simulations are used for the latter, simulation input data must be available.

4 **Synthesis of hardware components:** For each leaf node, application-specific hardware is synthesized. Since quite a number of hardware components may have to be synthesized, hardware synthesis should not be too slow. It was found that commercial synthesis tools focusing on gate level synthesis were too slow to be useful for COOL. However, high-level synthesis (HLS) tools working at the register-transfer-level (using adders, registers, and multiplexer as components, rather than gates) provided sufficient synthesis speed. Also, such tools could provide sufficiently precise values for delay times and required silicon area. In the actual implementation, the OSCAR high-level synthesis tool [Landwehr and Marwedel, 1997] was used.

5 **Flattening the hierarchy:** The next step is to extract a flat task graph from the hierarchical flow graph. Since no merging or splitting of nodes is performed, the granularity used by the designer is maintained. Cost and performance information gained from compilation and from hardware

---

[5] See page 43 for a definition of this term.

[6] In retrospect, we now know that C should have been used for this, as this choice would have made the partitioning for many standards described in C easier.

synthesis are added to the nodes. This is actually one of the key ideas of
COOL: the **information required for hardware/software partitioning is
precomputed and it is computed with good precision**. This information
forms the basis for generating cost-minimized designs meeting the design
constraints.

6 **Generating and solving a mathematical model of the optimization problem:** COOL uses integer linear programming (ILP) to solve the optimization problem. A commercial ILP solver is used to find values for decision variables minimizing the cost. The solution is optimal with respect
to the cost function derived from the available information. However, this
cost includes only a coarse approximation of the communication time. The
communication time between any two nodes of the task graph depends on
the mapping of those nodes to processors and hardware. If both nodes are
mapped to the same processor, communication will be local and thus quite
fast. If the nodes are mapped to different hardware components, communication will be non-local and may be slower. Modeling communication
costs for all possible mappings of task graph nodes would make the model
very complex and is therefore replaced by iterative improvements of the
initial solution. More details on this step will be presented below.

7 **Iterative improvements:** In order to work with good estimates of the communication time, adjacent nodes mapped to the same hardware component
are now merged. This merging is shown in fig. 6.28.

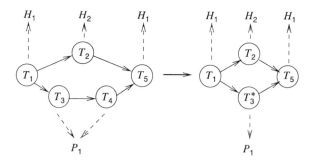

Figure 6.28.   Merging of task nodes mapped to the same hardware component

We assume that tasks T_1, T_2 and T_5 are mapped to hardware components
H_1 and H_2, whereas T_3 and T_4 are mapped to processor P_1. Accordingly,
communication between T_3 and T_4 is **local** communication. Therefore, we
merge T_3 and T_4, and assume that the communication between the two tasks
does not require a communication channel. Communication time can be
now estimated with improved precision. The resulting graph is then used

as new input for mathematical optimization. The previous and the current step are repeated until no more graph nodes are merged.

8 **Interface synthesis:** After partitioning, the glue logic required for interfacing processors, application-specific hardware and memories is created.

Next, we will describe how partitioning can be modeled using a 0/1-ILP model (see Appendix A, page 335). The following index sets will be used in the description of the ILP model:

- Index set V denotes task graph nodes. Each $v \in V$ corresponds to one task graph node.

- Index set L denotes task graph node **types**. Each $l \in L$ corresponds to one task graph node type. For example, there may be nodes describing square root, Discrete Cosine Transform (DCT) or Discrete Fast Fourier Transform (DFT) computations. Each of them is counted as one type.

- Index set M denotes hardware component **types**. Each $m \in M$ corresponds to one hardware component type. For example, there may be special hardware components for the DCT or the DFT. There is one index value for the DCT hardware component and one for the DFT hardware component.

- For each of the hardware components, there may be multiple copies, or "instances". Each instance is identified by an index $j \in J$.

- Index set KP denotes processors. Each $k \in KP$ identifies one of the processors (all of which are of the same type).

The following decision variables are required by the model:

- $X_{v,m}$: this variable will be 1, if node v is mapped to hardware component type $m \in M$ and 0 otherwise.

- $Y_{v,k}$: this variable will be 1, if node v is mapped to processor $k \in KP$ and 0 otherwise.

- $NY_{l,k}$: this variable will be 1, if at least one node of type l is mapped to processor $k \in KP$ and 0 otherwise.

- $Type$ is a mapping $V \rightarrow L$ from task graph nodes to their corresponding types.

In our particular case, the cost function accumulates the total cost of all hardware units:

C = processor costs + memory costs + cost of application specific hardware

We would obviously minimize the total cost if no processors, memory and application specific hardware were included in the "design". Due to the constraints, this is not a legal solution. We can now present a brief description of some of the constraints of the ILP model:

- **Operation assignment constraints:** These constraints guarantee that each operation is implemented either in hardware or in software. The corresponding constraints can be formulated as follows:

$$\forall v \in V \quad : \quad \sum_{m \in M} X_{v,m} + \sum_{k \in KP} Y_{v,k} = 1 \qquad (6.10)$$

In plain text, this means the following: for all task graph nodes v, the following must hold: v is implemented either in hardware (setting one of the $X_{v,m}$ variables to 1, for some m) or it is implemented in software (setting one of the $Y_{v,k}$ variables to 1, for some k).

All variables are assumed to be non-negative integer numbers:

$$X_{v,m} \quad \in \quad \mathbb{N}_0, \qquad (6.11)$$
$$Y_{v,k} \quad \in \quad \mathbb{N}_0 \qquad (6.12)$$

Additional constraints ensure that decision variables $X_{v,m}$ and $Y_{v,k}$ have 1 as an upper bound and, hence, are in fact 0/1-valued variables:

$$\forall v \in V : \forall m \in M \quad : \quad X_{v,m} \leq 1 \qquad (6.13)$$
$$\forall v \in V : \forall k \in KP \quad : \quad Y_{v,k} \leq 1 \qquad (6.14)$$

If the functionality of a certain node of type l is mapped to some processor k, then this processors' instruction memory must include a copy of the software for this function:

$$\forall l \in L, \forall v : Type(v) = c_l, \forall k \in KP : NY_{l,k} \quad \geq \quad Y_{v,k} \qquad (6.15)$$

In plain text, this means: for all types l of task graph nodes and for all nodes v of this type, the following must hold: if v is mapped to some processor k

(indicated by $Y_{v,k}$ being 1), then the software corresponding to functionality l must be provided by processor k, and the corresponding software must exist on that processor (indicated by $NY_{l,k}$ being 1).

Additional constraints ensure that decision variables $NY_{l,k}$ are also 0/1-valued variables:

$$\forall l \in L : \forall k \in KP : NY_{l,k} \leq 1 \qquad (6.16)$$

- **Resource constraints:** The next set of constraints ensures that "not too many" nodes are mapped to the same hardware component at the same time. We assume that, for every clock cycle, at most one operation can be performed per hardware component. Unfortunately, this means that the partitioning algorithm also has to generate a partial schedule for executing task graph nodes. Scheduling by itself is already an NP-complete problem for most of the relevant problem instances.

- **Precedence constraints:** These constraints ensure that the schedule for executing operations is consistent with the precedence constraints in the task graph.

- **Design constraints:** These constraints put a limit on the cost of certain hardware components, such as memories, processors or area of application-specific hardware.

- **Timing constraints:** Timing constraints, if present in the input to COOL, are converted into ILP constraints.

- Some additional, but less important constraints are not included in this list.

Example: In the following, we will show how these constraints can be generated for the task graph in fig. 6.29 (the same as the one in fig. 2.6).

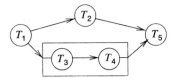

Figure 6.29. Task graph

Suppose that we have a hardware component library containing three components types H_1, H_2 and H_3 with costs of 20, 25 and 30 cost units, respectively. Furthermore, suppose that we can also use a processor P of cost 5. In addition,

T	H_1	H_2	H_3	P
1	20			100
2		20		100
3			12	10
4			12	10
5	20			100

Figure 6.30. Execution times of tasks T_1 to T_5 on components

we assume that the table in fig. 6.30 describes the execution times of our tasks on these components.

Tasks T_1 to T_5 can only be executed on the processor or on one application-specific hardware unit. Obviously, processors are assumed to be cheap but slow in executing tasks T_1, T_2, and T_5.

The following operation assignment constraints must be generated, assuming that a maximum of one processor (P_1) is to be used:

$$X_{1,1} + Y_{1,1} = 1 \text{ (Task 1 either mapped to } H_1 \text{ or to } P_1)$$
$$X_{2,2} + Y_{2,1} = 1 \text{ (Task 2 either mapped to } H_2 \text{ or to } P_1)$$
$$X_{3,3} + Y_{3,1} = 1 \text{ (Task 3 either mapped to } H_3 \text{ or to } P_1)$$
$$X_{4,3} + Y_{4,1} = 1 \text{ (Task 4 either mapped to } H_3 \text{ or to } P_1)$$
$$X_{5,1} + Y_{5,1} = 1 \text{ (Task 5 either mapped to } H_1 \text{ or to } P_1)$$

Furthermore, assume that the types of tasks T_1 to T_5 are $l = 1, 2, 3, 3$ and 1, respectively. Then, the following additional resource constraints are required:

$$NY_{1,1} \geq Y_{1,1} \qquad\qquad (6.17)$$
$$NY_{2,1} \geq Y_{2,1}$$
$$NY_{3,1} \geq Y_{3,1}$$
$$NY_{3,1} \geq Y_{4,1}$$
$$NY_{1,1} \geq Y_{5,1} \qquad\qquad (6.18)$$

Equation 6.17 means: if task 1 is mapped to the processor, then the function $l = 1$ must be implemented on that processor. The same function must also be implemented on the processor if task 5 is mapped to the processor (eq. 6.18).

We have not included timing constraints. However, it is obvious that the processor is slow in executing some of the tasks and that application-specific hardware is required for timing constraints below 100 time units.

The cost function is:

$$C = 20 * \#(H_1) + 25 * \#(H_2) + 30 * \#(H_3) + 5 * \#(P)$$

where $\#()$ denotes the number of instances of hardware components. This number can be computed from the variables introduced so far if the schedule is also taken into account. For a timing constraint of 100 time units, the minimum cost design comprises components H_1, H_2 and P. This means that tasks T_3 and T_4 are implemented in software and all others in hardware.

In general, due to the complexity of the combined partitioning and scheduling problem, only small problem instances of the combined problem can be solved in acceptable run-times. Therefore, the problem is heuristically split into the scheduling and the partitioning problem: an initial partitioning is based on estimated execution times and the final scheduling is done after partitioning. If it turns out that the schedule was too optimistic, the whole process has to be repeated with tighter timing constraints. Experiments for small examples have shown that the cost for heuristic solutions is only 1 or 2 % larger than the cost of optimal results.

Automatic partitioning can be used for analyzing the design space. In the following, we will present results for an audio lab, including mixer, fader, echo, equalizer and balance units. This example uses earlier target technologies in order to demonstrate the effect of partitioning. The target hardware consists of a (slow) SPARC processor, external memory, and application-specific hardware to be designed from an (outdated) 1μ ASIC library. The total allowable delay is set to 22675 ns, corresponding to a sample rate of 44.1 kHz, as used in CDs. Fig. 6.31 shows different design points which can be generated by changing the delay constraint.

The unit λ refers to a technology-dependent length unit. It is essentially one half of the closest distance between the centers of two metal wires on the chip (also called *half-pitch* [ITRS Organization, 2009]). The design point at the left corresponds to a solution implemented completely in hardware, the design point at the right to a software solution. Other design points use a mixture of hardware and software. The one corresponding to an area of 78.4 λ^2 is the cheapest meeting the deadline.

Obviously, technology has advanced to allow a 100% software-based audio lab design nowadays. Nevertheless, this example demonstrates the underlying design methodology which can also be used for more demanding applications, especially in the high-speed multimedia domain, such as MPEG-4.

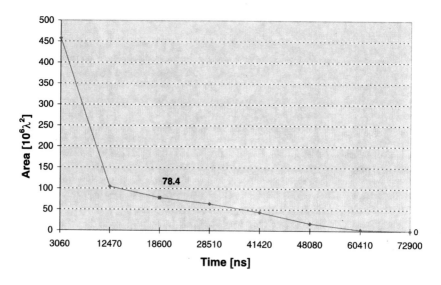

Figure 6.31. Design space for audio lab

6.4   Mapping to heterogeneous multi-processors

Currently (in 2010), mapping to heterogeneous multi-processors still is a research topic. Overviews of the state of art in this area are provided by the
Workshops on Mapping of Applications to MPSoCs, organized by the *Artist-
Design* European Network of Excellence. The following information is based
on the first [Marwedel, 2008a] and the second [Marwedel, 2009a] workshop in
this series as well as on a summary of the first workshop [Marwedel, 2009b].
The different approaches for this mapping can be classified by two criteria:
mapping tools may either assume a fixed execution platform or may design
such a platform during the mapping and they may or may not include automatic parallelization of the source codes. Fig. 6.32 contains a classification of
some of the available mapping tools by these two criteria.

The DOL tools from ETH Zürich [Thiele, L. et al., 2009] incorporate

- **Automatic selection of computation templates:** Processor types can be
  completely heterogeneous. Standard processors, micro-controllers, DSP
  processors, FPGAs etc. are all possible options.

- **Automatic selection of communication techniques:** Various interconnec-
  tion schemes like central buses, hierarchical buses, rings etc. are feasible.

Architecture fixed/ Auto parallelizing	Fixed Architecture	Architecture to be designed
Starting from a given model	HOPES, mapping to CELL proc., Q. Xu, T. Simunic	COOL, DOL, SystemCo-designer
Auto-parallelizing	Mnemee, O'Boyle and Franke	Daedalus
	MAPS	

Figure 6.32. Classification of mapping tools and authors' work

- **Automatic selection of scheduling and arbitration:** DOL design space exploration tools automatically choose between rate monotonic scheduling, EDF, TDMA- and priority-based schemes.

The input to DOL consists of a set of tasks together with use cases. The output describes the execution platform, the mapping of tasks to processors together with task schedules. This output is expected to meet constraints (like memory size and timing constraints) and to minimize objectives (like size, energy etc). Applications are represented by so-called problem graphs. Fig. 6.33 shows a simple DOL-problem graph. This graph models communication explicitly.

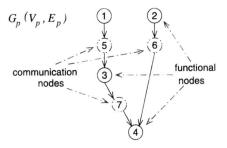

Figure 6.33. DOL problem graph

In addition, possible execution platforms are represented by so-called architecture graphs. Fig. 6.34 shows a simple hardware platform together with its architecture graph. Again, communication is modeled explicitly.

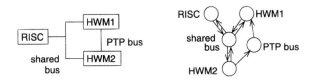

Figure 6.34. DOL architecture graph

The problem graph and the architecture graph are connected in the specification graph. Fig. 6.35 shows a DOL specification graph.

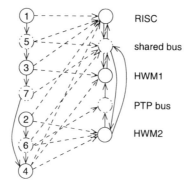

Figure 6.35. DOL specification graph

Such a specification graph consists of the problem graph and the architecture graph. Edges between the two subgraphs represent feasible implementations. In total, implementations are represented by a triple:

- **An allocation** α: α is a subset of the architecture graph, representing hardware components allocated (selected) for a particular design.

- **A binding** β: a selected subset of the edges between specification and architecture identifies a relation between the two. Selected edges are called **bindings**.

- **A schedule** τ: τ assigns start times to each node v in the problem graph.

Example: Fig. 6.36 shows how the specification of fig. 6.35 can be turned into an implementation.

In DOL, implementations are generated with evolutionary algorithms [Bäck and Schwefel, 1993], [Bäck et al., 1997], [Coello et al., 2007]. With such algorithms, solutions are represented as strings in chromosomes of "individuals". Using evolutionary algorithms, new sets of solutions can be derived from existing sets of solutions. The derivation is based on evolutionary operators such as mutation, selection and recombination. The selection of new sets of solutions is based on **fitness values**. Evolutionary algorithms are capable of solving complex optimization problems not tackable by other types of algorithms. Finding appropriate ways of encoding solutions in chromosomes is not easy. On one hand, the decoding should not require too much run-time. On the other hand, we must deal with the situation after the evolutionary transfor-

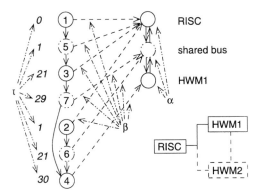

Figure 6.36. DOL implementation

mations. These transformations could generate infeasible solutions, except for some carefully designed encodings.

In DOL, chromosomes encode allocations and bindings. In order to evaluate the fitness of a certain solution, allocations and bindings must be decoded from the individuals (see fig. 6.37).

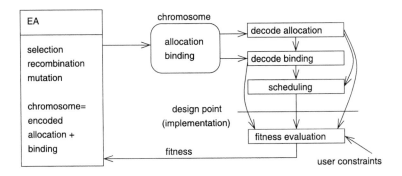

Figure 6.37. Decoding of solutions from chromosomes of individuals

In DOL, schedules are not encoded in the chromosomes. Rather, they are derived from the allocation and binding. This way overloading evolutionary algorithms with scheduling decisions is avoided. Once the schedule has been computed, the fitness of solutions can be evaluated.

The overall architecture of DOL is shown in fig. 6.38.

Initially, the task graph, use cases and available resources are defined. This can be done with a specialized editor called MOSES. This initial information is evaluated in the evaluation framework EXPO. Performance values computed by EXPO are then sent to SPEA2, an evolutionary algorithm-based optimiza-

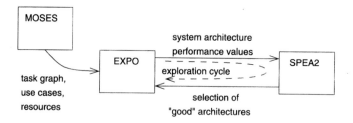

Figure 6.38. DOL tool

tion framework. SPEA2 selects good candidate architectures. These are sent
back to EXPO for an evaluation. Evaluation results are then communicated
again to SPEA2 for another round of evolutionary optimizations. This kind
of ping-pong game between EXPO and SPEA2 continues until good solutions
have been found. The selection of solutions is based on the principle of Pareto-
optimality. A set of Pareto-optimal designs is returned to the designer, who can
then analyze the trade-off between the different objectives. Fig. 6.39 shows the
resulting visualization of the Pareto-front.

The functionality of the SystemCodesigner [Keinert et al., 2009] is somewhat
similar to that of DOL. However, it differs in the way specifications are described (they can be represented in SystemC) and in the way, the optimizations
are performed. The mapping of applications is modeled as an ILP model. A
first solution is generated using an ILP optimizer. This solution is then improved by switching to evolutionary algorithms[7].

Daedalus [Nikolov et al., 2008] incorporates automatic parallelization. For
this purpose, sequential applications are mapped to Kahn process networks.
Design space exploration is then performed using Kahn process networks as
an intermediate representation.

Other approaches start from a given task graph and map to a fixed architecture. For example, Ruggiero maps applications to cell processors [Ruggiero
and Benini, 2008]. The HOPES-system is able to map to various processors [Ha, 2007], using models of computation supported by the Ptolemy tools.
Some tools take additional objectives into account. For example, Xu considers
the optimization of the dependable lifetime of the resulting system [Xu et al.,
2009]. Simunic incorporates thermal analysis into her work and tries to avoid
too hot spots on the MPSoC [Simunic-Rosing et al., 2007]. Further work includes that of Popovici et al. [Popovici et al., 2010]. This work uses several
levels of modeling, employing Simulink and SystemC as languages.

---

[7] A more recent version uses a satisfiability (SAT) solver for the same purpose.

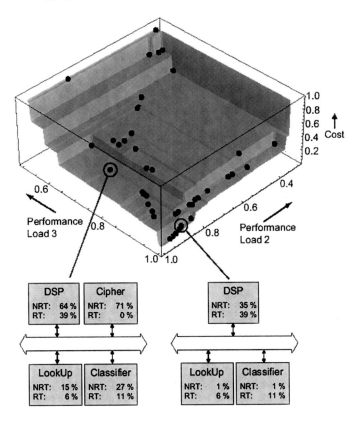

Figure 6.39. Pareto front of solutions for a design problem, ©ETHZ

Auto-parallelizing approaches for fixed architectures include the Mnemee tool set [Mnemee project, 2010] and work at the University of Edinburgh [Franke and O'Boyle, 2005]. MAPS tools [Ceng et al., 2008] combine automatic parallelization with a limited DSE.

6.5 Assignments

1 Suppose that we have a set of 4 tasks. Arrival times A_i, deadlines d_i and execution times c_i are as follows:

- T_1: $A_1=10$, $d_1=18$, $c_1=4$

- T_2: $A_2=0$, $d_2=28$, $c_2=12$

- T_3: $A_3=6$, $d_3=17$, $c_3=3$

- T_4: $A_4=3$, $d_4=13$, $c_4=6$

Generate a graphical representation of schedules for this task set, using *Earliest Deadline First* (EDF) and *Least Laxity* (LL) scheduling algorithms! For LL scheduling, indicate laxities for all tasks at all context switch times. Will any task miss its deadline?

2 Suppose that we have a task set of six tasks T_1 to T_6. Their execution times and their deadlines are as follows:

- T_1: $d_1=15$, $c_1=3$

- T_2: $d_2=13$, $c_2=5$

- T_3: $d_3=14$, $c_3=4$

- T_4: $d_4=16$, $c_4=2$

- T_5: $d_3=20$, $c_3=4$

- T_6: $d_4=22$, $c_4=3$

Task dependencies are as shown in fig. 6.40. Tasks T_1 and T_2 are available immediately.

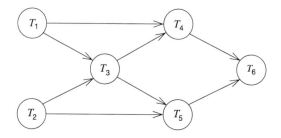

Figure 6.40.   Task dependencies

Generate a graphical representation of schedules for this task set, using the *Latest Deadline First* (LDF) algorithm!

3 Suppose that we have a system comprising two tasks. Task 1 has a period of 5 and and execution time of 2. The second task has a period of 7 and an execution time of 4. Let the deadlines be equal to the periods. Assume that we are using *Rate Monotonic Scheduling* (RMS). Could any of the two tasks miss its deadline, due to a too high processor utilization? Compute this utilization and compare it to a bound which would guarantee schedulability! Generate a graphical representation of the resulting schedule! Suppose that tasks will always run to their completion, even if they missed their deadline.

4 Consider the same task set as in the previous assignment. Use *Earliest Deadline First* (EDF) for scheduling. Can any of the tasks miss its deadline? If not, why not? Generate a graphical representation of the resulting schedule! Suppose that tasks will always run to their completion.

5 Consider a set of tasks. Let $V = \{v\}$ be the index set for tasks. Let $L = \{l\}$ be the set of task types and let $Type : V \rightarrow L$ be a mapping from tasks to their types. Assume that $M = \{m\}$ and $KP = \{k\}$ denote the set of hardware component types and processors, respectively. Describe the following elements of the hardware/software partitioning model used by COOL:

(a) Which decision variables are required?

(b) Which variables model whether or not tasks of type l are mapped to processor k?

(c) What does the objective function look like?

(d) Which equations are required to ensure that each task will be implemented in either hardware or in software?

(e) Which equations are required to ensure that tasks are mapped to a certain processor only if the software for the type of task is available on that processor?

Chapter 7

OPTIMIZATION

In order to make embedded systems as efficient as required, many optimizations have been developed. Only a small subset of those can be mentioned in this book. In this chapter, we will present a selected set of such optimizations. As indicated in our design flow, these optimizations complement the tools mapping applications to the final systems, as described in Chapter 6 and as shown in fig. 7.1.

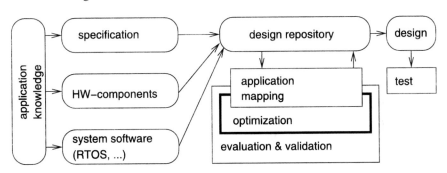

Figure 7.1. Context of the current Chapter

7.1 Task level concurrency management

As mentioned on page 31, the task graphs' **granularity** is one of their most important properties. Even for hierarchical task graphs, it may be useful to change the granularity of the nodes. The partitioning of specifications into tasks or processes does not necessarily aim at the maximum implementation efficiency. Rather, during the specification phase, a clear separation of concerns and a clean software model are more important than caring about the im-

P. Marwedel, *Embedded System Design*, Embedded Systems,
DOI 10.1007/978-94-007-0257-8_7, © Springer Science+Business Media B.V. 2011

plementation too much. For example, a clear separation of concerns includes
a clear separation of the implementation of abstract data types from their use.
Also, we might be using several tasks in a pipelined fashion in our specification, where merging some of them might reduce context switching overhead.
Hence, there will not necessarily be a one-to-one correspondence between the
tasks in the specification and those in the implementation. This means that a
regrouping of tasks may be advisable. Such a regrouping is indeed feasible by
merging and splitting of tasks.

Merging of task graphs can be performed whenever some task T_i is the immediate predecessor of some other task T_j and if T_j does not have any other
immediate predecessor (see fig. 7.2 with $T_i = T_3$ and $T_j = T_4$). This transformation can lead to a reduced overhead of context-switches if the node is
implemented in software, and it can lead to a larger potential for optimizations
in general.

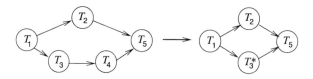

Figure 7.2.   Merging of tasks

On the other hand, splitting of tasks may be advantageous for the following
reasons:

Tasks may be holding resources (like large amounts of memory) while they
are waiting for some input. In order to maximize the use of these resources, it
may be best to constrain the use of these resources to the time intervals during
which these resources are actually needed. In fig. 7.3, we are assuming that
task T_2 requires some input somewhere in its code.

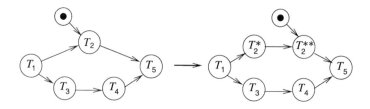

Figure 7.3.   Splitting of tasks

In the initial version, the execution of task T_2 can only start if this input is
available. We can split the node into T_2^* and T_2^{**} such that the input is only
required for the execution of T_2^{**}. Now, T_2^* can start earlier, resulting in more

scheduling freedom. This improved scheduling freedom might improve resource utilization and could even enable meeting some deadline. It may also have an impact on the memory required for data storage, since T_2^* could release some of its memory before terminating and this memory could be used by other tasks while T_2^{**} is waiting for input.

One might argue that the tasks should release resources like large amounts of memory anyway before waiting for input. However, the readability of the original specification could suffer from caring about implementation issues in an early design phase.

Quite complex transformations of the specifications can be performed with a Petri net-based technique described by Cortadella et al. [Cortadella et al., 2000]. Their technique starts with a specification consisting of a set of tasks described in a language called *FlowC*. FlowC extends C with process headers and intertask communication specified in the form of READ- and WRITE-function calls. Fig. 7.4 shows an input specification using FlowC.

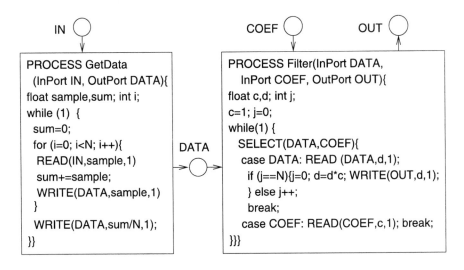

Figure 7.4. System specification

The example uses input ports IN and COEF, as well as output port OUT. Point-to-point interprocess communication between processes is realized through a uni-directional buffered channel DATA. Task GetData reads data from the environment and sends it to channel DATA. Each time N samples have been sent, their average value is also sent via the same channel. Task Filter reads N values from the channel (and ignores them) and then reads the average value, multiplies the average value by c (c can be read in from port COEF) and writes the result to port OUT. The third parameter in READ and WRITE calls is the num-

ber of items to be read or written. READ calls are blocking, WRITE calls are
blocking if the number of items in the channel exceeds a predefined threshold.
The SELECT statement has the same semantics as the statement with the same
name in ADA (see page 104): execution of this task is suspended until input
arrives from one of the ports. This example meets all criteria for splitting tasks
that were mentioned in the context of fig. 7.3. Both tasks will be waiting for
input while occupying resources. Efficiency could be improved by restructur-
ing these tasks. However, the simple splitting of fig. 7.3 is not sufficient. The
technique proposed by Cortadella et al. is a more comprehensive one. Using
their technique, FlowC-programs are first translated into (extended) Petri nets.
Petri nets for each of the tasks are then merged into a single Petri net. Using
results from Petri net theory, new tasks are then generated. Fig. 7.5 shows a
possible new task structure.

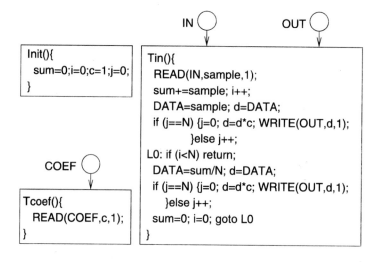

Figure 7.5.   Generated software tasks

In this new task structure, there is one task which performs all initializations:
In addition, there is one task for each of the input ports. An efficient implementation would raise interrupts each time new input is received for a port.
There should be a unique interrupt per port. The tasks could then be started
directly by those interrupts, and there would be no need to invoke the operating
system for that. Communication can be implemented as a single shared global
variable (assuming a shared address space for all tasks). The overall operating
system overhead would be quite small, if required at all.

The code for task Tin shown in fig. 7.5 is the one that is generated by the Petri
net-based inter-task optimization of the task structure. It should be further
optimized by intra-task optimizations, since the test performed for the first if-

statement is always false (j is equal to i-1 in this case, and i and j are reset to 0 whenever i becomes equal to N). For the third if-statement, the test is always true, since this point of control is only reached if i is equal to N and i is equal to j whenever label L0 is reached. Also, the number of variables can be reduced. The following is an optimized version Tin:

```
Tin () {
    READ (IN, sample, 1);
    sum += sample; i++;
    DATA = sample; d = DATA;
L0: if (i < N) return;
    DATA = sum/N; d = DATA;
    d = d*c; WRITE(OUT,d,1);
    sum = 0; i = 0;
    return;
}
```

The optimized version of Tin could be generated by a very clever compiler. Unfortunately, hardly any of today's compilers will perform this optimization. Nevertheless, the example shows the type of transformations required for generating "good" task structures. For more details about the task generation, refer to Cortadella et al. [Cortadella et al., 2000].

Optimizations similar to the one just presented are described in the book by Thoen [Thoen and Catthoor, 2000] and in a publication by Meijer et al. [Meijer et al., 2010].

7.2 High-level optimizations

There are many high-level optimizations which can potentially improve the efficiency of embedded software.

7.2.1 Floating-point to fixed-point conversion

Floating-point to fixed-point conversion is a commonly used technique. This conversion is motivated by the fact that many signal processing standards (such as MPEG-2 or MPEG-4) are specified in the form of C-programs using floating-point data types. It is left to the designer to find an efficient implementation of these standards.

For many signal processing applications, it is possible to replace floating-point numbers with fixed-point numbers (see page 144). The benefits may be significant. For example, a reduction of the cycle count by 75% and of the energy consumption by 76% has been reported for an MPEG-2 video compression algorithm [Hüls, 2002]. However, some loss of precision is normally incurred. More precisely, there is a trade-off between the cost of the implementation and the quality of the algorithm (evaluated e.g. in terms of the signal-to-noise ratio (SNR), see page 132). For small word-lengths, the quality may be seriously affected. Consequently, floating-point data types may be replaced by fixed-point data types, but the quality loss has to be analyzed. This replacement was initially performed manually. However, it is a very tedious and error-prone process.

Therefore, researchers have tried to support this replacement with tools. One of such tools is FRIDGE (fixed-point programming design environment) [Willems et al., 1997], [Keding et al., 1998]. FRIDGE tools have been made available commercially as part of the Synopsys CoCentric tool suite [Synopsys, 2010].

In FRIDGE, the design process starts with an algorithm described in C, including floating-point numbers. This algorithm is then converted to an algorithm described in **fixed-C**. Fixed-C extends C by two fixed-point data types, using the type definition features of C++. Fixed-C is a subset of C++ and provides two data types fixed and Fixed. Fixed-point data types can be declared very much like other variables. The following declaration declares a scalar variable, a pointer, and an array to be fixed-point data types.

```
fixed a,*b,c[8]
```

Providing parameters of fixed-point data types can (but does not have to) be delayed until assignment time:

```
a=fixed(5,4,s,wt,*b)
```

This assignment sets the word-length parameter of a to 5 bits, the fractional word-length to 4 bits, sign to present (s), overflow handling to wrap-around (w), and the rounding mode to truncation (t). The parameters for variables that are read in an assignment are determined by the assignment(s) to those variables. The data type Fixed is similar to fixed, except that a consistency check between parameters used in the declaration and those used in the assignment is performed. For every assignment to a variable, parameters (including the word-length) can be different. This parameter information can be added to the original C-program before the application is simulated. Simulation provides value ranges for all assignments. Based on that information, FRIDGE adds parameter information to all assignments. FRIDGE also infers parameter information from the context. For example, the maximum value of additions is considered to be the sum of the arguments. Added parameter information can be either based on simulations or on worst case considerations. Being based on simulations, FRIDGE does not necessarily assume the worst case values that would result from a formal analysis. The resulting C++-program is simulated again to check for the quality loss. The Synopsys version of FRIDGE uses SystemC fixed-point data types to express generated data type information. Accordingly, SystemC can be used for simulating fixed-point data types.

An analysis of the trade-offs between the additional noise introduced and the word-length needed was proposed by Shi and Brodersen [Shi and Brodersen, 2003] and also by Menard et al. [Menard and Sentieys, 2002].

7.2.2 Simple loop transformations

There is a number of loop transformations that can be applied to specifications. The following is a list of standard loop transformations:

- **Loop permutation:** Consider a two-dimensional array. According to the C standard [Kernighan and Ritchie, 1988], two-dimensional arrays are laid out in memory as shown in fig. 7.6. Adjacent index values of the second index are mapped to a contiguous block of locations in memory. This layout is called **row-major order** [Muchnick, 1997]. Note that the layout for arrays is different for FORTRAN: Adjacent values of the first index are mapped to a contiguous block of locations in memory (**column major order**). Publications describing optimizations for FORTRAN can therefore be confusing.

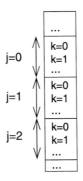

Figure 7.6. Memory layout for two-dimensional array p[j][k] in C

For row-major layout, it is usually beneficial to organize loops such that the last index corresponds to the innermost loop. A corresponding loop permutation is shown in the following example:

```
for (k=0; k<=m; k++)          for (j=0; j<=n; j++)
   for (j=0; j<=n; j++)    ⇒     for (k=0; k<=m; k++)
      p[j][k] = ...                 p[j][k] = ...
```

Such permutations may have a positive effect on the reuse of array elements
in the cache, since the next iteration of the loop body will access an adjacent location in memory. Caches are normally organized such that adjacent
locations can be accessed significantly faster than locations that are further
away from the previously accessed location.

- **Loop fusion, loop fission:** There may be cases in which two separate loops
  can be merged, and there may be cases in which a single loop is split into
  two. The following is an example:

```
for (j=0; j<=n; j++)          for (j=0; j<=n; j++)
   p[j]= ... ;                   {p[j]= ... ;
for (j=0; j<=n; j++)    ⇔          p[j]= p[j] + ...}
   p[j]= p[j] + ...
```

The left version may be advantageous if the target processor provides a
zero-overhead loop instruction which can only be used for small loops.
The right version might lead to an improved cache behavior (due to the
improved locality of references to array p), and also increases the potential
for parallel computations within the loop body. As with many other transformations, it is difficult to know which of the transformations leads to the
best code.

- **Loop unrolling:** Loop unrolling is a standard transformation creating several instances of the loop body. The following is an example in which the
  loop is being unrolled once:

```
for (j=0; j<=n; j++)          for (j=0; j<=n; j+=2)
   p[j]= ... ;          ⇒        {p[j]= ... ;
                                  p[j+1]= ...}
```

The number of copies of the loop is called the **unrolling factor**. Unrolling
factors larger than two are possible. Unrolling reduces the loop overhead
(less branches per execution of the original loop body) and therefore typically improves the speed. As an extreme case, loops can be completely
unrolled, removing control overhead and branches altogether. Unrolling
typically enables a number of following transformations and may therefore be beneficial even in cases where just unrolling the program does not

give any advantages. However, unrolling increases code size. Unrolling is normally restricted to loops with a constant number of iterations.

7.2.3   Loop tiling/blocking

It can be observed that the speed of memories is increasing at a slower rate than that of processors. Since small memories are faster than large memories (see page 155), the use of memory hierarchies may be beneficial. Possible "small" memories include caches and scratch-pad memories. A significant reuse factor for the information in those memories is required. Otherwise the memory hierarchy cannot be efficiently exploited.

Reuse effects can be demonstrated by an analysis of the following example. Let us consider matrix multiplication for arrays of size $N \times N$ [Lam et al., 1991]:

```
for (i=1; i<=N; i++)
    for(k=1; k<=N; k++){
        r=X[i,k]; /* to be allocated to a register*/
        for (j=1; j<=N; j++)
        Z[i,j] += r* Y[k,j]
    }
```

Let us consider access patterns for this code. The same element X[i,k] is used by all iterations of the innermost loop. Compilers will typically be capable of allocating this element to a register and reuse it for every execution of the innermost loop. We assume that array elements are allocated in row major order (as it is standard for C). This means that array elements with adjacent row (right most) index values are stored in adjacent memory locations. Accordingly, adjacent locations of Z and Y are fetched during the iterations of the innermost loop. This property is beneficial if the memory system uses prefetching (whenever a word is loaded into the cache, loading of the next word is started as well). Fig. 7.7 shows access patterns for this code.

For one iteration of the innermost loop, the black areas of arrays Z and Y are accessed (and loaded into the cache). Whether or not the same information is still in the cache for the next iteration of the middle or outermost loops depends on the size of the cache. In the worst case (if N is large or the cache is small), the information has to be reloaded for every execution of the innermost loop and cache elements are not reused. The total number of memory references may be as large as $2 N^3$ (for references to Z), N^3 (for references to Y), and N^2 (for references to X).

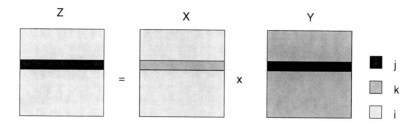

Figure 7.7. Access pattern for unblocked matrix multiplication

Research on scientific computing led to the design of **blocked** or **tiled algorithms** [Xue, 2000], which improve the **locality of references**. The following is a tiled version of the above algorithm:

```
for (kk=1; kk<= N; kk+=B)
  for (jj=1; jj<= N; jj+=B)
   for (i=1; i<= N; i++)
    for (k=kk; k<= min(kk+B-1,N); k++){
     r=X[i][k]; /* to be allocated to a register*/
     for (j=jj; j<= min(jj+B-1, N); j++)
      Z[i][j] += r* Y[k][j]
    }
```

Fig. 7.8 shows the corresponding access pattern.

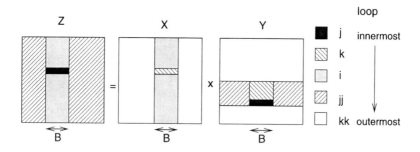

Figure 7.8. Access pattern for tiled/blocked matrix multiplication

The innermost loop is now restricted so that it accesses less array elements (those shown in black). Like above, references to X are replaced by references to r. If a proper blocking factor is selected, elements of Z and Y are still in the cache when the next iteration of the innermost loop starts. The blocking factor B can be chosen such that the elements of the innermost loops fit into the cache. In particular, it can be chosen such that a B × B sub-matrix of Y fits into the cache. This corresponds to a reuse factor of B for Y, since the elements in the sub-matrix are accessed B times for each iteration of i. Also, a block of B row elements of Z should fit into the cache. These will then be reused during the iterations of k, resulting in a reuse factor of B for Z as well. This reduces the overall number of memory references to at most $2 N^3/B$ (for references to Z) and N^2 (for references to X). In practice, the reuse factor may be less than B. Optimizing the reuse factor has been an area of comprehensive research. Initial research focused on the performance improvements that can be obtained by tiling. Performance improvements for matrix multiplication by a factor between 3 and 4.3 was reported by Lam [Lam et al., 1991]. Possible improvements are expected to increase with the increasing gap between processor and memory speeds. Tiling can also reduce the energy consumption of memory systems [Chung et al., 2001].

7.2.4 Loop splitting

Next, we discuss loop splitting as another optimization that can be applied before compiling the program. Potentially, this optimization could also be added to compilers.

Many image processing algorithms perform some kind of filtering. This filtering consists of considering the information about a certain pixel as well as that of some of its neighbors. Corresponding computations are typically quite regular. However, if the considered pixel is close to the boundary of the image, not all neighboring pixels exist and the computations must be modified. In a straightforward description of the filtering algorithm, these modifications may result in tests being performed in the innermost loop of the algorithm. A more efficient version of the algorithm can be generated by splitting the loops such that one loop body handles the regular cases and a second loop body handles the exceptions. Figure 7.9 is a graphical representation of this transformation.

Figure 7.9. Splitting image processing into regular and special cases

Performing this loop splitting manually is a very difficult and error-prone procedure. Falk et al. have published an algorithm [Falk and Marwedel, 2003] to perform a procedure which also works for larger dimensions automatically. It is based on a sophisticated analysis of accesses to array elements in loops. Op-

timized solutions are generated using genetic algorithms. The following code
shows a loop nest from the MPEG-4 standard performing motion estimation:

```
for (z=0; z<20; z++)
  for (x=0; x<36; x++) {x1=4*x;
   for (y=0; y<49; y++) {y1=4*y;
    for (k=0; k<9; k++) {x2=x1+k-4;
     for (l=0; l<9; ) {y2=y1+l-4;
      for (i=0; i<4; i++) {x3=x1+i; x4=x2+i;
       for (j=0; j<4;j++) {y3=y1+j; y4=y2+j;
        if (x3<0 || 35<x3||y3<0||48<y3)
          then_block_1; else else_block_1;
        if (x4<0|| 35<x4||y4<0||48<y4)
          then_block_2; else else_block_2;
}}}}}}
```

Using Falk's algorithm, this loop nest is transformed into the following one:

```
for (z=0; z<20; z++)
  for (x=0; x<36; x++) {x1=4*x;
   for (y=0; y<49; y++)
    if (x>=10||y>=14)
     for (; y<49; y++)
      for (k=0; k<9; k++)
       for (l=0; l<9;l++ )
        for (i=0; i<4; i++)
         for (j=0; j<4;j++) {
           then_block_1; then_block_2}
    else {y1=4*y;
     for (k=0; k<9; k++) {x2=x1+k-4;
      for (l=0; l<9; ) {y2=y1+l-4;
       for (i=0; i<4; i++) {x3=x1+i; x4=x2+i;
        for (j=0; j<4;j++) {y3=y1+j; y4=y2+j;
         if (0 || 35<x3 ||0|| 48<y3)
           then_block_1; else else_block_1;
```

```
        if (x4<0|| 35<x4||y4<0||48<y4)
          then_block_2; else else_block_2;
}}}}}}
```

Instead of complicated tests in the innermost loop, we now have a splitting if-statement after the third for-loop statement. All regular cases are handled in the then-part of this statement. The else-part handles the relatively small number of remaining cases.

Fig. 7.10 shows the number of cycles that can be saved by loop nest splitting for various applications and target processors.

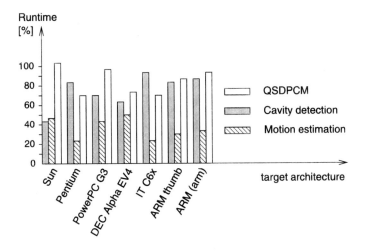

Figure 7.10. Results for loop splitting

For the motion estimation algorithm, cycle counts can be reduced by up to about 75 % (to 25 % of the original value). Obviously, substantial savings are possible. This potential should certainly not be ignored.

7.2.5 Array folding

Some embedded applications, especially in the multimedia domain, include large arrays. Since memory space in embedded systems is limited, options for reducing the storage requirements of arrays should be explored. Fig. 7.11 represents the addresses used by five arrays as a function of time. At any particular time only a subset of array elements is needed. The maximum number of elements needed is called the **address reference window** [De Greef et al., 1997b]. In fig. 7.11, this maximum is indicated by a double-headed arrow.

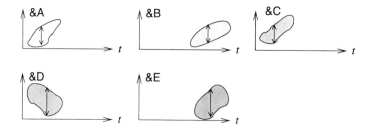

Figure 7.11.   Reference patterns for arrays

A classical memory allocation for arrays is shown in fig. 7.12 (left). Each array is allocated the maximum of the space it requires during the entire execution time (if we consider global arrays).

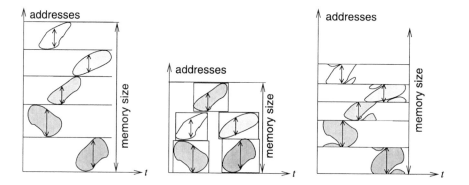

Figure 7.12.   Unfolded (left), inter-array folded (center), and intra-array folded (right) arrays

One of the possible improvements, **inter-array folding**, is shown in fig. 7.12 (center). Arrays which are not needed at overlapping time intervals can share the same memory space. A second improvement, intra-array folding [De Greef et al., 1997a], is shown in fig. 7.12 (right). It takes advantage of the limited sets of components needed **within** an array. Storage can be saved at the expense of more complex address computations. The two kinds of foldings can also be combined.

Other forms of high-level transformations have been analyzed by Chung, Benini and De Micheli [Chung et al., 2001], [Tan et al., 2003]. There are many additional contributions in this domain in the compiler community.

In particular, function inlining replaces function calls by the code of the called function. This transformation improves the speed of the code, but results in an increase in the code size. Increased code sizes may be a problem in SoC technologies. Traditional inlining techniques rely on the user identifying functions to be inlined. This is a problem in systems on a chip, since the size of the instruction memory is very critical for such systems. Hence, it is important to be able to constrain the size of the instruction memory and to let design tools find out automatically which of the functions should be inlined for a certain size of the memory. Known approaches for this include techniques by Teich [Teich et al., 1999], Leupers et al. [Leupers and Marwedel, 1999], Palkovic [Palkovic et al., 2002], and Lokuciejewski [Lokuciejewski et al., 2009]. These techniques can be either integrated into a compiler or can be applied as a source-to-source transformation before using any compiler.

7.3   Compilers for embedded systems

7.3.1   Introduction

Obviously, optimizations and compilers are available for the processors used in PCs and compiler generation for commonly used 32-bit processors is well understood. For embedded systems, standard compilers are also used in many cases, since they are typically cheap or even freely available.

However, there are several reasons for designing special optimizations and compilers for embedded systems:

- Processor architectures in embedded systems exhibit special features (see page 135). These features should be exploited by compilers in order to generate efficient code. Compilation techniques might also have to support compression techniques described on pages 138 to 140.

- A high efficiency of the code is more important than a high compilation speed.

- Compilers could potentially help to meet and prove real-time constraints. First of all, it would be nice if compilers contained explicit timing models. These could be used for optimizations which really improve the timing behavior. For example, it may be beneficial to freeze certain cache lines in order to prevent frequently executed code from being evicted and reloaded several times.

- Compilers may help to reduce the energy consumption of embedded systems. Compilers performing energy optimizations should be available.

- For embedded systems, there is a larger variety of instruction sets. Hence, there are more processors for which compilers should be available. Sometimes there is even the request to support the optimization of instruction sets with **retargetable** compilers. For such compilers, the instruction set can be specified as an input to a compiler generation system. Such systems can be used for experimentally modifying instruction sets and then observing the resulting changes for the generated machine code. This is one particular case of **design space exploration** and is supported, for example, by Tensilica tools [Tensilica Inc., 2010].

Some first approaches for retargetable compilers are described in the first book on this topic [Marwedel and Goossens, 1995]. Optimizations can be found in books by Leupers et al. [Leupers, 1997], [Leupers, 2000a]. In this section, we will present examples of compilation techniques for embedded processors.

7.3.2   Energy-aware compilation

Many embedded systems are mobile systems which must run on batteries. While computational demands on mobile systems are increasing, battery technology is expected to improve only slowly [ITRS Organization, 2009]. Hence, the availability of energy is a serious bottleneck for new applications.

Saving energy can be done at various levels, including the fabrication process technology, the device technology, circuit design, the operating system and the application algorithms. Adequate translation from algorithms to machine code can also help. High-level optimization techniques such as those presented on pages 285 to 295 can also help to reduce the energy consumption. In this section, we will look at compiler optimizations which can reduce the energy consumption (frequently called low *power* optimizations). **Energy models** are very essential ingredients of all energy optimizations. Energy models were presented in Chapter 5. Using models like those, the following compiler optimizations have been used for reducing the energy consumption:

- **Energy-aware scheduling:** the order of instructions can be changed as long as the meaning of the program does not change. The order can be changed such that the number of transitions on the instruction bus is minimized. This optimization can be performed on the output generated by a compiler and therefore does not require any change to the compiler.

- **Energy-aware instruction selection:** typically, there are different instruction sequences for implementing the same source code. In a standard compiler, the number of instructions or the number of cycles is used as a criterion (cost function) for selecting a good sequence. This criterion can be replaced by the energy consumed by that sequence. Steinke and others found that energy-aware instruction selection reduces the energy consumption by some percent [Steinke, 2003].

- **Replacing the cost function** is also possible for other standard compiler optimizations, such as register pipelining, loop invariant code motion etc. Possible improvements are also in the order of a few percent.

- **Exploitation of the memory hierarchy:** As already explained on page 155, smaller memories provide faster access and consume less energy per access. Therefore, a significant amount of energy can be saved if memory hierarchies are exploited. Of all the compiler optimizations analyzed by Steinke [Steinke et al., 2002b], [Steinke et al., 2002a], the energy savings enabled by memory hierarchies are the largest. It is therefore beneficial to use small scratch-pad memories (SPMs) in addition to large background memories. All accesses to the corresponding address range will then require less energy and are faster than accesses to the larger memory. The compiler should be responsible for allocating variables and instructions to the scratch pad. This approach does, however, require that frequently accessed variables and code sequences are identified and mapped to that address range.

7.3.3 Memory-architecture aware compilation

7.3.3.1 Compilation techniques for scratch-pads

The advantages of using scratch-pad memories (SPMs) have been very clearly demonstrated [Banakar et al., 2002]. Therefore, the exploitation of scratch-pad memories (SPMs) is the most prominent case of memory hierarchy exploitation. Available compilers are usually capable of mapping memory objects to certain address ranges in the memory. Towards this end, the source code typically has to be annotated. For example, memory segments can be introduced in the source code by using pragmas like

```
# pragma arm section rwdata = "foo", rodata = "bar"
```

Variables declared after this pragma would be mapped to read-write segment "foo" and constants would be mapped to read-only segment "bar". Linker commands can then map these segments to particular address ranges, including those belonging to the SPM. This is the approach taken in compilers for ARM processors [ARM Ltd., 2009b]. This is not a very comfortable approach and it would be nice if compilers could perform such a mapping automatically for frequently accessed objects. Therefore, optimization algorithms have been designed. A survey has been presented at the HiPEAC summer school [Marwedel, 2007]. Available SPM optimizations can be classified into two categories:

- **Non-overlaying** (or "static") memory allocation strategies: For these strategies, memory objects will stay in the SPM while the corresponding application is executed.

- **Overlaying** (or "dynamic") memory allocation strategies: For these strategies, memory objects are moved in and out of the SPM at run-time. This is a kind of "compiler-controlled paging", except the migration of objects happens between the SPM and some slower memory and does not involve any disks.

7.3.3.2   Non-overlaying allocation

For non-overlaying allocation, we can start by considering the allocation of functions and global variables to the SPM. For this purpose, each function and each global variable can be modeled as a memory object. Let

- S be the size of the SPM,

- sf_i and sv_i be the sizes of function i and variable i, respectively,

- g be the energy consumption saved per access to the SPM (that is, the difference between the energy required per access to the slow main memory and the one required per access to the SPM),

- nf_i and nv_i be the number of accesses to function i and variable i, respectively,

- xf_i and xv_i be defined as

$$xf_i = \begin{cases} 1 & \text{if function } i \text{ is mapped to the SPM} \\ 0 & \text{otherwise} \end{cases} \tag{7.1}$$

$$xv_i = \begin{cases} 1 & \text{if variable } i \text{ is mapped to the SPM} \\ 0 & \text{otherwise} \end{cases} \tag{7.2}$$

Then, the goal is to maximize the gain

$$G \; = \; g \left(\sum_i nf_i \cdot xf_i + \sum_i nv_i \cdot xv_i \right) \tag{7.3}$$

while respecting the size constraint

$$\sum_i sf_i \cdot xf_i + \sum_i sv_i \cdot xv_i \le S \qquad (7.4)$$

The problem is known as a **knapsack** problem. Standard knapsack algorithms can be used for selecting the objects to be allocated to the SPM. However, equations 7.3 and 7.4 also have the form of an integer linear programming (ILP) problem (see Appendix A) and ILP-solvers can be used as well. g is a constant factor in the objective function and is not needed for the solution of the ILP problem. The corresponding optimization can be implemented as a pre-pass optimization (see fig. 7.13).

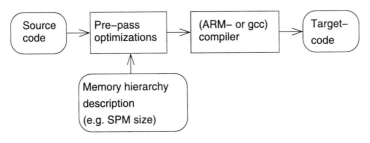

Figure 7.13. Pre-pass optimization

The optimization impacts addresses of functions and global variables. Compilers typically allow a manual specification of these addresses in the source code. Hence, no change to the compiler itself is required. The advantage of such a pre-pass optimization is that it can be used with compilers for many different target processors. There is no need to modify a large number of target-specific compilers.

This model can be extended into various directions:

- **Allocation of basic blocks:** The approach just described only allows the allocation of entire functions or variables to the SPM. As a result, a major fraction of the SPM may remain empty if functions and variables are large. Therefore, we try to reduce the granularity of the objects which are allocated to the SPM. The natural choice is to consider **basic blocks** as memory objects. In addition, we do also consider sets of adjacent basic blocks, where adjacency is defined as being placed next to each other in the instruction address space by the compiler. We call such sets of adjacent blocks **multi-blocks**. Fig. 7.14 shows the three multi-blocks M12, M23 and M123 for basic blocks BB1, BB2 and BB3.

The ILP model can be extended accordingly:

Let

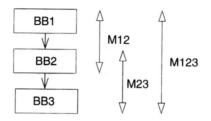

Figure 7.14.   Basic blocks and multi-blocks

 -   sb_i and sm_i be the sizes of basic blocks i and multi-blocks i, respectively,

 -   nb_i and nm_i be the number of accesses to basic block i and multi-blocks i, respectively,

 -   xb_i and xm_i be defined as

$$xb_i = \begin{cases} 1 & \text{if basic block } i \text{ is mapped to the SPM} \\ 0 & \text{otherwise} \end{cases} \quad (7.5)$$

$$xm_i = \begin{cases} 1 & \text{if multi-block } i \text{ is mapped to the SPM} \\ 0 & \text{otherwise} \end{cases} \quad (7.6)$$

Then, the goal is to maximize the gain

$$G = g\left(\sum_i nf_i \cdot xf_i + \sum_i nb_i \cdot xb_i + \sum_i nm_i \cdot xm_i + \sum_i nv_i \cdot xv_i\right) \quad (7.7)$$

while respecting the constraints

$$\sum_i sf_i \cdot xf_i + \sum_i sb_i \cdot xb_i + \sum_i sm_i \cdot xm_i + \sum_i sv_i \cdot xv_i \leq S \quad (7.8)$$

$$\forall \text{ basic blocks } i : xb_i + xf_{fct(i)} + \sum_{i' \in multiblock(i)} xm_{i'} \leq 1 \quad (7.9)$$

where $fct(i)$ is the function containing basic block i

and $multiblock(i)$ is the set of multi-blocks containing basic block i

The second constraint ensures that a basic block is mapped to the SPM only once, instead of potentially being mapped as a member of the enclosing function and a member of a multi-block.

Experiments using this model were performed by Steinke et al. [Steinke et al., 2002b]. For some benchmark applications, energy reductions of up to about 80% were found, even though the size of the SPM was just a small fraction of the total code size of the application. Results for the bubble sort program are shown in fig. 7.15.

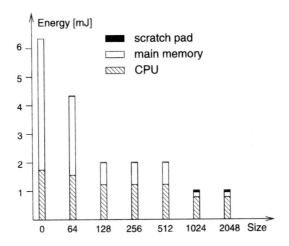

Figure 7.15.   Energy reduction by compiler-based mapping to scratch-pad for bubble sort

Obviously, larger SPMs lead to a reduced energy consumption in the main memory. The energy required in the processor is also reduced, since less wait cycles are required. Supply voltages have been assumed to be constant.

- **Partitioned memories** [Wehmeyer and Marwedel, 2006]: Small memories are faster and require less energy per access. Therefore, it makes sense to partition memories into several smaller memories. The ILP model can be extended easily to also model several memories. We do not distinguish between the various types of memory objects (functions, basic blocks, variables etc.), in this case. An index i represents any memory object. Let

  - S_j be the size of the memory j,
  - s_i be the size of object i (as before),
  - e_j be the energy consumption per access to memory j,
  - n_i the number of accesses to object i (as before),
  - $x_{i,j}$ be defined as

$$x_{i,j} = \begin{cases} 1 & \text{if object } i \text{ is mapped to memory } j \\ 0 & \text{otherwise} \end{cases} \qquad (7.10)$$

Instead of maximizing the energy saving, we are now minimizing the overall energy consumption. Hence, the goal is now to minimize

$$C \; = \; \sum_j e_j \sum_i x_{i,j} \cdot n_i \qquad (7.11)$$

while respecting the constraints

$$\forall j \; : \; \sum_i s_i \cdot x_{i,j} \leq S_j \qquad (7.12)$$

$$\forall i \; : \; \sum_j x_{i,j} = 1 \qquad (7.13)$$

Partitioned memories are advantageous especially for varying memory requirements. Storage locations accessed frequently are called the **working
set** of an application. Applications with a small working set could use a
very small fast memory, whereas applications requiring a larger working
set could be allocated to a somewhat larger memory. Therefore, a key advantage of partitioned memories is their ability to adapt to the size of the
current working set.

Furthermore, unused memories can be shut down to save additional energy. However, we are considering only the "dynamic" energy consumption caused by accesses to the memory. In addition, there may be some
energy consumption even if the memory is idle. This consumption is not
considered here. Therefore, savings from shutting down memories are not
reflected in equations 7.11 and 7.12.

- **Link/load-time allocation of memory** [Nguyen et al., 2005]: Optimizing
  code at compile time for a certain SPM size has a disadvantage: the code
  might perform badly if we run it on different variants of some processor, if
  these variants have differently sized SPMs. We would like to avoid requiring different executable files for the different variants of the processor. As
  a result, we are interested in executables which are independent of the SPM
  size. This is feasible, if we perform the optimization at link-time. The proposed approach computes the ratio of the number of accesses divided by
  the size of a variable at compile-time and stores this value together with
  other information about variables in the executable. At load time, the OS
  is queried for the size of the SPM. Then, the code is patched such that as
  many profitable variables as possible are allocated to the SPM.

- **Allocation of the stack**: In order to really reduce the energy consumption,
  all frequently accessed memory objects must be allocated to some small

memory. The stack must be included in this consideration. Otherwise, stack accesses will limit the overall improvements that are feasible. There are at least two approaches for this: Steinke [Steinke et al., 2002b] computed the worst case stack size using a **stack size analyzer**. Stacks which are small enough can then be allocated to the SPM. Avissar et al. [Avissar et al., 2002] proposed to partition the stack into frequently and less frequently accessed elements. Infrequently accessed elements will stay in the slower main memory, while frequently accessed variables will be allocated to the SPM. This scheme requires two stack pointers, one for each memory. In order to prevent the overhead of updating two pointers for each function call, splitting of the stack is avoided for "short" functions.

- **Allocation of the heap** [Dominguez et al., 2005]: Remarks regarding frequent accesses to the stack also apply to the heap. Heap elements which are frequently accessed should also be allocated to efficient memory layers. We could use a heap size analyzer in order to compute bounds on the heap size. Small, frequently accessed heaps can be allocated to the SPM entirely. However, heaps are frequently too large for this approach. Dominguez et al. propose a second approach. In their approach, programs are partitioned into **regions**.Regions are delimited by so-called **program points**. The selection of program points is crucial for this approach. Program points can be defined as [Udayakumaran et al., 2006]: *"(i) the start and end of each procedure; (ii) just before and just after each loop (even inner loops of nested loops); (iii) the start and end of each if statement's then part and else part, as well as the start and end of the entire if statement; and (iv) the start and end of each case in all switch statements in the program, as well as the start and end of the entire switch statement"*.

In [Dominguez et al., 2005], some space is kept available in the SPM and each time a code region is entered, heap elements to be moved in and out of the SPM are copied as needed. The copying is done such that pointers to heap elements will always remain valid.

- Consideration of the **impact on timing predictability** [Wehmeyer and Marwedel, 2006]: Most of the SPM allocation algorithms allocate memory such that we know at compile time, whether a memory access will be to a fast or to a slow memory. Therefore, it is possible to predict the speed of memory accesses more precisely than for caches. As a result, the worst case execution times of SPM-based systems are typically better that those of cache-based systems.

7.3.3.3   Overlaying allocation

Large applications may have multiple hot spots (multiple areas of code containing compute-intensive loops). Non-overlaying approaches fail to provide the best possible results in this context. For such applications, the SPM should be exploited for each of the hot spots. This requires an automatic migration between the layers in the memory hierarchy. There are several approaches for overlaying allocation:

- **Tiling of large arrays** [Kandemir et al., 2001], [Chen et al., 2006]: SPMs can be problematic when large arrays are used, which do not completely fit into the SPM. Algorithms presented so far will not allow subsets of arrays to be copied into the SPM. This has been changed with Kandemir's proposal to combine tiling with SPM allocation. His technique allows copying slices of arrays into the SPM. In [Kandemir et al., 2001], tiling was unconditionally applied. In [Chen et al., 2006], the authors propose to suppress tiling for irregular array accesses for which tiling would be inefficient.

- **Multiple hierarchy levels** [Brockmeyer et al., 2003]: The speed difference between large and small memories is increasing. Therefore, it makes sense to introduce multiple memory hierarchy levels. The MHLA (memory hierarchy layer assignment) tool of IMEC tries to find an appropriate allocation of variables to the different memory layers. MHLA automatically selects subsets of arrays which can be copied to faster memory layers before loops are entered. A new version of this tool is currently being designed in the Mnemee project [Mnemee project, 2010].

- **Region-based memory object migration** [Udayakumaran et al., 2006]: This approach is also based on regions (see page 303) and program points. For each program point, it is considered which variables should be moved out of the SPM and which variables should be moved into the SPM (code is modeled as a kind of variable).

- Verma's approach [Verma and Marwedel, 2004] is similar to the one by Udayakumaran. However, the selection of the memory objects to be copied is based on a global ILP model, instead of a more local, heuristic optimization.

7.3.3.4   Multiple threads/processes

The above approaches are still limited to handling a single process or thread. For multiple threads, moving objects into and out of the SPM at context switch time has to be considered. Verma [Verma et al., 2005] proposed three different approaches:

1 For the first approach, only a single process owns space in the SPM at any given time. At each context switch, the information of the preempted process in the occupied space is saved and the information for the process to be executed is restored. This approach is called the **saving/restoring approach**. This approach does not work well with large SPMs, since the copying would consume a significant amount of time and energy.

2 For the second approach, the space in the SPM is partitioned into areas for the various processes. The size of the partitions is determined in a special optimization. The SPM is filled during initialization. No further compiler-controlled copying is required. Therefore, this approach is called the **non-saving approach**. This approach makes sense only for SPMs large enough to contain areas for several processes.

3 The third approach is a **hybrid** approach: The SPM is split into an area jointly used by processes and a second area, in which processes obtain some exclusively allocated space. The size of the two areas is determined in an optimization.

Verma's approaches require a fixed set of processes to be known at compile time. The next step is to allow processes to enter and to leave the system. Pyka et al. [Pyka et al., 2007] describe run-time SPM allocation performed by an SPM memory manager (SPMM) to be integrated into the operating system. Pyka's approach allows space for pre-compiled libraries to be allocated in the SPM, in contrast to earlier algorithms. Unfortunately, Pyka's algorithm requires an additional level of indirection. Despite the overhead of this additional level of indirection, a reduction of the energy consumption by 25 % to 35 % with respect to a four-way set associative cache has been obtained.

This additional level of indirection can be avoided if a memory management unit (MMU) is available. Egger et al. [Egger et al., 2006] developed a technique exploiting MMUs: At compile time, sections of code are classified as either benefiting or not benefiting from an allocation to the SPM. The code benefiting is stored in a certain area in the virtual address space. Initially, this area is not mapped to physical memory. Therefore a page fault occurs when the code is accessed for the very first time. Page fault handling then invokes the SPMM and the SPMM allocates (and deallocates) space in the SPM, always updating the virtual-to-real addresses translation tables as needed.

In general, exploitation of SPM requires tool support, but leads to efficient designs. Caches can be taken advantage of without such support. Future systems might contain mixtures of caches and SPMs.

7.3.4   Reconciling compilers and timing analysis

Almost all compilers which are available today do not include a timing model. Therefore, the development of real-time software typically has to follow an iterative approach: software is compiled by a compiler which is unaware of any timing information. The resulting code is then analyzed using a timing analyzer such as aiT [Absint, 2010]. If the timing constraints are not met, some of the inputs to the compiler run must be changed and the procedure has to be repeated. We call this **"trial-and-error"-based development of real-time software**. This approach suffers from several problems. First of all, the number of required design iterations is initially unknown. Furthermore, the compiler used in this approach is "optimizing", but a precise evaluation of objectives apart from the code size is impossible. Hence, compiler writers can only **hope** that their "optimizations" have a positive impact of the quality of the code in terms of relevant objectives. Due to the complex timing behavior of modern processors, this hope is hardly supported by evidence. Finally, the "trial-and-error"-based development of real-time software requires the designer to find appropriate modifications of the input to the compiler such that the real-time constraints will eventually be met.

This "trial-and-error"-based approach can be avoided, if timing analysis is integrated into the compiler. The is the aim of the development of the worst case execution time aware compiler WCC at TU Dortmund. Developing a completely new timing analyzer independent of the existing ones would be a waste of efforts. Therefore, WCC is based on the integration of the timing analyzer aiT into an experimental compiler for the TriCore architecture. Fig. 7.16 shows the resulting overall structure.

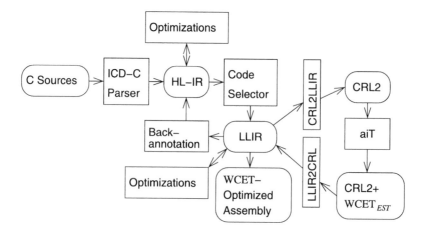

Figure 7.16.   Worst case execution time aware compiler WCC

WCC uses the ICD-C compiler infrastructure [ICD Staff, 2010] to read and parse C source code. The source is then converted into a "high-level intermediate representation" (HL-IR). The HL-IR is an abstract representation of the source code. Various optimizations can be applied to the HL-IR. The optimized HL-IR is passed to the code selector. The code selector maps source code operations to machine instructions. WCC so far focuses on the support of the Infineon TriCore architecture. TriCore instructions are represented in the low-level intermediate representation LLIR. In oder to estimate the $WCET_{EST}$, the LLIR is converted into the CRL2 representation used by aiT (using the converter LLIR2CRL). aiT is then able to generate $WCET_{EST}$ for the given machine code. This information is converted back into the LLIR representation (using the converter CRL2LLIR). WCC uses this information to consider $WCET_{EST}$ as the objective function during optimizations. This can be done in a straightforward manner for optimizations at the LLIR-level. However, many optimizations are performed at the HL-IR-level. $WCET_{EST}$-directed optimizations at this level require using back-annotation from the LLIR-level to the HR-IR-Level. ICD-C includes this back-annotation.

WCC has been used to study the impact of optimizing for a reduced $WCET_{EST}$ in the compiler. The numerous results include a study of the impact of employing this objective for register allocation [Falk, 2009]. Results indicate a dramatic impact, as can be seen from fig. 7.17.

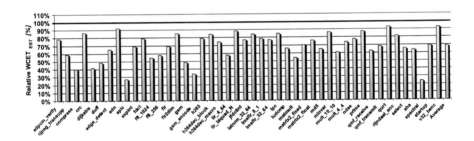

Figure 7.17. Reduction of $WCET_{EST}$ by WCET-aware register allocation

$WCET_{EST}$ can be reduced down to 68.8% of the original $WCET_{EST}$ on the average by just using WCET-aware register allocation in WCC. The largest reduction yields a $WCET_{EST}$ of only 24.1% of the original $WCET_{EST}$. The combined effect of several such optimizations has been analyzed by Lokuciejewski et al. [Lokuciejewski and Marwedel, 2010]. For the considered benchmarks, Lokuciejewski found a reduction of down to 57.1% of the original $WCET_{EST}$.

7.3.5   Compilation for digital signal processors

Features of DSP processors are described on page 143. Compilers should exploit these in order to optimize code with respect to the objectives mentioned in Chapter 5. Techniques for this can be demonstrated using address generation units as examples. The possibility of generating addresses "for free" has an important impact on how variables should be laid out in memory. Fig. 7.18 shows an example.

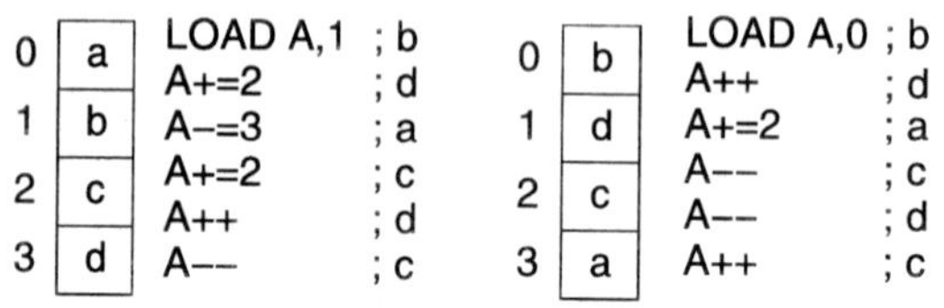

Figure 7.18.   Comparison of memory layouts

We assume that in some basic block, variables a to d are accessed in the sequence (b,d,a,c,d,c). Accessing these variables with register-indirect addressing requires, first of all, loading the address of b into an address register (see fig. 7.18, left). The instruction referring to variable b is not shown in fig. 7.18, since the current focus is on address generation. Therefore, the generation of the address for the access to the next variable (d) is considered next. Assuming that there is just a single address register A, A has to be updated to point to variable d. This requires adding 2 to the register. Again, we ignore the instruction loading the variable, and we immediately consider the access to a. For this, we must subtract 3, and for the next access we must add 2. Assuming that the auto-increment and -decrement range is restricted to ± 1, only the last two accesses shown in fig. 7.18 can be implemented with these operations. In total, 4 instructions for calculating addresses are needed.

In contrast, for the layout in fig. 7.18 (right), 4 address calculations are auto-increment and -decrement operations which will be executed in parallel with some operation in the main data path. Only 2 cycles are needed for address calculations with an offset larger than 1. Again, the instructions actually using the variables are not shown.

How do we generate such clever memory layouts? Algorithms doing this typically start from an access graph (see fig. 7.19).

Such access graphs have one node for each of the variables and have an edge for every pair of variables for which there are adjacent accesses. The weight of such edges corresponds to the number of adjacent accesses to the variables connected by that edge.

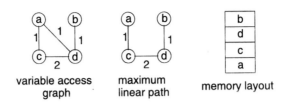

Figure 7.19. Memory allocation for access sequence (b, d, a, c, d, c) for a single address register A

Variables connected by an edge of a high weight should preferably be allocated to adjacent memory locations. The number of address calculations saved in this way is equal to the weight of the corresponding edge. For example, if c and d are allocated to adjacent locations, then the last two accesses in the sequence can be implemented with auto-increment and -decrement operations.

The overall goal of memory allocation is to find a linear order of variables in memory maximizing the use of auto-increment and -decrement operations. This corresponds to finding a linear path of maximum weight in the variable access graph. Unfortunately, the maximum weighted path problem in graphs is NP-complete. Hence, it is common to use heuristics for generating such paths [Liao et al., 1995b], [Sudarsanam et al., 1997]. Most of them are based on Kruskal's spanning tree heuristic. This is Liao's algorithm:

1 Sort edges of access graph $G = (V, E)$ according to their weight.

2 Construct a new graph $G' = (V', E')$, starting with $G' = G$ and $E' = 0$.

3 Select an edge e of G of highest weight; If this edge does not cause a cycle in G' and does not cause any node in G' to have a degree > 2 then add this node to E' otherwise discard e.

4 Goto 3 as long as not all edges from G have been selected and as long as G' has less than $(|V| - 1)$ edges.

Implicitly, all nodes are assumed to be connected by an edge of weight 0. This ensures that the algorithm continues even if parts of the graph become disconnected. The order of the variables in memory corresponds to the order of the variables along the generated linear path.

An application of this algorithm to the example of fig. 7.19 is shown in fig. 7.20.

Edge (c,d), due to its weight, is the first edge added to the empty graph G'. Among all edges of weight 1, the sequence is arbitrary. Suppose (a,c) is added

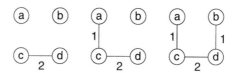

Figure 7.20.   Sequence of steps in Liao's algorithm

next (see fig. 7.20 (center)). (a,d) may be the next edge considered. Its inclusion in G' would cause a cycle and it is discarded. Finally, (b,d) is added. The algorithm stops with 3 edges added to a graph of 4 nodes.

The algorithm just sketched only covers a simple case. A tie-break heuristic for edges of equal weight was published by Leupers and Marwedel [Leupers and Marwedel, 1996]. Extensions of the basic algorithm cover more complex situations, such as:

- $n > 1$ address registers [Leupers and Marwedel, 1996],

- also using modify registers present in the AGU [Leupers and Marwedel, 1996], [Leupers and David, 1998],

- extension to arrays [Basu et al., 1999],

- larger auto-increment and -decrement ranges [Sudarsanam et al., 1997].

Memory allocation, as described above, improves both the code-size and the run-time of the generated code. Other proposed optimization algorithms exploit further architectural features of DSP processors, such as:

- multiple memory banks [Sudarsanam and Malik, 1995],

- heterogeneous register files [Araujo and Malik, 1995],

- modulo addressing [Quilleré and Rajopadhye, 2000],

- instruction level parallelism [Leupers and Marwedel, 1995],

- multiple operation modes [Liao et al., 1995a].

Other optimization techniques are described by Leupers [Leupers, 2000a].

7.3.6   Compilation for multimedia processors

In order to fully support packed data types as described on page 145, compilers must be able to automatically convert operations in loops to operations on packed data types. Taking advantage of this potential is necessary for generating efficient software. A very challenging task is to use this feature in compilers. Compiler algorithms exploiting operations on packed data types are extensions of vectorizing algorithms originally developed for supercomputers. Some algorithms for multimedia and SIMD short vector extensions have been described [Fisher and Dietz, 1998], [Fisher and Dietz, 1999], [Leupers, 2000b], [Krall, 2000], [Larsen and Amarasinghe, 2000].

Automatic parallelization of loops for the M3-DSP (see page 148) requires the use of vectorization techniques, which achieve significant speedups (compared to the case of sequential operations, see fig. 7.21) [Lorenz et al., 2002], [Lorenz et al., 2004]. For application dot_product_2, the size of the vectors was too small to lead to a speedup and no vectorization should be performed. The number of cycles can be reduced by 94 % for benchmark example if vectorization is combined with an exploitation of zero-overhead-loop instructions.

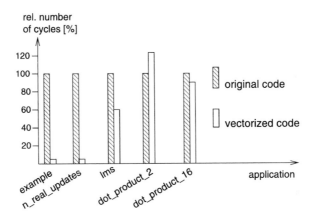

Figure 7.21. Reduction of the cycle count by vectorization for the M3-DSP

Due to the increased number of processors with SIMD extensions, compilation for SIMD instructions has received significant attention [Ren et al., 2006], [Nuzman et al., 2006]. In particular, compilation for the CELL processor has revived interest in such compilation techniques (see, for example, Eichenberger et al. [Eichenberger et al., 2005]). Furthermore, compiler writers addressed implications of the availability of short vector instructions on Pentium®-compatible processors [Gerber et al., 2005]. It is not possible to provide a full overview of this dynamic research area.

7.3.7   Compilation for VLIW processors

VLIW architectures (see page 146) require special compiler optimizations:

- A key optimization required for TMS 320C6xx compilers is to allocate, at compile time, the functional unit that should execute a certain operation. Due to the two data paths (see fig. 3.28), this implies a partitioning of the operations into two sets [Jacome and de Veciana, 1999], [Jacome et al., 2000], [Leupers, 2000c] and also includes an allocation to one of the register files.

- VLIW processors frequently have branch delay slots. For VLIW processors, the branch delay penalty is significantly larger than for other processors, because each of the branch delay slots could hold a full instruction packet, not just a single instruction. For example, for the TMS 320C6xx, the branch delay penalty is $5 \times 8 = 40$ instructions. In order to avoid this large penalty, most VLIW processors support predicated execution for a large number of condition code registers. Predicated execution can be employed to efficiently implement small if-statements. For large if-statements, however, conditional branches are more efficient, since these allow mutual exclusion of then- and else-branches to be exploited in hardware allocation. The precise trade-off between the two methods for implementing if-statements can be found with proper optimization techniques [Mahlke et al., 1992], [August et al., 1997], [Leupers, 1999].

- Due to the large branch delay penalty, inlining (see page 294) is another optimization that is very useful for VLIW processors.

- Significant effort has been invested into the design of compilers for the Intel IA-64 EPIC architecture (see [Dulong et al., 2001] as an example). Due to the peculiarities of the architecture, special optimization techniques are required.

- The Trimaran compiler infrastructure [Trimaran, 2010] is a platform for research on compilation techniques for instruction level parallelism and VLIW as well as EPIC architectures.

7.3.8   Compilation for network processors

Network processors are a new type of processors. They are optimized for high-speed Internet applications. Their instruction sets comprise numerous instructions for accessing and processing bit fields in streams of information. Typically, they are programmed in assembly languages, since their throughput is of utmost importance. Nevertheless, network protocols are becoming more and more complex and designing compilers for such processors supports the design of network components. The necessary bit-level details have been analyzed by Falk, Wagner et al. [Falk et al., 2006].

7.3.9 Compiler generation, retargetable compilers and design space exploration

When the first compilers were designed, compiler design was a totally manual process. In the meantime, some of the steps involved in generating a compiler have been automated or supported by tools. For example, lex and yacc and more recent versions of these tools (see [Johnson, 2010]) provide a standard means for parsing the source code. Generating machine instructions is another step which is now supported by tools. For example, **tree pattern matchers** such as **olive** [Tjiang, 1993] can be used for this task. Despite the use of such tools, compiler design is typically not a fully automated process.

However, there have been many attempts to design retargetable compilers. We distinguish between different kinds of retargetability:

- **Developer retargetability:** In this case, compiler specialists are responsible for retargeting compilers to new instruction sets.

- **User retargetability:** In this case, users are responsible for retargeting the compiler. This approach is much more challenging.

More information about retargetable compilers and their use for design space exploration can be found in a book by Leupers and Marwedel [Leupers and Marwedel, 2001]. Commercial products include those that are available from Tensilica Inc. [Tensilica Inc., 2010].

7.4 Power Management and Thermal Management

7.4.1 Dynamic voltage scaling (DVS)

Some embedded processors support dynamic power management (see page 136) and dynamic voltage scaling (see page 136). An additional optimization step can be used to exploit these features. Typically, such an optimization step follows code generation by the compiler. Optimizations in this step require a global view of all tasks of the system, including their dependencies, slack times etc.

The potential of dynamic voltage scheduling is demonstrated by the following example [Ishihara and Yasuura, 1998]. We assume that we have a processor which runs at three different voltages, 2.5 V, 4.0 V, and 5.0 V. Assuming an energy consumption of 40 nJ per cycle at 5.0 V, equation 3.14 can be used to compute the energy consumption at the other voltages (see table 7.22, where 25 nJ is a rounded value).

V_{dd} [V]	5.0	4.0	2.5
Energy per cycle [nJ]	40	25	10
f_{max} [MHz]	50	40	25
cycle time [ns]	20	25	40

Figure 7.22.   Characteristics of processor with DVS

Furthermore, we assume that our task needs to execute 10^9 cycles within 25 seconds. There are several ways of doing this, as can be seen from figures 7.23 to 7.25. Using the maximum voltage (case a), see fig. 7.23), it is possible to shut down the processor during the slack time of 5 seconds (we assume the power consumption to be zero during this time).

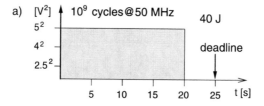

Figure 7.23.   Possible voltage schedule

Another option (case b)) is to initially run the processor at full speed and then reduce the voltage when the remaining cycles can be completed at the lowest voltage (see fig. 7.24).

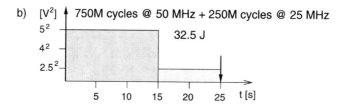

Figure 7.24.   Second voltage schedule

Finally, we can run the processor at a clock rate just large enough to complete the cycles within the available time (case c), see fig. 7.25).

The corresponding energy consumptions can be calculated as

$$E_a = 10^9 \times 40 \cdot 10^{-9} = 40 \; [J] \qquad (7.14)$$
$$E_b = 750 \cdot 10^6 \times 40 \cdot 10^{-9} + 250 \cdot 10^6 \times 10 \cdot 10^{-9} = 32.5 \; [J] \quad (7.15)$$

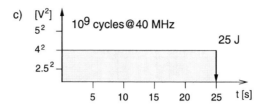

Figure 7.25. Third voltage schedule

$$E_c = 10^9 \times 25 \cdot 10^{-9} = 25 \, [J] \qquad (7.16)$$

A minimum energy consumption is achieved for the ideal supply voltage of 4 Volts. In the following, we use the term **variable voltage processor** only for processors that allow **any** supply voltage up to a certain maximum. It is expensive to support truly variable voltages, and therefore, actual processors support only a few fixed voltages.

The observations made for the above example can be generalized into the following statements. The proofs of these statements are given in the paper by Ishihara and Yasuura.

- If a variable voltage processor completes a task before the deadline, the energy consumption can be reduced[1].

- If a processor uses a single supply voltage V_s and completes a task T just at its deadline, then V_s is the unique supply voltage which minimizes the energy consumption of T.

If a processor can only use a number of discrete voltage levels, then a voltage schedule using the two voltages which are the two immediate neighbors of the ideal voltage V_{ideal} can be chosen. These two voltages lead to the minimum energy consumption except if the need to use an integer number of cycles results in a small deviation from the minimum[2].

The statements can be used for allocating voltages to tasks. Next, we will consider the allocation of voltages to a set of tasks. We will use the following notation:

---

[1] This formulation makes an implicit assumption in lemma 1 of the paper by Ishihara and Yasuura explicit.
[2] This need is not considered in the original paper.

N :   the number of tasks

EC_j :   the number of executed cycles of task j

L :   the number of voltages of the target processor

V_i :   the ith voltage, with $1 \leq i \leq L$

F_i :   the clock frequency for supply voltage V_i

D :   the global deadline at which all tasks must have been completed

SC_j :   the average switching capacitance during the execution of task j (SC_i comprises the actual capacitance C_L and the switching activity α (see eq. 3.14 on page 136))

The voltage scaling problem can then be formulated as an integer linear programming (ILP) problem (see page 335). Towards this end, we introduce variables $X_{i,j}$ denoting the number of cycles executed at a particular voltage:

$X_{i,j}$:   the number of clock cycles task j is executed at voltage V_i

Simplifying assumptions of the ILP-model include the following:

- There is one target processor that can be operated at a limited number of discrete voltages.

- The time for voltage and frequency switches is negligible.

- The worst case number of cycles for each task are known.

Using these assumptions, the ILP-problem can be formulated as follows:

Minimize

$$E = \sum_{j=1}^{N} \sum_{i=1}^{L} SC_j \cdot X_{i,j} \cdot V_i^2 \tag{7.17}$$

subject to

$$\forall j : \sum_{i=1}^{L} X_{i,j} = EC_j \tag{7.18}$$

and

$$\sum_{j=1}^{N} \sum_{i=1}^{L} \frac{X_{i,j}}{F_i} \leq D \tag{7.19}$$

The goal is to find the number $X_{i,j}$ of cycles that each task j is executed at a certain voltage V_i. According to the statements made above, no task will ever need more than two voltages. Using this model, Ishihara and Yasuura show that efficiency is typically improved if tasks have a larger number of voltages to choose from. If large amounts of slack time are available, many voltage levels help to find close to optimal voltage levels. However, four voltage levels do already give good results quite frequently.

There are many cases in which tasks actually run faster than predicted by their worst case execution times. This cannot be exploited by the above algorithm. This limitation can be removed by using checkpoints at which actual and worst case execution times are compared, and then to use this information to potentially scale down the voltage [Azevedo et al., 2002]. Also, voltage scaling in multi-rate task graphs was proposed [Schmitz et al., 2002]. DVS can be combined with other optimizations such as body biasing [Martin et al., 2002]. Body biasing is a technique for reducing leakage currents.

7.4.2   Dynamic power management (DPM)

In order to reduce the energy consumption, we can also take advantage of power saving states, as introduced on page 136. The essential question for exploiting DPM is: when should we go to a power-saving state? Straight-forward approaches just use a simple timer to transition into a power-saving state. More sophisticated approaches model the idle times by stochastic processes and use these to predict the use of subsystems with more accuracy. Models based on exponential distributions have been shown to be inaccurate. Sufficiently accurate models include those based on renewal theory [Simunic et al., 2000].

A comprehensive discussion of power management was published (see, for example, [Benini and De Micheli, 1998], [Lu et al., 2000]). There are also advanced algorithms which integrate DVS and DPM into a single optimization approach for saving energy [Simunic et al., 2001].

Allocating voltages and computing transition times for DPM may be two of the last steps of optimizing embedded software.

Power management is also linked to thermal management. Thermal management relies on temperature information being available at run-time. This information is then used to control the generation of additional heat, and possibly has an impact on cooling mechanisms as well. Controlling fans can be considered as a very simple case of thermal management. Also, systems may be shutting down completely, if temperatures are exceeding maximum thresholds. More advanced systems may be reducing the clock frequencies and voltages. For multiprocessor systems, tasks may be automatically migrated between various processors. In all of these cases, the objective "temperature" is evaluated at run-time and used to have an impact at run-time. Avoiding overheating is the goal of the work reported by Merkel et al. [Merkel and Bellosa, 2005] and by Donald et al. [Donald and Martonosi, 2006].

7.5   Assignments

1 Consider the following program:

```
1    #include <stdio.h>
2    #define DATALEN 15
3    #define FILTERTAPS 5
4    double x[DATALEN] = { 128.0, 130.0, 180.0, 140.0, 120.0,
5                                110.0, 107.0, 103.5, 102.0, 90.0,
6                                84.0, 70.0, 30.0, 77.3, 95.7 };
7    const double h[FILTERTAPS]={0.125,-0.25,0.5,-0.25,0.125};
8    double y[DATALEN]; // result;
9    int main(void)
10   {int i,n;
11   for(i=0;i<DATALEN;++i)
12   {y[i]=0;
13      for(n=0;n<FILTERTAPS;++n)
14        if ((i-n)>=0) y[i]+=h[n]*x[i-n];
15      }
16   for(i=0;i<DATALEN;++i) printf("%.2f ",y[i]);
17   return 0;
18 }
```

Perform at least the following optimizations:

- Removal of the if in the innermost loop (line 14)
- Loop unrolling (line 13)
- Constant propagation
- Floating-point to fixed-point conversion
- Avoidance of all accesses to arrays

Please provide the optimized version of the program after each of the transformations and do also check for consistent results!

2 Suppose that variables {a, b, c, d, e, f} are accessed in the sequence

(c a e d f a d a d e c b f d e d f b a d a).

Also, assume that our processor has the following characteristics:

- There is a single address register AR.
- All accesses to the memory must be via AR.
- Post-increment and post-decrement by 1 can be encoded in all load- and store-instructions.
- All other changes of AR need an extra instruction and an extra cycle.

Using Liao's algorithm, compute a variable order minimizing the total number of explicit address calculations! Include a graphical representation of the effect of each of the steps of the algorithm.

Create an assembly language program generating the indicated sequence of variable references. All references are assumed to be reading (not writing) the memory. Use the following assembly instructions (semantics indicated on the right):

ld r,(AR);	register[r]:= memory[AR]
ld r,(AR)++;	register[r]:= memory[AR]; AR++;
ld r,(AR)- -;	register[r]:= memory[AR]; AR- -;
li AR,constant;	AR:=constant;
addi AR,constant;	AR:=AR+constant; //constant can be negative

3 Suppose that your computer is equipped with a main memory and a scratch pad memory. Sizes and the required energy per access are shown in the table in fig. 7.26.

Memory	Size in bytes	Energy per access
Scratch pad	4096 (4k)	1.3 nJ
Main memory	262,144 (256 k)	31 nJ

Figure 7.26. Memory characteristics

Also, let us assume that we are accessing variables as shown in the table in fig. 7.27.

Which of those variables should be allocated to the scratch pad memory, provided that we use a static, non-overlaying allocation of variables? Use integer the linear problem (ILP) model to select the variables. Your result should include the ILP model as well as the results. You may use the *lp_solve* program [Anonymous, 2010a] to solve your ILP problem.

Variable	Size in bytes	Number of accesses
a	1024	16
b	2048	1024
c	512	2048
d	256	512
e	128	256
f	1024	512
g	512	64
h	256	512

Figure 7.27. Variable characteristics

4 Loop unrolling is one of the potentially useful optimizations. Please name two potential benefits and two potential problems!

Chapter 8

TEST

8.1 Scope

The purpose of testing is to make sure that a manufactured embedded system behaves as intended. Testing can be done during or after the fabrication (fabrication testing) and also after the system has been delivered to the customer (field testing). Testing of embedded systems needs special attention for several reasons:

- Embedded/cyber-physical systems integrated into a physical environment may be safety-critical. Therefore, their malfunctioning can be much more dangerous than, say, the malfunctioning of office equipment. As a result, expectations for the product quality are higher than for non-safety critical systems.

- Testing of timing-critical systems has to validate the correct timing behavior. This means that just testing the functional behavior is not sufficient.

- Testing embedded/cyber-physical systems in their real environment may be dangerous. For example, testing control software in a nuclear power plant can be a source of serious, far-reaching problems.

Preparations for testing should be done no later than at the end of the design phase. Preferably, necessary support for testing should even be considered earlier, intertwined with the design process and using testability as one of the objectives for evaluating designs. In order not to overload Chapter 5, we have moved all aspects of testing into this separate Chapter. The presentation corresponds to considering testing only at the very end of the design flow (see fig. 8.1), even though an earlier consideration during an actual design would

P. Marwedel, *Embedded System Design*, Embedded Systems,
DOI 10.1007/978-94-007-0257-8_8, © Springer Science+Business Media B.V. 2011

be advisable. However, an early consideration is not always common practice, and, therefore, fig. 8.1 might also correspond to an actual design flow.

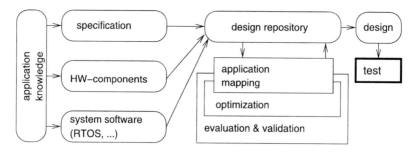

Figure 8.1.   Design flow with testing at its very end

In testing, we are typically denoting the system under design (SUD) as the **device under test** (DUT). To the DUT, we are applying a set of specially selected input patterns, so-called **test patterns** to the input of the system, observe its behavior and compare this behavior with the expected behavior. Test patterns are normally applied to the real, already manufactured system. The main purpose of testing is to identify systems that have not been correctly manufactured (manufacturing test) and to identify systems that fail later (field test).

Testing includes a number of different actions:

1 **test pattern generation,**

2 **test pattern application,**

3 **response observation,** and

4 **result comparison**.

8.2   Test procedures

8.2.1   Test pattern generation for gate level models

In test pattern generation, we try to identify a set of test patterns which distinguishes a correctly working from an incorrectly working system. Test pattern generation is usually based on **fault models**. Such fault models are models of possible faults. Test pattern generation tries to generate tests for all faults that are possible according to a certain fault model.

The stuck-at-fault model is a frequently used fault model. It is based on the assumption, that any internal wire of an electronic circuit is either permanently connected to '0' or '1'. It has been observed that many faults actually behave as if some wire was permanently connected that way.

As an example, consider the circuit shown in fig. 8.2[1].

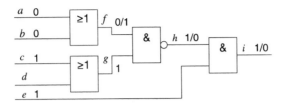

Figure 8.2. Test pattern at the gate level

Suppose that we would like to check if there is a stuck-at-1 fault for signal f. Toward this end, we try to set f to '0' by setting $a = b = $'0'. As a result, f should be '1' if there is this fault, and otherwise it should be '0'. In order to observe this difference, we must propagate it to the output signal i. For this to happen, we must set e to '1' and to set either c or d to '1'. h and i will be '1' if there is no fault and '0' otherwise. The test pattern comprises all values of inputs a to e. The D-algorithm can be used to generate this test pattern [Lala, 1985].

Many techniques for test pattern generation are based on the stuck-at-fault model. However, CMOS technologies require more comprehensive fault models. In CMOS technologies, faults can turn combinatorial devices into devices having internal states. This problem can occur, if wires are broken (this case is known as **stuck-at-open fault**). As a result of this, gates of transistors can become disconnected. Such transistors will be conducting or non-conducting, depending on the charge stored on the gate before the wire was broken. In this way, the gate "remembers" the input signal due to stored charges. Furthermore, there may be transient faults and delay faults (faults changing the delay of a circuit). Delay faults may be the result of cross-talk between adjacent wires. Fault models exist which take such hardware faults into account [Krstić and Cheng, 1998].

While good fault models exist for hardware testing, the same is not true for software testing.

---

[1]Please remember: consistent with standard ANSI/IEEE 91, the symbols ≥ 1 and & denote or-, and and-gates, respectively.

8.2.2   Self-test programs

One of the key problems of testing modern integrated circuits is their limited
number of pins, making it more and more difficult to access internal components. Also, it is getting very difficult to test these circuits at full speed, since
testers must be at least as fast as the circuits themselves. The fact that many
embedded systems are based on processors provides a way out of this dilemma:
processors are capable of running test programs or *diagnostics*. Such diagnostics have been used to test main frame machines for decades. Fig. 8.3 shows
some components that might be contained in some processor.

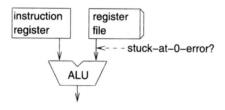

Figure 8.3.   Segment from processor hardware

In order to test for stuck-at-faults at the input of the ALU, we can execute a
small test program:

> store pattern of all '1's in the register file;
>
> perform xor between constant "0000...00" and register,
>
> test if result contains '0' bit,
>
> if yes, report error;
>
> otherwise start test for next fault

Similar small programs can be generated for other faults. Unfortunately, the
process of generating diagnostics for main frames has mostly been a manual
one. Some researchers have proposed to generate diagnostics automatically
[Brahme and Abraham, 1984], [Krüger, 1986], [Bieker and Marwedel, 1995],
[Krstic and Dey, 2002], [Kranitis et al., 2003], [Bernardi et al., 2005].

8.3   Evaluation of test pattern sets and system robustness

8.3.1   Fault coverage

The quality of test pattern sets can be evaluated using **fault coverage** as a
metric. Fault coverage is the percentage of potential faults that can be found
for a given test pattern set:

$$Coverage = \frac{\text{Number of detectable faults for a given test pattern set}}{\text{Number of faults possible due to the fault model}}$$

In practice, achieving a good product quality requires fault coverages in the area of at least 98 to 99 %. The requirements may be higher for particular systems. Also, special fault models may be necessary for certain hardware components (e.g. for batteries).

In addition to achieving a high coverage, we must also achieve a high **correctness coverage**. This means that a fault-free system must be recognized as such. Otherwise, it would be possible to achieve a 100% coverage by classifying all systems as faulty.

In order to increase the number of options that exist for system validation, it has been proposed to use test methods already during the design phase. For example, test pattern sets can be applied to software models of systems in order to check if two software models behave in the same way. More time-consuming formal methods need to be applied only to those cases in which the system passed this test-based equivalence check.

8.3.2 Fault simulation

It is currently not feasible (and it will probably not be feasible) to completely predict the behavior of systems in the presence of faults or to analytically compute the coverage. Therefore, the behavior of systems in the presence of faults is frequently simulated. This type of simulation is called **fault simulation**. In fault simulation, system models are modified to reflect the behavior of the system in the presence of a certain fault.

The goals of fault simulation include:

- to know the effect of a fault of the components at the system level. Faults are called **redundant** if they do not affect the observable behavior of the system, and

- to know whether or not mechanisms for improving fault tolerance actually help.

Fault simulation requires the simulation of the system for all faults feasible for the fault model and also for a possibly large number of different input patterns. Accordingly, fault simulation is an extremely time-consuming process. Different techniques have been proposed to speed up fault simulation.

One such technique applies to fault simulation at the gate level. In this case, internal signals are single bit signals. This fact enables the mapping of a signal to a single bit of some machine word of a simulating host machine. AND- and OR-machine instructions can then be used to simulate Boolean networks. However, only a single bit would be used per machine word. Efficiency is improved with **parallel fault simulation**. In parallel fault simulation, n different test patterns are simulated at the same time, if n is the machine word size. The values of each of the n test patterns are mapped to a different bit position in the machine word. Executing the same set of AND- and OR-instructions will then simulate the behavior of the Boolean network for n test patterns instead of for just one.

8.3.3   Fault injection

Fault simulation may be too time-consuming for real systems. If actual systems are available, fault injection can be used instead. In fault injection, real existing systems are modified and the overall effect on the system behavior is checked. Fault injection does not rely on fault models (even though they can be used). Hence, fault injection has the potential of generating faults that would not have been predicted by a fault model.

We can distinguish between two types of fault injection:

- local faults within the system, and

- faults in the environment (behaviors which do not correspond to the specification). For example, we can check how the system behaves if it is operated outside the specified temperature or radiation ranges.

Several methods can be used for fault injection:

- fault injection at the hardware level: Examples include pin-manipulation, electromagnetic and nuclear radiation, and

- fault injection at the software level: Examples include toggling some memory bits.

The quality of fault injection depends on the "probe effect": probing might have an impact on the behavior of the system. This impact should be as small as possible and essentially be negligible.

According to experiments reported by Kopetz [Kopetz, 1997], software-based fault injection was essentially as effective as hardware-based fault injection. Nuclear radiation was a noticeable exception in that it generated errors which were not generated with other methods.

8.4 Design for testability

8.4.1 Motivation

Ideas for test pattern generation for Boolean circuits have been presented in section 8.2.1. For circuits implementing state machines (automata), test pattern generation is more difficult. Verifying whether or not two finite state machines are equivalent may require complex input sequences [Kohavi, 1987]. For example, consider the state chart of fig. 2.27, shown again in fig. 8.4 for convenience:

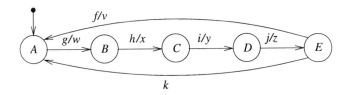

Figure 8.4. Finite state machine to be tested

Suppose that we would like to test the transition from state C to state D. This requires us to get into state C first, by applying an appropriate sequence of input patterns. Next, we must generate input event i and check, if output y is generated. Also, we need to check if we reached state D. This procedure is rather complicated, takes a lot of time and is susceptible to interference with other errors[2].

This example demonstrates: If testing comes in only as an afterthought, it may be very difficult to test a system. In order to simplify tests, special hardware can be added such that testing becomes easier. The process of designing for better testability is called **design for testability**, or **DfT**. Special purpose hardware for testing finite state machines is a prominent example of this.

8.4.2 Scan design

Reaching certain states and observing states resulting from the application of input patterns is very much simplified with **scan design**. In scan design, all flip-flops storing states are connected to form serial shift registers (see fig. 8.5).

The circuit contains three D-type flip-flops DFF and one multiplexer at each of the flip-flop inputs. Using the control input of the multiplexers (shown at the bottom of the multiplexer inputs), we can either connect the flip-flops to the

---

[2]The overall test of this FSM is simplified by the fact that this FSM contains a linear chain of transitions (c.f. to the assignments of this chapter).

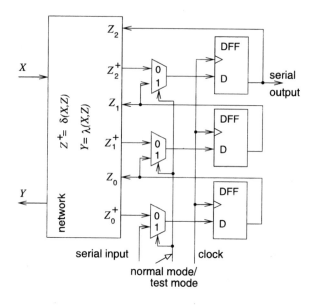

Figure 8.5.   Scan path design

network generating the next state from the current state and the current input or we can connect flip-flops to form a serial chain. Setting the multiplexers to scan mode, we can load state bit after state bit into the scan chain (one bit at every clock tick). This way, we can load any state into the three flip-flops serially. In a second phase, we can apply an input pattern to the FSM while the multiplexers are set to normal mode. After the next clock tick, the FSM will be in a new state. This new state can be serially shifted out in the third and final phase, using the serial mode again (one bit per clock tick). The net effect is that we do not need to worry about how to get into certain states and how to observe whether or not the Boolean function δ for computing the next state has been correctly implemented while we are generating tests for the FSM. Effectively, the fact that we are dealing with state-based systems has an impact only on the two (simple) shift phases, and test pattern generation for (stateless) Boolean networks can be used for checking for correct outputs. This means that it is sufficient to use test pattern generation methods for Boolean functions (stateless networks) instead of caring about complex input sequences etc.

Scan design is a technique which works well for single chips. For board-level integration it is necessary to have some technique for connecting scan chains of several chips. **JTAG** is a standard which does exactly this. The standard defines registers at the boundaries of all chips and a number of test pins and control commands such that all chips can be connected in scan chains. JTAG is also known as boundary scan [Parker, 1992].

8.4.3   Signature analysis

In order to also avoid shifting out the response of the device under test (DUT), responses can be compacted. A setup like the one shown in fig. 8.6 can be used.

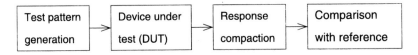

Figure 8.6.   Testing a device under test (DUT)

Generated test patterns are used as inputs (or so-called stimuli) to the DUT. The response of the DUT is then compacted to form a **signature**, which characterizes the response. This response is later compared to the expected response. The expected response can be computed by simulation.

The compaction is typically performed with linear feedback shift registers (LFSRs), shift registers with an XOR-feedback. Fig. 8.7 shows a 4-bit LFSR (left) and the associated state diagram (right) [Lala, 1985].

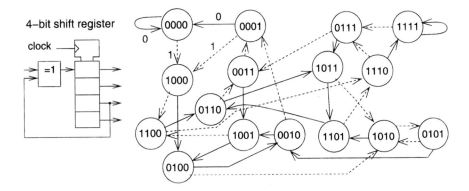

Figure 8.7.   Linear feedback shift register for response compaction

Dashed lines denote an input of '1', uninterrupted lines denote an input of '0'. The selected feedback yields all possible signatures.

During testing, the response of the system tested is sent to the input of the LFSR. The LFSR will then generate a signature reflecting the response. Due to storing the signature instead of the full response, several response patterns can be mapped to the same signature. What is the probability of obtaining a correct signature from an incorrect response?

In general, an n-bit signature generator can generate 2^n signatures. For an m-bit response of the DUT, the best that we can do is to evenly map $2^{(m-n)}$

responses to the same signature. Suppose that we expect a certain signature to be generated for the correct response of the system. Then, $2^{(m-n)} - 1$ incorrect responses would also map to the same signature. There is a total of $2^m - 1$ incorrect responses if responses are m-bit long. Hence, the probability of an incorrect response to map to the correct signature (provided patterns map evenly to signatures) is

$$P = Pr\left(\frac{\text{other patterns mapping to the same signature}}{\text{total number of other patterns}}\right) \quad (8.1)$$

$$= \frac{2^{(m-n)} - 1}{2^m - 1} \quad (8.2)$$

$$\approx \frac{1}{2^n} \text{ for } m \gg n \quad (8.3)$$

This means that the probability of generating correct signatures from an incorrect test response is very small if the shift register is long.

8.4.4   Pseudo-random test pattern generation

For chips with a large number of flip-flops, it can take quite some time to shift-in the test patterns. In order to speed up the process of generating patterns on the chip, it has been proposed to also integrate hardware for generating test patterns on the chip.

For example, pseudo-random patterns (also generated by LFSRs) can be used as test patterns. For example, we can modify the circuit of fig. 8.7 as shown in fig. 8.8.

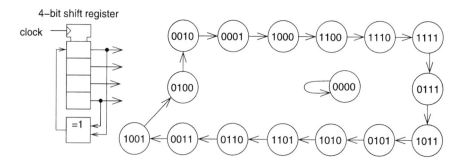

Figure 8.8.   Linear feedback shift register for test pattern generation

The circuit generates all possible test patterns, except the pattern consisting of all zeros. The pattern consisting of all zeros has to be avoided, since the generator would get stuck once it arrives at this pattern. The generated patterns are typically exercising systems to be tested much better than simple counters.

8.4.5   The built-in logic block observer (BILBO)

The **built-in logic block observer** (BILBO) [Könemann et al., 1979] has been proposed as a circuit combining test pattern generation, test response compaction and serial scan capabilities. A BILBO with three D-type flip-flops is shown in fig. 8.9.

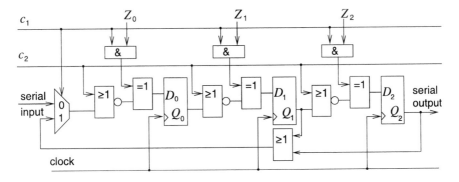

Figure 8.9.   BILBO

Modes of BILBO registers are shown in table 8.10. The 3-bit register shown in fig. 8.9 can be in scan path, reset, linear-feedback shift register (LFSR) and normal mode. In LFSR mode, it can be used for either generating pseudo-random patterns or for compacting responses from inputs (Z_0 to Z_2). In this case, compaction is based on parallel inputs instead of the sequential inputs we have considered so far. Purpose and behavior of compaction from parallel inputs is similar to that for serial inputs.

c_1	c_2	D_i	
'0'	'0'	$'0' \oplus \overline{Q_{i-1}} = \overline{Q_{i-1}}$	scan path mode
'0'	'1'	$'0' \oplus '1' = '0'$	reset
'1'	'0'	$Z_i \oplus \overline{Q_{i-1}}$	LFSR mode
'1'	'1'	$Z_i \oplus '1' = Z_i$	normal mode

Figure 8.10.   Modes of BILBO registers

Typically, BILBOs are used in pairs (see fig. 8.11).

One BILBO generates pseudo-random test patterns, feeding some Boolean network with these patterns. The response of the Boolean network is then compressed by a second BILBO connected to the output of the network. At the end

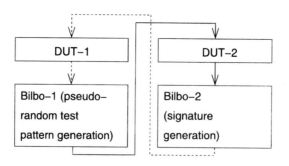

Figure 8.11.   Cross-coupled BILBOs

of the test sequence, the compacted response is serially shifted out and compared with the expected response. The expected response can be computed by
simulation.

During a second phase, the roles of the two BILBOs can be swapped. During
this phase, the connection shown as a dashed line in fig. 8.11 is used. In normal
mode, BILBOs can be used as state registers.

DfT hardware is of great help during the prototyping and debugging of hardware. It is also useful to have DfT hardware in the final product, since hardware
fabrication never has a zero defect rate. Testing fabricated hardware significantly contributes to the overall cost of a product and mechanisms that reduce
this cost are highly appreciated by all companies.

8.5   Assignments

1 Consider the circuit shown in fig. 8.2. Generate a test pattern for a stuck-
  at-0 fault at signal *h*!

2 Which state diagram corresponds to the LFSR shown in fig. 8.12?

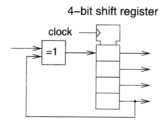

Figure 8.12.   LFSR

3 Specify test patterns and expected responses for the FSM shown in fig.
  8.4. These patterns must be specified as a sequence of pairs (test pattern,

expected response). Events shown in fig. 8.4 can be used as test patterns. We assume that the FSM will be in the default state after power-on. Provide a complete test for all transitions! Note that the special chain-like structure of the FSM simplifies testing.

Appendix A
Integer linear programming

Integer linear programming (ILP) is a mathematical optimization technique applicable to a large number of optimization problems.

ILP models provide a general approach for modeling optimization problems. ILP models consist of two parts: a cost function and a set of constraints. Both parts involve references to a set $X = \{x_i\}$ of integer-valued variables. Cost functions must be linear functions of those variables. So, they must be of the general form

$$C = \sum_i a_i x_i, \text{ with } a_i \in I\!R, x_i \in I\!N_0 \tag{A.1}$$

The set J of constraints must also consist of linear functions of integer-valued variables. They must be of the form

$$\forall j \in J : \sum_i b_{i,j} x_i \geq c_j \text{ with } b_{i,j}, c_j \in I\!R \tag{A.2}$$

Def.: The **integer linear programming (ILP-) problem** is the problem of minimizing the cost function of eq. (A.1) subject to the constraints given in eq. A.2. If all variables are constrained to being either 0 or 1, the corresponding model is called a **0/1-integer linear programming model**. In this case, variables are also denoted as **(binary) decision variables**.

Note that $\geq$ can be replaced by $\leq$ in equation (A.2) if constants $b_{i,j}$ are modified accordingly. Also, the case of negative variables x_i (that is, allowing x_i to have any integer value) can be transformed into the case of non-negative

P. Marwedel, *Embedded System Design*, Embedded Systems,
DOI 10.1007/978-94-007-0257-8, © Springer Science+Business Media B.V. 2011

variables shown above by multiplying constants by -1. Applications requiring **maximizing** some **gain** function C' can be changed into the above form by setting $C = -C'$.

For example, assuming that x_1, x_2 and x_3 must be integers, the following set of equations represent a 0/1-IP model:

$$C = 5x_1 + 6x_2 + 4x_3 \tag{A.3}$$
$$x_1 + x_2 + x_3 \geq 2 \tag{A.4}$$
$$x_1 \leq 1 \tag{A.5}$$
$$x_2 \leq 1 \tag{A.6}$$
$$x_3 \leq 1 \tag{A.7}$$

Due to the constraints, all variables are either 0 or 1. There are four possible solutions. These are listed in fig. A.1. The solution with a cost of 9 is optimal.

x_1	x_2	x_3	C
0	1	1	10
1	0	1	9
1	1	0	11
1	1	1	15

Figure A.1.   Possible solutions of the presented ILP-problem

ILP is a variant of linear programming (LP). For linear programming, variables can take any real values. ILP and LP models can be solved optimally using mathematical programming techniques. Unfortunately, ILP is NP-complete (but LP is not) and ILP execution times may become very large.

Nevertheless, ILP models are useful for modeling optimization problems as long as the model sizes are not extremely large. Modeling optimization problems as integer linear programming problems makes sense despite the complexity of the problem: many problems can be solved in acceptable execution times and if they cannot, ILP models provide a good starting point for heuristics. Execution times depend on the number of variables and on the number and structure of the constraints. Good ILP solvers (like lp_solve [Anonymous, 2010a] or CPLEX) can solve well-structured problems containing a few thousand variables in acceptable computation times (e.g. minutes). For more information on ILP and LP, refer to books on the topic (e.g. to Wolsey [Wolsey, 1998]).

Appendix B
Kirchhoff's laws and operational amplifiers

Our presentation of D/A-converters on page 164 assumes some basic knowledge about operational amplifiers. This knowledge is frequently lacking among computer science students and therefore the necessary fundamentals are presented in this appendix. These fundamentals require an understanding of Kirchhoff's laws, of which students will also be reminded in this appendix.

Kirchhoff's laws

Kirchoff's laws provide a means for analyzing electrical circuits. The first rule is Kirchhoff's Current Law, also called Kirchhoff's Junction Rule, or Kirchhoff's First Law. The rule applies to junctions such as the one shown in fig. B.1.

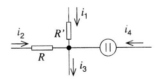

Figure B.1. Junction in an electrical circuit

Kirchhoff's Current Law: At any point in an electrical circuit, the sum of currents flowing towards that point is equal to the sum of currents flowing away from that point [Jewett and Serway, 2007]. Formally, for any node in a circuit we have:

P. Marwedel, *Embedded System Design*, Embedded Systems,
DOI 10.1007/978-94-007-0257-8, © Springer Science+Business Media B.V. 2011

$$\sum_k i_k = 0 \tag{B.1}$$

If Kirchhoff's law is used in the form of equation B.1, currents denoted by arrows pointing away from the node must be counted as negative, and this counting is independent of the direction into which electrons are actually flowing. Example: for the currents of fig. B.1, we have

$$i_1 + i_2 - i_3 + i_4 \quad = \quad 0 \tag{B.2}$$
$$i_1 + i_2 + i_4 \quad = \quad i_3 \tag{B.3}$$

This invariance exists due to the conservation of electrical charge. Without this rule, the total electrical charge would not remain constant, and the voltage would increase.

Kirchhoff's second rule applies to loops in a circuit. It is known as Kirchhof's Voltage Law, Kirchhoff's Loop Rule or Kirchhoff's Second Law. Fig. B.4 shows an example.

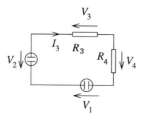

Figure B.2.   Loop in an electrical circuit

Kirchhoff's Voltage Law: The sum of the potential differences (voltages) across all elements around any closed circuit must be zero [Jewett and Serway, 2007]. Formally, for any loop in a circuit we have:

$$\sum_k V_k \quad = \quad 0 \tag{B.4}$$

If we traverse voltages against the arrow direction, we have to count them as negative. Example: for the schematic of fig. B.2, we have

$$V_1 - V_2 - V_3 + V_4 \quad = \quad 0 \tag{B.5}$$

The underlying reason for this invariance is the conservation of energy. Without this rule, we could accelerate charge in the loop and the charge would accumulate energy without any energy consumption elsewhere.

In general, it is not relevant into which direction electrons are actually flowing and which of two terminals is actually positive with respect to some other terminal. Arrows can be selected in an arbitrary way. We just have to make sure that we respect the direction of the arrows when we apply Kirchhoff's laws. If arrows for voltages and currents across components are pointing in opposite directions, the equation for that component has to take that into account. For example, Ohm's law for resistor R_3 in fig. B.2 reads as follows, due to the opposite directions of voltage and current arrows:

$$I_3 = -\frac{V_3}{R_3} \tag{B.6}$$

Of course, we will typically try to define the direction of voltages and currents such that we avoid having too many minus signs.

Operational amplifiers

In electronics, there is frequently the need to amplify some signal $x(t)$ in order to obtain some amplified signal $y(t) = a \cdot x(t)$, with $a > 1$. a is called the **gain**. Designing different circuits for each and every gain would be a laborious task. Therefore, designers are frequently using a general amplifier which can be easily configured to have the required gain. Such a general amplifier is called **operational amplifier**, or op-amp for short. Op-amps are designed for a very large maximum gain. The required actual gain can be adjusted with a proper selection of a few hardware components in the circuit surrounding the op-amp.

More precisely, an operational amplifier is a component having two signal inputs and one signal output. In addition, there are at least two power supply inputs (see fig. B.3).

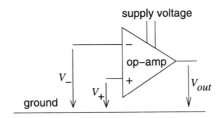

Figure B.3. Operational amplifier

Op-amps amplify the difference between the voltages at the two signal inputs with respect to ground by a gain g:

$$V_{out} \;=\; g \cdot (V_+ - V_-) \tag{B.7}$$

g is called the **open loop gain** and is typically very large ($10^4 < g < 10^6$). For an ideal op-amp, g would approach infinity. Furthermore, op-amps usually come with a very high input impedance ($> 1M\Omega$). Hence, we can frequently ignore signal input currents. For an ideal op-amp, the input impedance would be infinity and input currents would be zero.

Op-amps have been commercially available for decades, both as separate integrated circuits and within other circuits. They differ by their speed, their voltage ranges, their current drive capability, and other characteristics. The actual gain of the circuit is selected with external resistors. Fig. B.4 shows how this can be done.

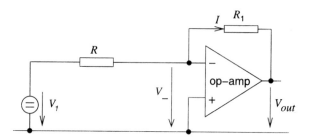

Figure B.4.   Operational amplifier with feed back

Any small voltage between the two signal inputs is amplified by a large factor. Via resistor R_1, the resulting output voltage is feed back. Feed back is to the inverting input and therefore, any positive voltage V_- results in a negative voltage V_{out} and vice versa. This means that the feed back will work against the input voltage, and it does so very strongly, due to the large amplification. Therefore, the feed back will reduce the voltage at the input pin. The question is: by how much? We can use Kirchhoff's rules to find the resulting voltage V_- (see fig. B.5).

Due to the characteristics of op-amps, we have

$$V_{out} \;=\; -g \cdot V_- \tag{B.8}$$

Due to Kirchhoff's law for the loop shown by a dashed line in fig. B.5, we have

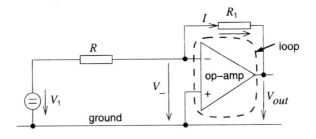

Figure B.5. Op-amp with feed back (loop highlighted)

$$I \cdot R_1 + V_{out} - V_- = 0 \tag{B.9}$$

Note that we include a minus sign for V_- since we are traversing a segment of the loop against the direction of the arrow. From equations B.8 and B.9, we get

$$
\begin{align}
I \cdot R_1 + (-g) \cdot V_- - V_- &= 0 \tag{B.10} \\
(1+g) \cdot V_- &= I \cdot R_1 \tag{B.11} \\
V_- &= \frac{I \cdot R_1}{1+g} \tag{B.12} \\
V_{-,ideal} &= \lim_{g \to \infty} \frac{I \cdot R_1}{1+g} \tag{B.13} \\
&= 0 \tag{B.14}
\end{align}
$$

This means that, for an ideal op-amp, V_- is 0. Due to this, the inverting signal input is called **virtual ground**. Nevertheless, this input cannot be connected to ground, since this would change the currents.

Computing the actual gain of the circuit in fig. B.4 is left as an exercise for Chapter 3.

References

[Aamodt and Chow, 2000] Aamodt, T. and Chow, P. (2000). Embedded ISA support for enhanced floating-point to fixed-point ANSI C compilation. *3rd ACM Intern. Conf. on Compilers, Architectures and Synthesis for Embedded Systems (CASES)*, pages 128–137.

[Absint, 2002] Absint (2002). Absint: WCET analyses. *http://www.absint.de/wcet.htm*.

[Absint, 2010] Absint (2010). aiT worst-case execution time analyzers. *http://www.absint.de/ait*.

[Accellera Inc., 2003] Accellera Inc. (2003). SystemVerilog 3.1 - Accellera's extensions to Verilog®. *http://www.eda.org/sv/SystemVerilog_3.1_final.pdf*.

[ACM SIGBED, 2010] ACM SIGBED (2010). Home page. *http://www.sigbed.org*.

[ACM/IEEE, 2008] ACM/IEEE (Dec. 2008). Computer science curriculum 2008: An interim revision of CS 2001. *Association for Computing Machinery, IEEE Computer Society, http://www.acm.org/education/curricula/ComputerScience2008.pdf*.

[Ambler, 2003] Ambler, S. (2003). The diagrams of UML 2.0. *http://www.agilemodeling.com/essays/umlDiagrams.htm*.

[Analog Devices Inc. Eng., 2004] Analog Devices Inc. Eng. (2004). *Data Conversion Handbook (Analog Devices)*. Newnes.

[Anonymous, 2010a] Anonymous (2010a). Introduction to lp_solve 5.5.0.14. *http://lpsolve.sourceforge.net*.

[Anonymous, 2010b] Anonymous (2010b). RTJS home page. *http://www.rtsj.org*.

[Araujo and Malik, 1995] Araujo, G. and Malik, S. (1995). Optimal code generation for embedded memory non-homogenous register architectures. *8th Int. Symp. on System Synthesis (ISSS)*, pages 36–41.

[ARM Ltd., 2009a] ARM Ltd. (2009a). AMBA 2 specification. *http://www.arm.com/products/solutions/AMBA_Spec.html*.

P. Marwedel, *Embedded System Design*, Embedded Systems,
DOI 10.1007/978-94-007-0257-8, © Springer Science+Business Media B.V. 2011

[ARM Ltd., 2009b] ARM Ltd. (2009b). Realview compilation tools compiler reference guide. *http://infocenter.arm.com/help/index.jsp?topic=/com.arm.doc.set.swdev/index.html.*

[ARTEMIS Joint Undertaking, 2010] ARTEMIS Joint Undertaking (2010). Home page. *https://www.artemis-ju.eu/organisation.*

[Artist Consortium, 2010] Artist Consortium (2010). Home page. *http://www.artist-embedded.org.*

[Atienza et al., 2007] Atienza, D., Baloukas, C., Papadopoulos, L., Poucet, C., Mamagkakis, S., Hidalgo, J. I., Catthoor, F., Soudris, D., and Lanchares, J. (2007). Optimization of dynamic data structures in multimedia embedded systems using evolutionary computation. *In 10th Int. Workshop on Software & Compilers for Embedded Systems (SCOPES)*, pages 31–40.

[August et al., 1997] August, D. I., Hwu, W. W., and Mahlke, S. (1997). A framework for balancing control flow and predication. *Ann. Workshop on Microprogramming and Microarchitecture (MICRO)*, pages 92–103.

[AUTOSAR, 2010] AUTOSAR (2010). Automotive open system architecture. *http://www.autosar.org.*

[Avissar et al., 2002] Avissar, O., Barua, R., and Stewart, D. (2002). An optimal memory allocation scheme for scratch-pad-based embedded systems. *Transactions on Embedded Computing Systems.*

[Avižienis et al., 2004] Avižienis, A., Laprie, J.-C., Randell, B., and Landwehr, C. (2004). Basic concepts and taxonomy of dependable and secure computing. *IEEE Transactions on Dependable and Secure Computing*, 1(1):11–33.

[Azevedo et al., 2002] Azevedo, A., Issenin, I., Cornea, R., Gupta, R., Dutt, N., Veidenbaum, A., and Nicolau, A. (2002). Profile-based dynamic voltage scheduling using program checkpoints. *Design, Automation and Test in Europe (DATE)*, pages 168–175.

[Bäck et al., 1997] Bäck, T., Fogel, D., and Michalewicz, Z. (1997). *Handbook of Evolutionary Computation.* Oxford Univ. Press.

[Bäck and Schwefel, 1993] Bäck, T. and Schwefel, H.-P. (1993). An overview of evolutionary algorithms for parameter optimization. *Evolutionary computation*, pages 1–23.

[Balarin et al., 1998] Balarin, F., Lavagno, L., Murthy, P., and Sangiovanni-Vincentelli, A. (1998). Scheduling for embedded real-time systems. *IEEE Design & Test of Computers*, pages 71–82.

[Ball, 1996] Ball, S. R. (1996). *Embedded Microprocessor Systems - Real world designs.* Newnes.

[Ball, 1998] Ball, S. R. (1998). *Debugging Embedded Microprocessor Systems.* Newnes.

[Banakar et al., 2002] Banakar, R., Steinke, S., Lee, B.-S., Balakrishnan, M., and Marwedel, P. (2002). Scratchpad memory: a design alternative for cache on-chip memory in embedded systems. *10th Intern. Symp. on Hardware/software Codesign (CODES)*, pages 73–78.

[Barney, 2010] Barney, B. (2010). POSIX threads programming. *https://computing.llnl.gov/tutorials/pthreads.*

[Barr, 1999] Barr, M. (1999). *Programming Embedded Systems*. O'Reilly.

[Barrett and Pack, 2005] Barrett, S. and Pack, D. (2005). *Embedded Systems - Design and Applications with the 68HC12 and HCS12*. Prentice Hall.

[Basten, 2008] Basten, T. (2008). Opening remarks, 2nd Artist workshop on models of computation and communication. *Eindhoven*, *http://www.es.ele.tue.nl/~tbasten/mocc2008/presentations/mocc.pdf*.

[Basu et al., 1999] Basu, A., Leupers, R., and Marwedel, P. (1999). Array index allocation under register constraints in dsp programs. *Int. Conf. on VLSI Design*, pages 330–335.

[Belbachir, 2010] Belbachir, A. N., editor (2010). *Smart cameras*. Springer.

[Bengtsson and Yi, 2004] Bengtsson, J. and Yi, W. (2004). Timed automata: Semantics, algorithms and tools. *In: J. Desel, W. Reisig and G. Rozenberg (eds.): ACPN 2003, Springer LNCS*, 3098:87–124.

[Benini et al., 2000] Benini, L., Bogliolo, A., and De Micheli, G. (2000). A survey of design techniques for system-level dynamic power management. *IEEE Trans. Very Large Scale Integr. Syst.*, 8(3):299–316.

[Benini and De Micheli, 1998] Benini, L. and De Micheli, G. (1998). *Dynamic Power Management – Design Techniques and CAD Tools*. Kluwer Academic Publishers.

[Bergé et al., 1995] Bergé, J.-M., Levia, O., and Rouillard, J. (1995). *High-Level System Modeling*. Kluwer Academic Publishers.

[Bernardi et al., 2005] Bernardi, P., Rebaudengo, M., and Reorda, S. (2005). Using infrastructure IPs to support SW-based self-test of processor cores. *Workshop on Fibres and Optical Passive Components*, pages 22–27.

[Bertolotti, 2006] Bertolotti, I. C. (2006). Real-time embedded operating systems: Standards and perspectives. *In: R. Zurawski (ed.): Embedded Systems Handbook, CRC Press*.

[Beszedes, 2003] Beszedes, A. (2003). Survey of code size reduction methods. *ACM Computing Surveys*, pages 223–267.

[Bieker and Marwedel, 1995] Bieker, U. and Marwedel, P. (1995). Retargetable self-test program generation using constraint logic programming. *32nd annual Design Automation Conference (DAC)*, pages 605–611.

[Bini et al., 2001] Bini, E., Buttazzo, G., and Buttazzo, G. (2001). A hyperbolic bound for the rate monotonic algorithm. *13th Euromicro Conference on Real-Time Systems (ECRTS)*, pages 59–73.

[Boussinot and de Simone, 1991] Boussinot, F. and de Simone, R. (1991). The Esterel language. *Proc. of the IEEE*, Vol. 79, No. 9, pages 1293–1304.

[Bouwmeester et al., 2000] Bouwmeester, D., Ekert, A. and Zeilinger, A. (eds.) (2000). *The Physics of Quantum Information: Quantum Cryptography, Quantum Teleportation, Quantum Computation*. Springer.

[Bouyssounouse and Sifakis, 2005] Bouyssounouse, B. and Sifakis, J., editors (2005). *Embedded Systems Design, The ARTIST Roadmap for Research and Development*. Lecture Notes in Computer Science, Vol. 3436, Springer.

[Brahme and Abraham, 1984] Brahme, D. and Abraham, J. A. (1984). Functional testing of microprocessors. *IEEE Trans. on Computers*, pages 475–485.

[Braun et al., 2010] Braun, A., Bringmann, O., Lettnin, D., and Rosenstiel, W. (2010). Simulation-based verification of the MOST netinterface specification revision 3.0. *Design, Automation and Test in Europe (DATE)*.

[Bremaud, 1999] Bremaud, P. (1999). *Markov Chains*. Springer Verlag.

[Brockmeyer et al., 2003] Brockmeyer, E., Miranda, M., and Catthoor, F. (2003). Layer assignment techniques for low energy in multi-layered memory organisations. *Design, Automation and Test in Europe (DATE)*, pages 1070–1075.

[Broesma, 2004] Broesma, M. (Sep. 2004). Microsoft server crash nearly causes 800-plane pile-up. *Techworld, http://www.techworld.com/opsys/news/index.cfm?newsid=2275*.

[Brooks et al., 2000] Brooks, D., Tiwari, V., and Martonosi, M. (2000). Wattch: a framework for architectural-level power analysis and optimizations. *27th Int. Symp. on Computer Architecture (ISCA)*, pages 83–94.

[Bruno and Bollella, 2009] Bruno, E. and Bollella, G. (2009). *Real-Time Java Programming: With Java RTS*. Prentice Hall.

[Budkowski and Dembinski, 1987] Budkowski, S. and Dembinski, P. (1987). An introduction to Estelle: A specification language for distributed systems. *Computer Networks and ISDN Systems*, 14(1):3 – 23.

[Burd and Brodersen, 2000] Burd, T. and Brodersen, R. (2000). Design issues for dynamic voltage scaling. *Int. Symp. on Low Power Electronics and Design (ISLPED)*, pages 9–14.

[Burd and Brodersen, 2003] Burd, T. and Brodersen, R. W. (2003). *Energy efficient microprocessor design*. Kluwer Academic Publishers.

[Burkhardt, 2001] Burkhardt, J. (2001). *Pervasive Computing*. Addison-Wesley.

[Burns and Wellings, 1990] Burns, A. and Wellings, A. (1990). *Real-Time Systems and Their Programming Languages*. Addison-Wesley.

[Burns and Wellings, 2001] Burns, A. and Wellings, A. (2001). *Real-Time Systems and Programming Languages (Third Edition)*. Addison Wesley.

[Buttazzo, 2002] Buttazzo, G. (2002). *Hard Real-time computing systems*. Kluwer Academic Publishers, 4th printing.

[Byteflight Consortium, 2003] Byteflight Consortium (2003). Home page. *http://www.byteflight.com*.

[Camposano and Wolf, 1996] Camposano, R. and Wolf, W. (1996). Message from the editors-in-chief. *Design Automation for Embedded Systems*.

[Caspi et al., 2005] Caspi, P., Sangiovanni-Vincentelli, A., Almeida, L., and et al. (2005). Guidelines for a graduate curriculum on embedded software and systems. *ACM Transactions on Embedded Computing Systems (TECS)*, pages 587–611.

[Cederqvist, 2006] Cederqvist, P. (2006). The CVS manual - version management with CVS. *Network Theory Ltd.*

[Ceng et al., 2008] Ceng, J., Castrillón, J., Sheng, W., Scharwächter, H., Leupers, R., Ascheid, G., Meyr, H., Isshiki, T., and Kunieda, H. (2008). MAPS: an integrated framework for MP-SoC application parallelization. In *45th annual Design Automation Conference (DAC)*, pages 754–759.

[Chandrakasan et al., 1992] Chandrakasan, A. P., Sheng, S., and Brodersen, R. W. (1992). Low-power CMOS digital design. *IEEE Journal of Solid-State Circuits*, 27(4):119–123.

[Chandrakasan et al., 1995] Chandrakasan, A. P., Sheng, S., and Brodersen, R. W. (1995). Low power CMOS digital design. *Kluwer Academic Publishers*.

[Chanet et al., 2007] Chanet, D., Sutter, B. D., Bus, B. D., Put, L. V., and Bosschere, K. D. (2007). Automated reduction of the memory footprint of the linux kernel. *ACM Trans. Embed. Comput. Syst.*, 6(4):23.

[Chen et al., 2006] Chen, G., Ozturk, O., Kandemir, M., and Karakoy, M. (2006). Dynamic scratch-pad memory management for irregular array access patterns. *Design, Automation and Test in Europe (DATE)*, pages 931–936.

[Chen et al., 2007] Chen, K., Sztipanovits, J., and Neema, S. (2007). Compositional specification of behavioral semantics. *Design, Automation and Test in Europe (DATE)*, pages 906–911.

[Chen et al., 2010] Chen, X., Dick, R., and Shang, L. (2010). Properties of and improvements to time-domain dynamic thermal analysis algorithms. *Design, Automation and Test in Europe (DATE)*.

[Chetto et al., 1990] Chetto, H., Silly, M., and Bouchentouf, T. (1990). Dynamic scheduling of real-time tasks under precedence constraints. *Journal of Real-Time Systems, 2*.

[Chung et al., 2001] Chung, E.-Y., Benini, L., and De Micheli, G. (2001). Source code transformation based on software cost analysis. *Int. Symp. on System Synthesis (ISSS)*, pages 153–158.

[Clarke and et al., 2003] Clarke, E. and et al. (2003). Model checking@CMU. *http://www-2.cs. cmu.edu/~modelcheck/index.html*.

[Clarke et al., 2005] Clarke, E. M., Grumberg, O., Hiraishi, H., Jha, S., Long, D. E., McMillan, K. L., and Ness, L. A. (2005). Verification of the futurebus+ cache coherence protocol. *Formal Methods in System Design*, 6(2):217–232.

[Clavier and Gaj, 2009] Clavier, C. and Gaj, K. (2009). Int. workshop on cryptographic hardware and embedded systems (CHES).

[Clouard et al., 2003] Clouard, A., Jain, K., Ghenassia, F., Maillet-Contoz, L., and Strassen, J. (2003). Using transactional models in SoC design flow. *In: [Müller et al., 2003]*, pages 29–64.

[Coelho, 1989] Coelho, D. R. (1989). *The VHDL handbook.* Kluwer Academic Publishers.

[Coello et al., 2007] Coello, C. A. C., Lamont, G. B., and Veldhuizen, D. A. v. (2007). *Evolutionary Algorithms for Solving Multi-Objective Problems.* Springer.

[Collins-Sussman et al., 2008] Collins-Sussman, B., Fitzpatrick, B., and Pilato, C. (2008). Version control with subversion – for subversion 1.5. *http://svnbook.red-bean.com/en/1.5/svn-book.pdf.*

[Cooling, 2003] Cooling, J. (2003). *Software Engineering for Real-Time Systems.* Addison Wesley.

[Cortadella et al., 2000] Cortadella, J., Kondratyev, A., Lavagno, L., Massot, M., Moral, S., Passerone, C., Watanabe, Y., and Sangiovanni-Vincentelli, A. (2000). Task generation and compile-time scheduling for mixed data-control embedded software. *37th Design Automation Conference (DAC)*, pages 489–494.

[Coussy and Morawiec, 2008] Coussy, P. and Morawiec, A. (2008). *High-Level Synthesis.* Springer.

[Craig, 2006] Craig, I. D. (2006). *Virtual Machines.* Springer.

[Damm and Harel, 2001] Damm, W. and Harel, D. (2001). LSCs: Breathing life into message sequence charts. *Formal Methods in System Design.*

[Dasgupta, 1979] Dasgupta, S. (1979). The organization of microprogram stores. *ACM Computing Surveys*, Vol. 11, pages 39–65.

[Davis et al., 2001] Davis, J., Hylands, C., Janneck, J., Lee, E. A., Liu, J., Liu, X., Neuendorffer, S., Sachs, S., Stewart, M., Vissers, K., Whitaker, P., and Xiong, Y. (2001). Overview of the Ptolemy project. *Technical Memorandum UCB/ERL M01/11; http://ptolemy.eecs.berkeley.edu.*

[De Greef et al., 1997a] De Greef, E., Catthoor, F., and Man, H. (1997a). Memory size reduction through storage order optimization for embedded parallel multimedia applications. *Proc. Workshop on Parallel Processing and Multimedia*, pages 84–98.

[De Greef et al., 1997b] De Greef, E., Catthoor, F., and Man, H. D. (1997b). Array placement for storage size reduction in embedded multimedia systems. *IEEE Int. Conf. on Application-Specific Systems, Architectures and Processors (ASAP)*, pages 66–75.

[De Micheli et al., 2002] De Micheli, G., Ernst, R., and Wolf, W. (2002). *Readings in Hardware/Software Co-Design.* Academic Press.

[Deutsches Institut für Normung, 1997] Deutsches Institut für Normung (1997). *DIN 66253, Programmiersprache PEARL, Teil 2 PEARL 90.* Beuth-Verlag; English version available through *http://www.din.de.*

[Dibble, 2008] Dibble, P. C. (2008). *Real-Time Java Platform Programming: Second Edition.* BookSurge Publishing.

[Diederichs et al., 2008] Diederichs, C., Margull, U., Slomka, F., and Wirrer, G. (2008). An application-based EDF scheduler for OSEK/VDX. *Design, Automation and Test in Europe (DATE)*, pages 1045–1050.

[Dierickx, 2000] Dierickx, B. (2000). CMOS image sensors - concepts, Photonics West 2000 short course. *http://www.cypress.com/?rID=14527*.

[Dill and Alur, 1994] Dill, D. and Alur, R. (1994). A theory of timed automata. *Theoretical Computer Science*, pages 183–235.

[Dominguez et al., 2005] Dominguez, A., Udayakumaran, S., and Barua, R. (2005). Heap data allocation to scratch-pad memory in embedded systems. *Journal of Embedded Computing*, 1(4):521–540.

[Donald and Martonosi, 2006] Donald, J. and Martonosi, M. (2006). Techniques for multicore thermal management: Classification and new exploration. *SIGARCH Comput. Archit. News*, 34(2):78–88.

[Douglass, 2000] Douglass, B. P. (2000). *Real-Time UML, 2nd edition*. Addison Wesley.

[Drusinsky and Harel, 1989] Drusinsky, D. and Harel, D. (1989). Using statecharts for hardware description and synthesis. *IEEE Trans. on Computer Design*, pages 798–807.

[Dulong et al., 2001] Dulong, C., Shrivastav, P., and Refah, A. (2001). The making of a compiler for the Intel® Itanium^{TM} processor. *Intel Technology Journal Q3, http://download. intel.com/technology/itj/q32001/pdf/art_4.pdf*.

[Dunn, 2002] Dunn, W. (2002). *Practical Design of Safety-Critical Computer Systems*. Reliability Press.

[Ecker et al., 2009] Ecker, W., Müller, W., and Dömer, R. (2009). *Hardware-dependent software - Principles and practice*. Springer.

[Edwards, 2001] Edwards, S. (2001). Dataflow languages. *http://www.cs.columbia.edu/ ˜sedwards/classes/2001/w4995-02/presentations/dataflow.ppt*.

[Edwards, 2006] Edwards, S. (2006). Languages for embedded systems. *In: R. Zurawski (ed.): Embedded Systems Handbook, CRC Press*.

[Egger et al., 2006] Egger, B., Lee, J., and Shin, H. (2006). Scratchpad memory management for portable systems with a memory management unit. *9rd ACM Intern. Conf. on Compilers, Architectures and Synthesis for Embedded Systems (CASES)*, pages 321–330.

[Eggermont, 2002] Eggermont, L. (2002). Embedded systems roadmap. *STW, http://www.stw. nl/NR/rdonlyres/3E59AA43-68B1-4E83-BA95-20376EB00560/0/ESRversion1.pdf*.

[Eichenberger et al., 2005] Eichenberger, A. E., O'Brien, K., O'Brien, Kevin, Wu, P. , Chen, T., Oden, P. H., Prener, D. A., Shepherd, J. C., So, B., Sura, Z., Wang, A., Zhang, T., Zhao, P., and Gschwind, M. (2005). Optimizing compiler for a CELL processor. *Proceedings of the 14th International Conference on Parallel Architectures and Compilation Techniques (PACT'05)*, pages 161–172.

[Elsevier B.V., 2010a] Elsevier B.V. (2010a). Sensors and actuators A: Physical. *An International Journal*.

[Elsevier B.V., 2010b] Elsevier B.V. (2010b). Sensors and actuators B: Chemical. *An International Journal*.

[Esterel Technologies, 2010] Esterel Technologies (2010). Scade suite^{TM} - the standard for the development of safety-critical embedded software in aerospace & defense, rail transportation, energy and heavy equipment industries. *http://www.esterel-technologies.com/products/scade-suite*.

[Esterel Technologies Inc., 2010] Esterel Technologies Inc. (2010). Homepage. *http://www.esterel-technologies.com*.

[European Commission Cordis, 2010] European Commission Cordis (2010). Seventh Framework Programme (FP7). *http://cordis.europa.eu/fp7*.

[Evidence, 2010] Evidence (2010). Erika enterprise. *http://erika.tuxfamily.org*.

[Falk, 2009] Falk, H. (2009). WCET-aware register allocation based on graph coloring. *Proceedings of the 46th Design Automation Conference (DAC)*, pages 726–731.

[Falk and Marwedel, 2003] Falk, H. and Marwedel, P. (2003). Control flow driven splitting of loop nests at the source code level. *Design, Automation and Test in Europe (DATE)*, pages 410–415.

[Falk et al., 2006] Falk, H., Wagner, J., and Schaefer, A. (2006). Use of a Bit-true Data Flow Analysis for Processor-Specific Source Code Optimization. In *4th IEEE Workshop on Embedded Systems for Real-Time Multimedia (ESTIMedia)*, pages 133–138, Seoul/Korea.

[Fettweis et al., 1998] Fettweis, G., Weiss, M., Drescher, W., Walther, U., Engel, F., Kobayashi, S., and Richter, T. (1998). Breaking new grounds over 3000 MMAC/s: a broadband mobile multimedia modem DSP. *Intern. Conf. on Signal Processing Application & Technology (IC-SPA)*, available at *http://citeseerx.ist.psu.edu/viewdoc/summary?doi=10.1.1.50.9340*.

[Fiorin et al., 2007] Fiorin, L., Palermo, G., Lukovic, S., and Silvano, C. (2007). A data protection unit for NoC-based architectures. In *5th IEEE/ACM Int. Conf. on Hardware/software Codesign and System Synthesis (CODES+ISSS)*, pages 167–172.

[Fisher and Dietz, 1998] Fisher, R. and Dietz, H. G. (1998). Compiling for SIMD within a single register. *Annual Workshop on Lang. & Compilers for Parallel Computing (LCPC)*, pages 290–304.

[Fisher and Dietz, 1999] Fisher, R. J. and Dietz, H. G. (1999). The Scc compiler: SWARing at MMX and 3DNow! *Annual Workshop on Lang. & Compilers for Parallel Computing (LCPC)*, pages 399–414.

[FlexRay Consortium, 2002] FlexRay Consortium (2002). Flexray® requirement specification. version 2.01. *http://www.flexray.de*.

[Fowler and Scott, 1998] Fowler, M. and Scott, K. (1998). *UML Distilled - Applying the Standard Object Modeling Language*. Addison-Wesley.

[Franke, 2008] Franke, B. (2008). Fast cycle-approximate instruction set simulation. In *10th Int. Workshop on Software & Compilers for Embedded Systems (SCOPES)*, pages 69–78.

[Franke and O'Boyle, 2005] Franke, B. and O'Boyle, M. F. (2005). A complete compiler approach to auto-parallelizing C programs for multi-DSP systems. *IEEE Transactions on Parallel and Distributed Systems*, 16:234–245.

[Freescale semiconductor, 2005] Freescale semiconductor (2005). ColdFire® family programmer's reference manual. *http://www.freescale.com/files/dsp/doc/ref_manual/CFPRM.pdf*.

[Fu et al., 1987] Fu, K., Gonzalez, R., and Lee, C. (1987). *Robotics*. McGraw-Hill.

[Gajski and Kuhn, 1983] Gajski, D. and Kuhn, R. (1983). New VLSI tools. *IEEE Computer*, pages 11–14.

[Gajski et al., 1994] Gajski, D., Vahid, F., Narayan, S., and Gong, J. (1994). *Specification and Design of Embedded Systems*. Prentice Hall.

[Gajski et al., 2000] Gajski, D., Zhu, J., Dömer, R., Gerstlauer, A., and Zhao, S. (2000). *SpecC: Specification Language Methodology*. Kluwer Academic Publishers.

[Gajski et al., 2009] Gajski, D. D., Abdi, S., Gerstlauer, A., and Schirner, G. (2009). *Embedded System Design*. Springer, Heidelberg.

[Ganssle, 2008] Ganssle, J., editor (2008). *Embedded Systems (World Class Designs)*. Newnes.

[Ganssle, 2000] Ganssle, J. G. (2000). *The Art of Designing Embedded Systems*. Newnes.

[Ganssle et al., 2008] Ganssle, J. G., Noergaard, T., Eady, F., Edwards, L., Katz, D. J., Gentile, R., Arnold, K., Hyder, K., and Perrin, B. (2008). *Embedded Hardware - Know it all*. Newnes.

[Garey and Johnson, 1979] Garey, M. R. and Johnson, D. S. (1979). *Computers and Intractability*. Bell Labaratories, Murray Hill, New Jersey.

[Garg and Khatri, 2009] Garg, R. and Khatri, S. (2009). *Analysis and Design of Resilient VLSI Circuits*. Springer.

[Gebotys, 2010] Gebotys, C. (2010). *Security in Embedded Devices*. Springer.

[Geffroy and Motet, 2002] Geffroy, J.-C. and Motet, G. (2002). *Design of Dependable computing Systems*. Kluwer Academic Publishers.

[Gelsen, 2003] Gelsen, O. (2003). Organic displays enter consumer electronics. *Opto & Laser Europe, June; availabe at http://optics.org/cws/article/articles/17598*.

[Gerber et al., 2005] Gerber, R., Bik, A. J. C., Smith, K., and Tian, X. (2005). *The Software Optimization Cookbook Second Edition. High Performance Recipes for IA 32 Platforms*. Intel Press.

[Gomez and Fernandes, 2010] Gomez, L. and Fernandes, J. (2010). *Behavioral Modeling for Embedded Systems and Technologies*. IGI Global.

[Grötker et al., 2002] Grötker, T., Liao, S., and Martin, G. (2002). *System design with SystemC*. Springer.

[Gupta, 2002] Gupta, R. (2002). Tasks and task management. *Course ICS 212, Winter 2002, UC Irvine, http://citeseerx.ist.psu.edu/viewdoc/download?doi=10.1.1.7.8704&rep=rep1&type=pdf*.

[Ha, 2007] Ha, S. (2007). Model-based programming environment of embedded software for mpsoc. *Asia and South Pacific Design Automation Conference (ASP-DAC)*, pages 330–335.

[Halbwachs, 1998] Halbwachs, N. (1998). Synchronous programming of reactive systems, a tutorial and commented bibliography. *Tenth International Conference on Computer-Aided Verification, CAV'98, LNCS 1427, Springer Verlag; see also: http://www.springerlink.com/content/5127074271136j71/fulltext.pdf.*

[Halbwachs, 2008] Halbwachs, N. (2008). Personal communication. *South American Artist School on Embedded Systems, Florianopolis.*

[Halbwachs et al., 1991] Halbwachs, N., Caspi, P., Raymond, P., and Pilaud, D. (1991). The synchronous dataflow language LUSTRE. *Proc. of the IEEE Trans. on Software Engineering,* 79:1305–1320.

[Hansmann, 2001] Hansmann, U. (2001). *Pervasive Computing.* Springer Verlag.

[Harbour, 1993] Harbour, M. G. (1993). RT-POSIX: An overview. *http://www.ctr.unican.es/publications/mgh-1993a.pdf.*

[Harel, 1987] Harel, D. (1987). StateCharts: A visual formalism for complex systems. *Science of Computer Programming,* pages 231–274.

[Hattori, 2007] Hattori, T. (2007). MPSoC approaches for low-power embedded SoC's. *Int. Forum on. Application Specific Multi Processor SoC, http://www.mpsoc-forum.org/2007/slides/Hattori.pdf.*

[Haugen and Moller-Pedersen, 2006] Haugen, O. and Moller-Pedersen, B. (2006). Introduction to UML and the modeling of embedded systems. *In: R. Zurawski (ed.): Embedded Systems Handbook, CRC Press.*

[Hayes, 1982] Hayes, J. (1982). A unified switching theory with applications to VLSI design. *Proceedings of the IEEE, Vol. 70,* pages 1140–1151.

[Heath, 2000] Heath, S. (2000). *Embedded System Design.* Newnes.

[Henia et al., 2005] Henia, R., Hamann, A., Jersak, M., Racu, R., Richter, K., and Ernst, R. (2005). System level performance analysis - the SymTA/S approach. *IEEE Computers and Digital Techniques,* pages 148–166.

[Hennessy and Patterson, 2002] Hennessy, J. L. and Patterson, D. A. (2002). *Computer Architecture – A Quantitative Approach.* Morgan Kaufmann Publishers Inc.

[Hennessy and Patterson, 2008] Hennessy, J. L. and Patterson, D. A. (2008). *Computer Organization – The Hardware/Software Interface.* Morgan Kaufmann Publishers Inc.

[Herken, 1995] Herken, R. (1995). *The Universal Turing Machine: A half-century survey.* Springer.

[Herrera et al., 2003a] Herrera, F., Fernández, V., Sánchez, P., and Villar, E. (2003a). Embedded software generation from SystemC for platform based design. *In: [Müller et al., 2003],* pages 247–272.

[Herrera et al., 2003b] Herrera, F., Posadas, H., Sánchez, P., and Villar, E. (2003b). Systemic embedded software generation from SystemC. *Design, Automation and Test in Europe (DATE),* pages 10142–10149.

[Hoare, 1985] Hoare, C. (1985). *Communicating Sequential Processes*. Prentice Hall International Series in Computer Science.

[Hopcroft et al., 2006] Hopcroft, J., Motwani, R., and Ullman, J. D. (2006). *Introduction to Automata Theory, Languages, and Computation*. Addison Wesley.

[Horn, 1974] Horn, W. (1974). Some simple scheduling algorithms. *Naval Research Logistics Quarterly, Vol. 21*, pages 177–185.

[Huang and Xu, 2010] Huang, L. and Xu, Q. (2010). AgeSim: A simulation framework for evaluating the lifetime reliability of processor-based SoCs. *Design, Automation and Test in Europe (DATE)*.

[Huerlimann, 2003] Huerlimann, D. (2003). Opentrack home page. *http://www.opentrack.ch*.

[Hüls, 2002] Hüls, T. (2002). Optimizing the energy consumption of an MPEG application (in German). *Master thesis, CS Dept., Univ. Dortmund, http://ls12-www.cs.uni-dortmund.de/publications/theses*.

[Hunt et al., 2007] Hunt, V. D., Puglia, A., and Puglia, M. (2007). *RFID: a guide to radio frequency identification*. Wiley.

[IBM, 2002] IBM (2002). Security: User authentication. *http://www.pc.ibm.com/us/security/userauth.html*.

[IBM, 2009] IBM (2009). What's new in Rational Rhapsody 7.5.1. *http://www.ibm.com/developerworks/rational/library/09/whatsnewinrationalrhapsody-7-5-1*.

[IBM, 2010a] IBM (2010a). IBM Rational StateMate. *http://www.ibm.com/developerworks/rational/products/statemate/*.

[IBM, 2010b] IBM (2010b). Rational DOORS. *http://www-01.ibm.com/software/awdtools/doors/*.

[ICD Staff, 2010] ICD Staff (2010). ICD-C compiler framework. *http://www.icd.de/es/icd-c*.

[IEC, 2002] IEC (2002). IEC 60848 – GRAFCET specification language for sequential function charts. *http://webstore.iec.ch/preview/info_iec60848{ed2.0}b.pdf*.

[IEEE, 1991] IEEE (1991). IEEE graphic symbols for logic functions std 91a-1991. *http://ieeexplore.ieee.org/stamp/stamp.jsp?tp=&arnumber=27895*.

[IEEE, 1997] IEEE (1997). *IEEE Standard VHDL Language Reference Manual (1076-1997)*. IEEE.

[IEEE, 2002] IEEE (2002). *IEEE Standard VHDL Language Reference Manual (1076-2002)*. IEEE.

[IEEE, 2009] IEEE (2009). IEEE Standard for SystemVerilog- unified hardware design, specification, and verification language. *http://www.ieee.org*.

[IMEC, 1997] IMEC (1997). LIC-SMARTpen identifies signer. *IMEC Newsletter, http://www2.imec.be/content/user/File/Newsletters/newsletter_18.pdf*.

[IMEC, 2010] IMEC (2010). ADRES multimedia processor & 3mf multimedia platform. *http://www2.imec.be/content/user/File/ADRES_3MF.pdf.*

[Intel, 2004] Intel (2004). Enhanced Intel® SpeedStep® Technology for the Intel® Pentium® M Processor - White paper. *ftp://download.intel.com/design/network/papers/30117401.pdf.*

[Intel, 2008] Intel (2008). Motion estimation with Intel® streaming SIMD extensions 4 (Intel® SSE4). *http://software.intel.com/en-us/articles/motion-estimation-with-intel-streaming-simd-extensions-4-intel-sse4.*

[Intel, 2010a] Intel (2010a). Intel® AVX. *http://software.intel.com/en-us/avx.*

[Intel, 2010b] Intel (2010b). Intel Itanium processor family. *http://www.intel.com/itcenter/products/itanium.*

[Ishihara and Yasuura, 1998] Ishihara, T. and Yasuura, H. (1998). Voltage scheduling problem for dynamically variable voltage processors. *Intern. Symp. on Low Power Electronics and Design (ISLPED)*, pages 197–202.

[Israr and Huss, 2008] Israr, A. and Huss, S. (2008). Specification and design considerations for reliable embedded systems. *Design, Automation and Test in Europe (DATE)*, pages 1111–1116.

[IT Facts, 2010] IT Facts (2010). Home page. *http://www.itfacts.biz.*

[ITRS Organization, 2009] ITRS Organization (2009). International technology roadmap for semiconductors (ITRS). *http://public.itrs.net.*

[Iyer and Marculescu, 2002] Iyer, A. and Marculescu, D. (2002). Power and performance evaluation of globally asynchronous locally synchronous processors. *Int. Symp. on Computer Architecture (ISCA)*, pages 158–168.

[Jackson, 1955] Jackson, J. (1955). Scheduling a production line to minimize maximum tardiness. *Management Science Research Project 43, University of California, Los Angeles.*

[Jackson et al., 2009] Jackson, J., Marwedel, P., and Ricks, K. (2009). Workshop on embedded system education. *http://www.artist-embedded.org/artist/WESE-09.html.*

[Jacome et al., 2000] Jacome, M., de Veciana, G., and Lapinskii, V. (2000). Exploring performance tradeoffs for clustered VLIW ASIPs. *IEEE Int. Conf. on Computer-Aided Design (ICCAD)*, pages 504–510.

[Jacome and de Veciana, 1999] Jacome, M. F. and de Veciana, G. (1999). Lower bound on latency for VLIW ASIP datapaths. *IEEE Int. Conf. on Computer-Aided Design (ICCAD)*, pages 261–269.

[Jain et al., 2001] Jain, M., Balakrishnan, M., and Kumar, A. (2001). ASIP design methodologies: Survey and issues. *14th Int. Conf. on VLSI Design*, pages 76–81.

[Janka, 2002] Janka, R. (2002). *Specification and Design Methodology for Real-Time Embedded Systems*. Kluwer Academic Publishers.

[Jantsch, 2004] Jantsch, A. (2004). *Modeling Embedded Systems and SoC's: Concurrency and Time in Models of Computation*. Morgan Kaufmann.

[Jantsch, 2006] Jantsch, A. (2006). Models of embedded computation. *In: R. Zurawski (ed.): Embedded Systems Handbook, CRC Press*.

[Java Community Process, 2002] Java Community Process (2002). JSR-1 – real-time specification for Java. *http://www.jcp.org/en/jsr/detail?id=1*.

[Jewett and Serway, 2007] Jewett, J. W. and Serway, R. A. (2007). *Physics for scientists and engineers with modern physics*. Thomson Higher Education.

[Jha and Dutt, 1993] Jha, P. and Dutt, N. (1993). Rapid estimation for parameterized components in high-level synthesis. *IEEE Transactions on VLSI Systems*, pages 296–303.

[Johnson, 2010] Johnson, S. C. (2010). The Lex & Yacc Page. *http://dinosaur.compilertools. net*.

[Jones, 1997] Jones, M. (1997). What really happened on Mars Rover Pathfinder. *In: P.G. Neumann (ed.): comp.risks, The Risks Digest, Vol. 19, Issue 49; available at http://research. microsoft.com/en-us/um/people/mbj/mars_pathfinder/Mars_Pathfinder.html*.

[Jones, 1996] Jones, N. D. (1996). An introduction to partial evaluation. *ACM Comput. Surv.*, 28(3):480–503.

[JXTA Community, 2010] JXTA Community (2010). Home page. *https://jxta.dev.java.net*.

[Kahn, 1974] Kahn, G. (1974). The semantics of a simple language for parallel programming. *Proc. of the Int. Federation for Information Processing (IFIP)*, pages 471–475.

[Kandemir et al., 2001] Kandemir, M., Ramanujam, J., Irwin, M. J., Vijaykrishnan, N., Kadayif, I., and Parikh, A. (2001). Dynamic management of scratch-pad memory space. *38th annual Design Automation Conference (DAC)*, pages 690–695.

[Karp and Miller, 1966] Karp, R. M. and Miller, R. E. (1966). Properties of a model for parallel computations: Determinancy, termination, queueing. *SIAM Journal of Applied Mathematics*, 14:1390–1411.

[Keding et al., 1998] Keding, H., Willems, M., Coors, M., and Meyr, H. (1998). FRIDGE: A fixed-point design and simulation environment. *Design, Automation and Test in Europe (DATE)*, pages 429–435.

[Keinert et al., 2009] Keinert, J., Streubühr, M., Schlichter, T., Falk, J., Gladigau, J., Haubelt, C., Teich, J., and Meredith, M. (2009). SystemCodesigner - an automatic ESL synthesis approach by design space exploration and behavioral synthesis for streaming applications. *ACM Transactions on Design Automation of Electronic Systems*, 14:1–23.

[Kempe, 1995] Kempe, M. (1995). Ada 95 reference manual, ISO/IEC standard 8652:1995. *(HTML-version), http://www.adahome.com/rm95/*.

[Kempe Software Capital Enterprises (KSCE), 2010] Kempe Software Capital Enterprises (KSCE) (2010). Ada home: The web site for Ada. *http://www.adahome.com*.

[Kernighan and Ritchie, 1988] Kernighan, B. W. and Ritchie, D. M. (1988). *The C Programming Language*. Prentice Hall.

[Kienhuis et al., 2000] Kienhuis, B., Rijjpkema, E., and Deprettere, E. (2000). Compaan: Deriving process networks from Matlab for embedded signal processing architectures. *Proc. 8th Int. Workshop on Hardware/Software Codesign (CODES)*, pages 29–40.

[Klaiber, 2000] Klaiber, A. (2000). The technology behind Crusoe^{TM} processors. *http://web.archive.org/web/20010602205826/www.transmeta.com/crusoe/download/pdf/crusoetechwp.pdf.*

[Könemann et al., 1979] Könemann, B., Mucha, J., and Zwiehoff, G. (1979). Built-in logic block observer. *IEEE Int. Test Conf.*, pages 261–266.

[Ko and Koo, 1996] Ko, M. and Koo, I. (1996). An overview of interactive video on demand system. *www.ece.ubc.ca/˜irenek/techpaps/vod/vod.html.*

[Kobryn, 2001] Kobryn, C. (2001). UML 2001: A standardization Odyssey. *Communications of the ACM (CACM), available at http://www.omg.org/attachments/pdf/UML_2001_CACM_Oct99_p29-Kobryn.pdf,* pages 29–36.

[Kohavi, 1987] Kohavi, Z. (1987). *Switching and Finite Automata Theory.* Tata McGraw-Hill Publishing Company, New Delhi, 9th reprint.

[Koninklijke Philips Electronics N.V., 2003] Koninklijke Philips Electronics N.V. (2003). Ambient intelligence. *http://www.research.philips.com/technologies/projects/ambintel.html.*

[Koopman and Upender, 1995] Koopman, P. J. and Upender, B. P. (1995). Time division multiple access without a bus master. *United Technologies Research Center, UTRC Technical Report RR-9500470, http://www.ece.cmu.edu/˜koopman/jtdma/jtdma.html.*

[Kopetz, 1997] Kopetz, H. (1997). *Real-Time Systems – Design Principles for Distributed Embedded Applications.* Kluwer Academic Publishers.

[Kopetz, 2003] Kopetz, H. (2003). Architecture of safety-critical distributed real-time systems. *Invited Talk; Design, Automation and Test in Europe (DATE).*

[Kopetz and Grunsteidl, 1994] Kopetz, H. and Grunsteidl, G. (1994). TTP –a protocol for fault-tolerant real-time systems. *IEEE Computer*, 27:14–23.

[Krall, 2000] Krall, A. (2000). Compilation techniques for multimedia extensions. *International Journal of Parallel Programming*, 28:347–361.

[Kranitis et al., 2003] Kranitis, N., Paschalis, A., Gizopoulos, D., and Zorian, Y. (2003). Instruction-based self-testing of processor cores. *Journal of Electronic Testing*, 19:103–112.

[Krhovjak and Matyas, 2006] Krhovjak, J. and Matyas, V. (2006). Secure hardware - pv018. *http://www.fi.muni.cz/˜xkrhovj/lectures/2006_PV018_Secure_Hardware_slides.pdf.*

[Krishna and Shin, 1997] Krishna, C. and Shin, K. G. (1997). *Real-Time Systems.* McGraw-Hill, Computer Science Series.

[Krstić and Cheng, 1998] Krstić, A. and Cheng, K. (1998). *Delay fault testing of VLSI circuits.* Kluwer Academic Publishers.

[Krstic and Dey, 2002] Krstic, A. and Dey, S. (2002). Embedded software-based self-test for programmable core-based designs. *IEEE Design & Test*, pages 18–27.

[Krüger, 1986] Krüger, G. (1986). Automatic generation of self-test programs: A new feature of the MIMOLA design system. *23rd annual Design Automation Conference (DAC)*, pages 378–384.

[Kuchcinski, 2002] Kuchcinski, K. (2002). System partitioning (course notes). *http://www.cs. lth.se/home/Krzysztof_Kuchcinski/DES/Lectures/Lecture7.pdf.*

[Kwok and Ahmad, 1999] Kwok, Y.-K. and Ahmad, I. (1999). Static scheduling algorithms for allocation directed task graphs to multiprocessors. *ACM Computing Surveys*, 31:406–471.

[Labrosse, 2000] Labrosse, J. (2000). *Embedded Systems Building Blocks - Complete and Ready-to-use Modules in C.* Elsevier.

[Lala, 1985] Lala, P. (1985). *Fault tolerant and Fault Testable Hardware Design.* Prentice Hall.

[Lam et al., 1991] Lam, M. S., Rothberg, E. E., and Wolf, M. E. (1991). The cache performance and optimizations of blocked algorithms. *Architectural Support for Programming Languages and Operating Systems (ASPLOS)*, pages 63–74.

[Landwehr and Marwedel, 1997] Landwehr, B. and Marwedel, P. (1997). A new optimization technique for improving resource exploitation and critical path minimization. *10th Int. Symp. on System Synthesis (ISSS)*, pages 65–72.

[Lapinskii et al., 2001] Lapinskii, V., Jacome, M. F., and de Veciana, G. (2001). Application-specific clustered VLIW datapaths: Early exploration on a parameterized design space. Technical Report UT-CERC-TR-MFJ/GDV-01-1, Computer Engineering Research Center, University of Texas at Austin.

[Laplante, 1997] Laplante, P. (1997). *Real-Time Systems: Design and Analysis - An Engineer's Handbook.* IEEE Press.

[Laprie, 1992] Laprie, J. C., editor (1992). *Dependability: basic concepts and terminology in English, French, German, Italian and Japanese.* IFIP WG 10.4, Dependable Computing and Fault Tolerance, In: volume 5 of Dependable computing and fault tolerant systems, Springer Verlag.

[Larsen and Amarasinghe, 2000] Larsen, S. and Amarasinghe, S. (2000). Exploiting superword parallelism with multimedia instructions sets. *Programming Language Design and Implementation (PLDI)*, pages 145–156.

[Latendresse, 2004] Latendresse, M. (2004). The code compression bibliography. *http://www. iro.umontreal.ca/~latendre/compactBib.*

[Law, 2006] Law, A. M. (2006). *Simulation Modeling & Analysis.* McGraw-Hill.

[Lawler, 1973] Lawler, E. L. (1973). Optimal sequencing of a single machine subject to precedence constraints. *Managements Science, Vol. 19*, pages 544–546.

[Le Boudec and Thiran, 2001] Le Boudec, J. and Thiran, P. (2001). *Network Calculus.* Springer, LNCS # 2050.

[Lee, 1999] Lee, E. A. (1999). Embedded software – an agenda for research. Technical report, UCB ERL Memorandum M99/63.

[Lee, 2006] Lee, E. A. (2006). The future of embedded software. *ARTEMIS Conference, Graz, http://ptolemy.eecs.berkeley.edu/presentations/06/FutureOfEmbeddedSoftware_Lee_ Graz.ppt.*

[Lee, 2007] Lee, E. A. (2007). Computing foundations and practice for cyber-physical systems: A preliminary report. Technical Report UCB/EECS-2007-72, EECS Department, University of California, Berkeley.

[Lee, 2005] Lee, E. A. (July, 2005). Absolutely positively on time. *IEEE Computer.*

[Lee and Messerschmitt, 1987] Lee, E. A. and Messerschmitt, D. (1987). Synchronous data flow. *Proc. of the IEEE, vol. 75*, pages 1235–1245.

[Lee et al., 2001] Lee, S., Ermedahl, A., and Min, S. (2001). An accurate instruction-level energy consumption model for embedded ROSC processors. *ACM SIGPLAN Conference on Languages, Compilers, and Tools for Embedded Systems (LCTES)*, pages 1–10.

[Leupers, 1997] Leupers, R. (1997). *Retargetable Code Generation for Digital Signal Processors.* Kluwer Academic Publishers.

[Leupers, 1999] Leupers, R. (1999). Exploiting conditional instructions in code generation for embedded VLIW processors. *Design, Automation and Test in Europe (DATE)*, pages 23–27.

[Leupers, 2000a] Leupers, R. (2000a). *Code Optimization Techniques for Embedded Processors - Methods, Algorithms, and Tools.* Kluwer Academic Publishers.

[Leupers, 2000b] Leupers, R. (2000b). Code selection for media processors with SIMD instructions. *Design, Automation and Test in Europe (DATE)*, pages 4–8.

[Leupers, 2000c] Leupers, R. (2000c). Instruction scheduling for clustered VLIW DSPs. *Int. Conf. on Parallel Architectures and Compilation Techniques (PACT)*, pages 291–300.

[Leupers and David, 1998] Leupers, R. and David, F. (1998). A uniform optimization technique for offset assignment problems. *Int. Symp. on System Synthesis (ISSS)*, pages 3–8.

[Leupers and Marwedel, 1995] Leupers, R. and Marwedel, P. (1995). Time-constrained code compaction for DSPs. *Int. Symp. on System Synthesis (ISSS)*, pages 54–59.

[Leupers and Marwedel, 1996] Leupers, R. and Marwedel, P. (1996). Algorithms for address assignment in DSP code generation. *IEEE Int. Conf. on Computer-Aided Design (ICCAD)*, pages 109–112.

[Leupers and Marwedel, 1999] Leupers, R. and Marwedel, P. (1999). Function inlining under code size constraints for embedded processors. *IEEE Int. Conf. on Computer-Aided Design (ICCAD)*, pages 253–256.

[Leupers and Marwedel, 2001] Leupers, R. and Marwedel, P. (2001). *Retargetable Compiler Technology for Embedded Systems – Tools and Applications.* Kluwer Academic Publishers.

[Leveson, 1995] Leveson, N. (1995). *Safeware, System Safety and Computers.* Addison Wesley.

[Lewis et al., 2007] Lewis, J., Rashba, E., and Brophy, D. (2007). VHDL-2006-D3.0 Tutorial. *Tutorial at Design, Automation, and Test in Europe (DATE), http://www.accellera.org/apps/ group_public/download.php/934/date_vhdl_tutorial.pdf.*

[Liao et al., 1995a] Liao, S., Devadas, S., Keutzer, K., and Tijang, S. (1995a). Code optimization techniques for embedded DSP microprocessors. *32nd Design Automation Conference (DAC)*, pages 599–604.

[Liao et al., 1995b] Liao, S., Devadas, S., Keutzer, K., Tijang, S., and Wang, A. (1995b). Storage assignment to decrease code size. *Programming Language Design and Implementation (PLDI)*, pages 186–195.

[Liebisch and Jain, 1992] Liebisch, D. C. and Jain, A. (1992). Jessi common framework design management: the means to configuration and execution of the design process. In *Conf. on European Design Automation (EURO-DAC)*, pages 552–557. IEEE Computer Society Press.

[LIN Administration, 2010] LIN Administration (2010). Home page. *http://www.lin-subbus. org/*.

[Liu and Layland, 1973] Liu, C. L. and Layland, J. W. (1973). Scheduling algorithms for multi-programming in a hard real-time environment. *Journal of the Association for Computing Machinery (JACM)*, pages 40–61.

[Liu, 2000] Liu, J. W. (2000). *Real-Time Systems*. Prentice Hall.

[Lohmann et al., 2009] Lohmann, D., Hofer, W., Schröder-Preikschat, W., and Spinczyk, O. (2009). CiAO: An aspect-oriented operating-system family for resource-constrained embedded systems. In *USENIX Annual Technical Conference*.

[Lohmann et al., 2006] Lohmann, D., Scheler, F., Schröder-Preikschat, W., and Spinczyk, O. (2006). PURE Embedded Operating Systems - CiAO. *Proc. International Workshop on Operating System Platforms for Embedded Real-Time Applications, (OSPERT)*.

[Lokuciejewski et al., 2009] Lokuciejewski, P., Gedikli, F., Marwedel, P., and Morik, K. (2009). Automatic WCET Reduction by Machine Learning Based Heuristics for Function Inlining. In *3rd Workshop on Statistical and Machine Learning Approaches to Architectures and Compilation (SMART)*, pages 1–15.

[Lokuciejewski and Marwedel, 2010] Lokuciejewski, P. and Marwedel, P. (2010). *WCET-aware Source Code and Assembly Level Optimization Techniques for Real-Time Systems*. Springer.

[Lorenz et al., 2004] Lorenz, M., Marwedel, P., Dräger, T., Fettweis, G., and Leupers, R. (2004). Compiler based exploration of DSP energy savings by SIMD operations. In *ASP-DAC '04: Proceedings of the 2004 Asia and South Pacific Design Automation Conference*, pages 838–841, Piscataway, NJ, USA. IEEE Press.

[Lorenz et al., 2002] Lorenz, M., Wehmeyer, L., Draeger, T., and Leupers, R. (2002). Energy aware compilation for DSPs with SIMD instructions. *LCTES/SCOPES*, pages 94–101.

[Lu et al., 2000] Lu, Y.-H., Chung, E.-Y., Šimunic, T., Benini, L., and De Micheli, G. (2000). Quantitative comparison of power management algorithms. In *Design, Automation and Test in Europe (DATE)*, pages 20–26.

[Machanik, 2002] Machanik, P. (2002). Approaches to addressing the memory wall. *Technical Report, November, Univ. Brisbane*.

[Macii et al., 2002] Macii, A., Benini, L., and Poncino, M. (2002). *Memory Design Techniques for Low Energy Embedded Systems*. Kluwer Academic Publishers.

[Macii, 2004] Macii, E., editor (2004). *Ultra low-power electronics and design.* Springer.

[Mahlke et al., 1992] Mahlke, S. A., Lin, D. C., Chen, W. Y., Hank, R. E., and Bringmann, R. A. (1992). Effective compiler support for predicated execution using the hyperblock. *25th annual Int. Symp. on Microarchitecture (MICRO)*, pages 45–54.

[Man, 2007] Man, H. D. (2007). From the heaven of software to the hell of nanoscale physics: an industry in transition... *Keynote, HiPEAC ACACES Summer School, L'Aquila.*

[Marian and Ma, 2007] Marian, N. and Ma, Y. (2007). Translation of Simulink models to component-based software models. *8th Int. Workshop on Research and Education in Mechatronics REM, http://seg.mci.sdu.dk/publications/Translation%20of%20Simulink%20Models%20to%20Component-based%20Software%20Models.pdf*, pages 262–267.

[Marongiu and Benini, 2009] Marongiu, A. and Benini, L. (2009). Efficient OpenMP support and extensions for MPSoCs with explicitly managed memory hierarchy. *Design, Automation and Test in Europe (DATE)*, pages 809–814.

[Martin and Müller, 2005] Martin, G. and Müller, W., editors (2005). *UML^{TM} for SoC Design.* Springer.

[Martin et al., 2002] Martin, S. M., Flautner, K., Mudge, T., and Blaauw, D. (2002). Combined dynamic voltage scaling and adaptive body biasing for lower power microprocessors under dynamic workloads. In *ICCAD '02: Proceedings of the 2002 IEEE/ACM international conference on Computer-aided design*, pages 721–725, New York, NY, USA. ACM.

[Marwedel, 1990] Marwedel, P. (1990). A software system for the synthesis of computer structures and the generation of microcode (in German). *habilitation thesis, Universität Kiel, 1985, Reprint: Report Nr.356, CS Dept., TU Dortmund.*

[Marwedel, 2003] Marwedel, P. (2003). *Embedded System Design.* Kluwer Academic Publishers.

[Marwedel, 2005] Marwedel, P. (2005). Towards laying common grounds for embedded system design education. *ACM SIGBED Review*, pages 25–28.

[Marwedel, 2007] Marwedel, P. (2007). Memory-architecture aware compilation. *Tutorial, HiPEAC ACACES Summer School, L'Aquila, http://ls12-www.cs.tu-dortmund.de/publications/papers/2007-marwedel-acaces.zip.*

[Marwedel, 2008a] Marwedel, P. (2008a). 1st workshop on mapping of applications to MPSoCs. *http://www.artist-embedded.org/artist/-map2mpsoc-2008-.html.*

[Marwedel, 2008b] Marwedel, P. (2008b). MIMOLA - a fully synthesizable language. *in: Prabhat Mishra, Nikil Dutt (Ed.): Processor Description Languages - Applications and Methodologies, Morgan Kaufmann*, pages 35–63.

[Marwedel, 2009a] Marwedel, P. (2009a). 2nd workshop on mapping of applications to MPSoCs. *http://www.artist-embedded.org/artist/-map2mpsoc-2009-.html.*

[Marwedel, 2009b] Marwedel, P. (2009b). Mapping of applications to MPSoCs. *IP-Embedded Systems Conference, Grenoble, http://ls12-www.cs.tu-dortmund.de/publications/papers/2009-ip-esc-marwedel.pdf.*

[Marwedel and Goossens, 1995] Marwedel, P. and Goossens, G., editors (1995). *Code Generation for Embedded Processors*. Kluwer Academic Publishers.

[Marwedel and Schenk, 1993] Marwedel, P. and Schenk, W. (1993). Cooperation of synthesis, retargetable code generation and test generation in the MSS. *European Design and Test Conf. (EDAC-EUROASIC)*, pages 63–69.

[Marzano and Aarts, 2003] Marzano, S. and Aarts, E. (2003). *The New Everyday*. 010 Publishers.

[Marzario et al., 2004] Marzario, L., Lipari, G., Balbastre, P., and Crespo, A. (2004). IRIS: a new reclaiming algorithm for server-based real-time systems. *Real-Time Application Symposium (RTAS 04)*.

[Massa, 2002] Massa, A. J. (2002). *Embedded Software Development with eCos*. Prentice Hall.

[MathWorks, 2010] MathWorks, T. (2010). Stateflow 7.3. *http://www.mathworks.com/products/stateflow*.

[McGregor, 2002] McGregor, I. (2002). The relationship between simulation and emulation. *Winter Simulation Conference*, pages 1683–1688.

[McLaughlin and Moore, 1998] McLaughlin, M. and Moore, A. (1998). Real-Time Extensions to UML. *http://www.ddj.com/184410749*.

[McNamee et al., 2001] McNamee, D., Walpole, J., Pu, C., Cowan, C., Krasic, C., Goel, A., Wagle, P., Consel, C., Muller, G., and Marlet, R. (2001). Specialization tools and techniques for systematic optimization of system software. *ACM Trans. Comput. Syst.*, 19(2):217–251.

[Meijer et al., 2010] Meijer, S., Nikolov, H., and Stefanov, T. (2010). Throughput modeling to evaluate process merging transformations in polyhedral process networks. *Design, Automation and Test in Europe (DATE)*.

[Menard and Sentieys, 2002] Menard, D. and Sentieys, O. (2002). Automatic evaluation of the accuracy of fixed-point algorithms. *Design, Automation and Test in Europe (DATE)*, pages 529–535.

[Merkel and Bellosa, 2005] Merkel, A. and Bellosa, F. (2005). Event-driven thermal management in SMP systems. *Proceedings of the Second Workshop on Temperature-Aware Computer Systems (TACS'05)*.

[Mermet et al., 1998] Mermet, J., Marwedel, P., Ramming, F. J., Newton, C., Borrione, D., and Lefaou, C. (1998). Three decades of hardware description languages in Europe. *Journal of Electrical Engineering and Information Science*, 3:106pp.

[Mesa-Martinez et al., 2010] Mesa-Martinez, F. J., Ardestani, E. K., and Renau, J. (2010). Characterizing processor thermal behavior. In *ASPLOS '10: Proceedings of the fifteenth edition of ASPLOS on Architectural support for programming languages and operating systems*, pages 193–204, New York, NY, USA. ACM.

[MHPCC, 2010] MHPCC, M. (2010). SP parallel programming workshop - message passing interface (MPI). *http://www.mhpcc.edu/training/workshop/mpi/MAIN.html*.

[Microsoft Inc., 2003] Microsoft Inc. (2003). Windows® embedded home. *http://www.microsoft.com/windowsembedded*.

[Mnemee project, 2010] Mnemee project (2010). Memory maNagEMEnt technology for adaptive and efficient design of Embedded systems. *http://www.mnemee.org*.

[Monteiro and van Leuken, 2010] Monteiro, J. and van Leuken, R., editors (2010). *Integrated circuit and system design: power and timing modeling, optimization and simulation : 19th international workshop, PATMOS 2009*. Springer LNCS 5953.

[MOST Cooperation, 2010] MOST Cooperation (2010). Home page. *http://www.mostcooperation.com/home*.

[MPI/RT forum, 2001] MPI/RT forum (2001). Document for the real-time message passing interface (MPI/RT-1.1). *http://www.mpirt.org/drafts/mpirt-report-18dec01.pdf*.

[Muchnick, 1997] Muchnick, S. S. (1997). *Advanced compiler design and implementation*. Morgan Kaufmann Publishers, Inc.

[Mukherjee, 2008] Mukherjee, S. (2008). *Architecture Design for Soft Errors*. Morgan Kaufmann Publishers Inc., San Francisco, CA, USA.

[Müller, 2007] Müller, W. (2007). UML^{TM} for SoC and embedded systems design. *DATE 2007 Friday Workshop, http://www.c-lab.de/uml-soc/uml-date07/date07-uml-workshop.pdf*.

[Müller et al., 2003] Müller, W., Rosenstiel, W., and Ruf, J. (2003). *SystemC – Methodologies and Applications*. Kluwer Academic Publications.

[National Research Council, 2001] National Research Council (2001). *Embedded, Everywhere*. National Academies Press.

[National Science Foundation, 2010] National Science Foundation (2010). Cyber-Physical Systems (CPS). *http://www.nsf.gov/pubs/2010/nsf10515/nsf10515.htm*.

[National Space-Based Positioning, Navigation, and Timing Coordination Office, 2010] National Space-Based Positioning, Navigation, and Timing Coordination Office (2010). Global positioning system. *http://www.gps.gov*.

[Neumann, 1995] Neumann, P. G. (1995). *Computer Related Risks*. Addison Wesley.

[Neumann, 2010] Neumann, P. G., editor (2010). *The risks digest, forum on the risks to the public in computers and related Systems. http://catless.ncl.ac.uk/risks*.

[Nguyen et al., 2005] Nguyen, N., Dominguez, A., and Barua, R. (2005). Memory allocation for embedded systems with a compile-time-unknown scratch-pad size. *Int. Conf. on Compilers, architectures and synthesis for embedded systems (CASES)*, pages 115–125.

[Niemann, 1998] Niemann, R. (1998). *Hardware/Software Co-Design for Data-Flow Dominated Embedded Systems*. Kluwer Academic Publishers.

[Nikolov et al., 2008] Nikolov, H., Thompson, M., Stefanov, T., Pimentel, A., Polstra, S., Bose, R., Zissulescu, C., and Deprettere, E. (2008). Daedalus: toward composable multimedia MP-SoC design. In *45th annual Design Automation Conference (DAC)*, pages 574–579.

[Nilsen, 1998] Nilsen, K. (1998). Adding real-time capabilities to Java. *Commun. ACM*, 41(6):49–56.

[Northeast Sustainable Energy Association, 2010] Northeast Sustainable Energy Association (2010). Zero-energy building award. *http://zeroenergybuilding.org.*

[Novosel, 2009] Novosel, D. (2009). Timing the power grid. *http://www.pserc.wisc.edu/ documents/general_information/presentations/smartr_grid_executive_forum/.*

[Nuzman et al., 2006] Nuzman, D., Rosen, I., and Zaks, A. (2006). Auto-vectorization of interleaved data for SIMD. In *PLDI '06: Proceedings of the 2006 ACM SIGPLAN conference on Programming language design and implementation*, pages 132–143, New York, NY, USA. ACM.

[Object Management Group (OMG), 2003] Object Management Group (OMG) (2003). CORBA® basics. *http://www.omg.org/gettingstarted/corbafaq.htm.*

[Object Management Group (OMG), 2005a] Object Management Group (OMG) (2005a). Real-time CORBA specification, version 1.2, jan. 2005. *http://www.omg.org/cgi-bin/doc? formal/05-01-04.ps.*

[Object Management Group (OMG), 2005b] Object Management Group (OMG) (2005b). UML^{TM} profile for schedulability, performance, and time specification, version 1.1. *http:// www.omg.org/cgi-bin/doc?formal/05-01-02.pdf.*

[Object Management Group (OMG), 2008] Object Management Group (OMG) (2008). OMG systems modeling language (OMG SysML^{TM}). *http://www.omg.org/spec/SysML/1.1/ changebar/PDF.*

[Object Management Group (OMG), 2009] Object Management Group (OMG) (2009). A UML^{TM} profile for MARTE: Modeling and analysis of real-time embedded systems - 1.0. *http://www.omg.org/spec/MARTE/1.0/PDF.*

[Object Management Group (OMG), 2010a] Object Management Group (OMG) (2010a). Catalog of UML^{TM} profile specifications. *http://www.omg.org/technology/documents/profile_ catalog.htm.*

[Object Management Group (OMG), 2010b] Object Management Group (OMG) (2010b). Unified modeling language (tm) resource page. *http://www.uml.org.*

[O'Neill, 2006] O'Neill, A. (2006). Analog to digital types. *IEEE tv (for members only), http://www.ieee.org/portal/ieeetv/viewer.html?progId=81045.*

[Open SystemC Initiative, 2005] Open SystemC Initiative (2005). IEEE 1666 LRM. *http://www.systemc.org/downloads/lrm.*

[OpenMP Architecture Review Board, 2008] OpenMP Architecture Review Board (2008). OpenMP application program interface. *http://www.openmp.org/mp-documents/spec30. pdf.*

[Oppenheim et al., 2009] Oppenheim, A. V., Schafer, R., and Buck, J. R. (2009). *Digital Signal Processing*. Pearson Higher Education.

[OSEK Group, 2004] OSEK Group (2004). OSEK/VDX - communication (version 3.0.3). *http://portal.osek-vdx.org/files/pdf/specs/osekcom303.pdf.*

[OSEK Group, 2010] OSEK Group (2010). Home page. *http://www.osek-vdx.org.*

[Palkovic et al., 2002] Palkovic, M., Miranda, M., and Catthoor, F. (2002). Systematic power-performance trade-off in MPEG-4 by means of selective function inlining steered by address optimisation opportunities. *Design, Automation and Test in Europe (DATE)*, pages 1072–1079.

[Pan et al., 2010] Pan, S., Hu, Y., and Li, X. (2010). IVF: Characterizing the vulnerability of microprocessor structures to intermittent faults. *Design, Automation and Test in Europe (DATE)*.

[Parker, 1992] Parker, K. P. (1992). *The Boundary Scan Handbook*. Kluwer Academic Press.

[Paulin and Knight, 1987] Paulin, P. and Knight, J. (1987). Force-directed scheduling in automatic data path synthesis. *24th annual Design Automation Conference (DAC)*.

[Petri, 1962] Petri, C. A. (1962). Kommunikation mit Automaten. *Schriften des Rheinisch-Westfälischen Institutes für instrumentelle Mathematik an der Universität Bonn*.

[Pino and Lee, 1995] Pino, J. L. and Lee, E. A. (1995). Hierarchical static scheduling of dataflow graphs onto multiple processors. *IEEE Int. Conf. on Acoustics, Speech, and Signal Processing*, pages 2643–2646.

[Pohl et al., 2005] Pohl, K., Böckle, G., and van der Linden, F. (2005). *Software Product Line Engineering*. Springer, ISBN-10: 3540289011.

[Popovici et al., 2010] Popovici, K., Rousseau, F., Jerraya, A. A., and Wolf, M. (2010). *Embedded Software Design and Programming of Multiprocessor System-on-Chip*. Springer.

[Potop-Butucaru et al., 2006] Potop-Butucaru, D., de Simone, R., and Talpin, J.-P. (2006). The synchronous hypothesis and synchronous languages. *In: R. Zurawski (ed.): Embedded Systems Handbook, CRC Press*.

[Press, 2003] Press, D. (2003). *Guidelines for Failure Mode and Effects Analysis for Automotive, Aerospace and General Manufacturing Industries*. CRC Press.

[Pyka et al., 2007] Pyka, R., Faßbach, C., Verma, M., Falk, H., and Marwedel, P. (2007). Operating system integrated energy aware scratchpad allocation strategies for multi-process applications. *Int. Workshop on Software & Compilers for Embedded Systems (SCOPES)*, pages 41–50.

[Quilleré and Rajopadhye, 2000] Quilleré, F. and Rajopadhye, S. (2000). Optimizing memory usage in the polyhedral model. *ACM Transactions on Programming Languages and Systems*, 22:773–815.

[Radetzki, 2009] Radetzki, M., editor (2009). *Languages for Embedded Systems and their Applications*. Springer.

[Ramamritham, 2002] Ramamritham, K. (2002). System support for real-time embedded systems. *In: Tutorial 1, 39th Design Automation Conference (DAC)*.

[Ramamritham et al., 1998] Ramamritham, K., Shen, C., Gonzalez, O., Sen, S., and Shirgurkar, S. B. (1998). Using Windows NT for real-time applications: Experimental observations and recommendations. *IEEE Real-Time Technology and Applications Symposium (RTAS)*, pages 102–111.

[Reisig, 1985] Reisig, W. (1985). *Petri nets*. Springer Verlag.

[Ren et al., 2006] Ren, G., Wu, P., and Padua, D. (2006). Optimizing data permutations for SIMD devices. *ACM SIGPLAN Notices*, 41(6):118–131.

[Riccobene et al., 2005] Riccobene, E., Scandurra, P., Rosti, A., and Bocchio, S. (2005). A UML^{TM} 2.0 profile for SystemC: toward high-level SoC design. In *5th ACM Int. Conf. on Embedded Software (EMSOFT)*, pages 138–141.

[Rixner et al., 2000] Rixner, S., Dally, W. J., Khailany, B. J., Mattson, P. J., and Kapasi, U. J. (2000). Register organization for media processing. *6th High-Performance Computer Architecture (HPCA-6)*, pages 375–386.

[Ruggiero and Benini, 2008] Ruggiero, M. and Benini, L. (2008). Mapping task graphs to the CELL BE processor. *http://www.artist-embedded.org/docs/Events/2008/Map2MPSoC/ Map2mpsoc-08-ruggiero.pdf*.

[Russell and Jacome, 1998] Russell, T. and Jacome, M. F. (1998). Software power estimation and optimization for high performance, 32-bit embedded processors. *Int. Conf. on Computer Design (ICCD)*, pages 328–333.

[Ryan, 1995] Ryan, M. (1995). Market focus – insight into markets that are making the news in EE Times. *EE Times (was available at http://eetimes.com/columns/mfocus95/mfocus11. html)*.

[Sangiovanni-Vincentelli, 2002] Sangiovanni-Vincentelli, A. (2002). The context for platform-based design. *IEEE Design & Test of Computers*, page 120.

[Schmitz et al., 2002] Schmitz, M., Al-Hashimi, B., and Eles, P. (2002). Energy-efficient mapping and scheduling for DVS enabled distributed embedded systems. *Design, Automation and Test in Europe (DATE)*, pages 514–521.

[SDL Forum Society, 2009] SDL Forum Society (2009). List of commercial tools. *http:// www.sdl-forum.org/Tools/Commercial.htm*.

[SDL Forum Society, 2010] SDL Forum Society (2010). Home page. *http://www.sdl-forum. org*.

[Sha et al., 1990] Sha, L., Rajkumar, R., and Lehoczky, J. (1990). Priority inheritance protocols: An approach to real-time synchronisation. *IEEE Trans. on Computers*, pages 1175–1185.

[Shi and Brodersen, 2003] Shi, C. and Brodersen, R. (2003). An automated floating-point to fixed-point conversion methodology. *Int. Conf. on Audio Speed and Signal Processing (ICASSP)*, pages 529–532.

[Siemens, 2010] Siemens (2010). Simatic step 7 programming software. *http://www. automation.siemens.com/simatic/industriesoftware/html_76/products/step7.htm*.

[Sifakis, 2008] Sifakis, J. (2008). A notion of expressiveness for component-based design. *Workshop on Foundations and Applications of Component-based Design, ES-Week, http://www.artist-embedded.org/docs/Events/2008/Components/SLIDES/12-JosephSifakis-WFCD-ArtistDesign-Oct192008.pdf*.

[Simple Scalar LLC, 2004] Simple Scalar LLC (2004). Home page. *http://www.simplescalar. com*.

[Simunic et al., 2000] Simunic, T., Benini, L., Acquaviva, A., Glynn, P., and De Micheli, G. (2000). Energy efficient design of portable wireless devices. *Intern. Symp. on Low Power Electronics and Design (ISLPED)*, pages 49–54.

[Simunic et al., 2001] Simunic, T., Benini, L., Acquaviva, A., Glynn, P., and De Micheli, G. (2001). Dynamic voltage scaling and power management for portable systems. *Design Automation Conference (DAC)*, pages 524–529.

[Simunic et al., 1999] Simunic, T., Benini, L., and De Micheli, G. (1999). Cycle-accurate simulation of energy consumption in embedded systems. *Design Automation Conference (DAC)*, pages 876–872.

[Simunic-Rosing et al., 2007] Simunic-Rosing, T., Coskun, A. K., and Whisnant, K. (2007). Temperature aware task scheduling in MPSoCs. *Design, Automation and Test in Europe (DATE)*, pages 1659–1664.

[Sipser, 2006] Sipser, M. (2006). *Introduction to the Theory of Computation*. Thomson Course Technology, Parts One and Two.

[Sirocic and Marwedel, 2007a] Sirocic, B. and Marwedel, P. (2007a). Levi Flexray® simulation software. *http://ls12-www.cs.tu-dortmund.de/teaching/download/levi/download/leviFRP.zip*.

[Sirocic and Marwedel, 2007b] Sirocic, B. and Marwedel, P. (2007b). Levi KPN simulation software. *http://ls12-www.cs.tu-dortmund.de/teaching/download/levi/download/leviKPN.zip*.

[Sirocic and Marwedel, 2007c] Sirocic, B. and Marwedel, P. (2007c). Levi RTS simulation software. *http://ls12-www.cs.tu-dortmund.de/teaching/download/levi/download/leviRTS.zip*.

[Sirocic and Marwedel, 2007d] Sirocic, B. and Marwedel, P. (2007d). Levi TDD simulation software. *http://ls12-www.cs.tu-dortmund.de/teaching/download/levi/download/leviTDD.zip*.

[Skadron et al., 2009] Skadron, K., Stan, M. R., Ribando, R. J., Gurumurthi, S., Huang, W., Sankaranarayanan, K., Tarjan, D., Burr, J., Ghosh, S., Velusamy, S., and Link, G. (2009). Hotspot 5.0. *http://lava.cs.virginia.edu/HotSpot/index.htm*.

[Smith and Nair, 2005] Smith, J. J. and Nair, R. (2005). *Virtual Machines: Versatile Platforms For Systems And Processes*. Morgan Kaufmann Publishers.

[Society for Display Technology, 2003] Society for Display Technology (2003). Home page. *http://www.sid.org*.

[Sprint Consortium, 2008] Sprint Consortium (2008). Open SoC design platform for reuse and integration of IPs. *http://www.sprint-project.net*.

[Stallings, 2009] Stallings, W. (2009). *Operating Systems: Internals and Design Principles*. Prentice Hall.

[Stankovic and Ramamritham, 1991] Stankovic, J. and Ramamritham, K. (1991). The Spring kernel: a new paradigm for real-time systems. *IEEE Software*, 8:62–72.

[Stankovic et al., 1998] Stankovic, J., Spuri, M., Ramamritham, K., and Buttazzo, G. (1998). *Deadline Scheduling for Real-Time Systems, EDF and related algorithms*. Kluwer Academic Publishers.

[Steinke, 2003] Steinke, S. (2003). *Analysis of the potential for saving energy in embedded systems through energy-aware compilation (in German)*. PhD thesis, TU Dortmund, *http://hdl.handle.net/2003/2769*.

[Steinke et al., 2002a] Steinke, S., Grunwald, N., Wehmeyer, L., Banakar, R., Balakrishnan, M., and Marwedel, P. (2002a). Reducing energy consumption by dynamic copying of instructions onto onchip memory. *Int. Symp. on System Synthesis (ISSS)*, pages 213–218.

[Steinke et al., 2001] Steinke, S., Knauer, M., Wehmeyer, L., and Marwedel, P. (2001). An accurate and fine grain instruction-level energy model supporting software optimizations. *Int. Workshop on Power and Timing Modeling, Optimization and Simulation (PATMOS)*.

[Steinke et al., 2002b] Steinke, S., Wehmeyer, L., Lee, B.-S., and Marwedel, P. (2002b). Assigning program and data objects to scratchpad for energy reduction. *Design, Automation and Test in Europe (DATE)*, pages 409–417.

[Stiller, 2000] Stiller, A. (2000). New processors (in German). *c't*, 22:52.

[Storey, 1996] Storey, N. (1996). *Safety-critical Computer Systems*. Addison Wesley.

[Stritter and Gunter, 1979] Stritter, E. and Gunter, T. (1979). Microprocessor architecture for a changing world: The Motorola 68000. *IEEE Computer*, 12:43–52.

[Stuijk, 2007] Stuijk, S. (2007). *Predictable Mapping of Streaming Applications on Multiprocessors*. Dissertation, TU Eindhoven.

[Sudarsanam et al., 1997] Sudarsanam, A., Liao, S., and Devadas, S. (1997). Analysis and evaluation of address arithmetic capabilities in custom DSP architectures. *Design Automation Conference (DAC)*, pages 287–292.

[Sudarsanam and Malik, 1995] Sudarsanam, A. and Malik, S. (1995). Memory bank and register allocation in software synthesis for ASIPs. *Intern. Conf. on Computer-Aided Design (ICCAD)*, pages 388–392.

[Sun, 2010] Sun (2010). Java technology concept map. *http://java.sun.com/new2java/javamap/Java_Technology_Concept_Map.pdf*.

[Sutherland, 2003] Sutherland, S. (2003). An overview of SystemVerilog 3.1. *EEdesign, May, available at http://www.eetimes.com/news/design/features/showArticle.jhtml?articleID=16501063*.

[Synopsys, 2010] Synopsys (2010). System studio. *http://www.synopsys.com/apps/docs/pdfs/ip/system_studio_ds.pdf*.

[SYSGO AG, 2010] SYSGO AG (2010). PikeOS RTOS and Virtualization Concept. *http://www.sysgo.com*.

[SystemC, 2010] SystemC (2010). Home page. *http://www.SystemC.org*.

[Takada, 2001] Takada, H. (2001). Real-time operating system for embedded systems. *In: M. Imai and N. Yoshida (eds.): Tutorial 2 – Software Development Methods for Embedded Systems, Asia South-Pacific Design Automation Conference (ASP-DAC)*.

[Tan et al., 2003] Tan, T. K., Raghunathan, A., and Jha, N. K. (2003). Software architectural transformations: A new approach to low energy embedded software. *Design, Automation and Test in Europe (DATE)*, pages 11046–11051.

[Tanenbaum, 2001] Tanenbaum, A. (2001). *Modern Operating Systems*. Prentice Hall.

[Teich, 1997] Teich, J. (1997). *Digitale Hardware/Software-Systeme*. Springer.

[Teich et al., 1999] Teich, J., Zitzler, E., and Bhattacharyya, S. (1999). 3D exploration of software schedules for DSP algorithms. *7th Int. Symp. on Hardware/Software Codesign (CODES)*, pages 168–172.

[Tensilica Inc., 2010] Tensilica Inc. (2010). Home page. *http://www.tensilica.com*.

[Tewari, 2001] Tewari, A. (2001). *Modern Control Design with MATLAB and SIMULINK*. John Wiley and Sons Ltd.

[The Dobelle Institute, 2003] The Dobelle Institute (2003). Home page. *http://www.dobelle. com (no longer accessible)*.

[The MathWorks Inc., 2010] The MathWorks Inc. (2010). Simulink - simulation and model-based design. *http://www.mathworks.com/products/simulink*.

[Thesing, 2004] Thesing, S. (2004). *Safe and Precise WCET Determination by Abstract Interpretation of Pipeline Models*. Pirrot Verlag.

[Thiébaut, 1995] Thiébaut, D. (1995). Parallel programming in C for the transputer. *http:// cs.smith.edu/~thiebaut/transputer/descript.html*.

[Thiele, 2006a] Thiele, L. (2006a). Design space exploration of embedded systems. *Artist Spring School on Embedded Systems, Xi-an, http://www.artist-embedded.org/docs/Events/ 2006/ChinaSchool/4_DesignSpaceExploration.pdf*.

[Thiele, 2006b] Thiele, L. (2006b). Performance analysis of distributed embedded systems. *In: R. Zurawski (Hrg.): Embedded Systems Handbook, CRC Press, 2006*.

[Thiele, L. et al., 2009] Thiele, L. et al. (2009). SHAPES @ TIK. *http://www.tik.ee.ethz.ch/~ shapes/dol.html*.

[Thoen and Catthoor, 2000] Thoen, F. and Catthoor, F. (2000). *Modelling, Verification and Exploration of Task-Level Concurrency in Real-Time Embedded Systems*. Kluwer Academic Publishers.

[Tiwari et al., 1994] Tiwari, V., Malik, S., and Wolfe, A. (1994). Power analysis of embedded software: A first step towards software power minimization. *IEEE Trans. On VLSI Systems*, pages 437–445.

[Tjiang, 1993] Tjiang, W.-K. (1993). An olive twig. *Technical Report, Synopsys*.

[Trimaran, 2010] Trimaran (2010). An infrastructure for research in backend compilation and architecture exploration. *http://www.trimaran.org*.

[TriQuint Semiconductor Inc., 2010] TriQuint Semiconductor Inc. (2010). FAQ 11: What is the MTBF for gallium arsenide devices? *http://www.triquint.com/company/quality/faqs/ faq_11.cfm.*

[Tsai and Yang, 1995] Tsai, J. and Yang, S. J. H. (1995). *Monitoring and Debugging of Distributed Real-Time Systems.* IEEE Computer Society Press.

[Udayakumaran et al., 2006] Udayakumaran, S., Dominguez, A., and Barua, R. (2006). Dynamic allocation for scratch-pad memory using compile-time decisions. *ACM Transactions in Embedded Computing Systems,* V:472–511.

[University of Cambridge, 2010] University of Cambridge (2010). HOL4. *http://hol. sourceforge.net.*

[UPnP Forum, 2010] UPnP Forum (2010). UPnP ^{TM} resources. *http://www.upnp.org/ resources/default.asp.*

[V-Modell XT Authors, 2010] V-Modell XT Authors (2010). V-Modell XT Gesamt 1.3. *http://v-modell.iabg.de/dmdocuments/V-Modell-XT-Gesamt-Englisch-V1.3.pdf.*

[Vaandrager, 1998] Vaandrager, F. (1998). Lectures on embedded systems. *in Rozenberg, Vaandrager (eds), LNCS, Vol. 1494.*

[Vahid, 1995] Vahid, F. (1995). Procedure exlining. *Int. Symp. on System Synthesis (ISSS),* pages 84–89.

[Vahid, 2002] Vahid, F. (2002). *Embedded System Design.* John Wiley& Sons.

[Verachtert, 2008] Verachtert, W. (2008). Introduction to parallelism. *Tutorial at Design, Automation, and Test in Europe (DATE).*

[Verma and Marwedel, 2004] Verma, M. and Marwedel, P. (2004). Dynamic overlay of scratch-pad memory for energy minimization. *8th IEEE/ACM Int. Conf. on Hardware/software Codesign and System Synthesis (CODES+ISSS),* pages 104–109.

[Verma et al., 2005] Verma, M., Petzold, K., Wehmeyer, L., Falk, H., and Marwedel, P. (2005). Scratchpad sharing strategies for multiprocess embedded systems: A first approach. In *IEEE 3rd Workshop on Embedded System for Real-Time Multimedia (ESTIMedia),* pages 115–120.

[Vladimirescu, 1987] Vladimirescu, A. (1987). SPICE user's guide. *Northwest Laboratory for Integrated Systems, Seattle.*

[Vogels and Gielen, 2003] Vogels, M. and Gielen, G. (2003). Figure of merit based selection of A/D converters. *Design, Automation and Test in Europe (DATE),* pages 1190–1191.

[Wandeler and Thiele, 2006] Wandeler, E. and Thiele, L. (2006). Real-Time Calculus (RTC) Toolbox.

[Wedde and Lind, 1998] Wedde, H. and Lind, J. (1998). Integration of task scheduling and file services in the safety-critical system MELODY. *EUROMICRO '98 Workshop on Real-Time Systems, IEEE Computer Society Press,* pages 18–25.

[Wegener, 2000] Wegener, I. (2000). *Branching programs and binary decision diagrams – Theory and Applications.* SIAM Monographs on Discrete Mathematics and Applications.

[Wehmeyer and Marwedel, 2006] Wehmeyer, L. and Marwedel, P. (2006). *Fast, Efficient and Predictable Memory Accesses*. Springer.

[Weiser, 2003] Weiser, M. (2003). Ubiquitous computing. *http://www.ubiq.com/hypertext/ weiser/UbiHome.html*.

[Wellings, 2004] Wellings, A. (2004). *Concurrent and Real-Time Programming in Java*. Wiley.

[Weste et al., 2000] Weste, N. H. H., Eshraghian, K., Michael, S., Michael, J. S., and Smith, J. S. (2000). *Principles of CMOS VLSI Design: A Systems Perspective*. Addision-Wesley.

[Wikipedia, 2010] Wikipedia (2010). Structured systems analysis and design method. *http:// en.wikipedia.org/wiki/Structured_Systems_Analysis_and_Design_Methodology*.

[Wilhelm, 2006] Wilhelm, R. (2006). Determining bounds on execution times. *In: R. Zurawski (Ed.): Embedded Systems Handbook, CRC Press, 2006*.

[Willems et al., 1997] Willems, M., Bürsgens, V., Keding, H., Grötker, T., and Meyr, H. (1997). System level fixed-point design based on an interpolative approach. *Design Automation Conference (DAC)*, pages 293–298.

[Wilton and Jouppi, 1996] Wilton, S. and Jouppi, N. (1996). CACTI: An enhanced access and cycle time model. *Int. Journal on Solid State Circuits*, 31(5):677–688.

[Wind River, 2010a] Wind River (2010a). VxWorks. *http://www.windriver.com/products/ vxworks*.

[Wind River, 2010b] Wind River (2010b). Web pages. *http://www.windriver.com*.

[Winkler, 2002] Winkler, J. (2002). The CHILL homepage. *http://psc.informatik.uni-jena.de/ languages/chill/chill.htm*.

[Wolf, 2001] Wolf, W. (2001). *Computers as Components*. Morgan Kaufmann Publishers.

[Wolsey, 1998] Wolsey, L. (1998). *Integer Programming*. Jon Wiley & Sons.

[Wong et al., 2001] Wong, C., Marchal, P., Yang, P., Prayati, A., Catthoor, F., Lauwereins, R., Verkest, D., and Man, H. D. (2001). Task concurrency management methodology to schedule the MPEG4 IM1 player on a highly parallel processor platform. *9th Int. Symp. on Hardware/Software Codesign (CODES)*, pages 170–177.

[ws4d, 2010] ws4d (2010). Web services for devices. *http://www.ws4d.org*.

[Xilinx, 2008] Xilinx (2008). MicroBlaze processor reference guide. *http://www.xilinx.com/ support/documentation/sw_manuals/mb_ref_guide.pdf*.

[Xilinx, 2009] Xilinx (2009). Device reliability report - second quarter 2009. *http://www. xilinx.com/support/documentation/user_guides/ug116.pdf*.

[Xilinx, 2009] Xilinx (May, 2009). Virtex-5 user guide, v 4.7. *http://www.xilinx.com/support/ documentation/user_guides/ug190.pdf*.

[Xilinx, 2007] Xilinx (Nov., 2007). Virtex-II Platform User Guide, V 2.2. *http://www.xilinx. com/support/documentation/user_guides/ug002.pdf*.

[XMOS Ltd., 2010] XMOS Ltd. (2010). Home page. *http://www.xmos.com/*.

[Xu and Parnas, 1993] Xu, J. and Parnas, D. L. (1993). On satisfying timing constraints in hard real-time systems. *IEEE Transactions on Software Engineering*, 19:70–84.

[Xu et al., 2009] Xu, Q., Huang, L., and Yuan, F. (2009). Lifetime reliability-aware task allocation and scheduling for MPSoC platforms. *Design, Automation and Test in Europe (DATE)*, pages 51–56.

[Xue, 2000] Xue, J. (2000). *Loop tiling for parallelism*. Kluwer Academic Publishers.

[Young, 1982] Young, S. (1982). *Real Time Languages – design and development*. Ellis Horwood.

[Zhuo et al., 2010] Zhuo, C., Sylvester, D., and Blaauw, D. (2010). Process variation and temperature-aware reliability management. *Design, Automation and Test in Europe (DATE)*.

[Zurawski, 2006] Zurawski, R., editor (2006). *Embedded Systems Handbook*. CRC Press.

About the Author

Peter Marwedel

Peter Marwedel was born in Hamburg, Germany. He received his PhD in Physics from the University of Kiel, Germany, in 1974. From 1974 to 1989, he was a faculty member of the Institute for Computer Science and Applied Mathematics at the same University. He has been a professor at TU Dortmund, Germany, since 1989. He holds a chair for embedded systems at the Computer Science Department and is also chairing ICD e.V., a local company specializing in technology transfer. Dr. Marwedel was a visiting professor of the University of Paderborn in 1985/1986 and of the University of California at Irvine in 1995. He served as Dean of the Computer Science Department from 1992 to 1995. Dr. Marwedel has been active in making the DATE conference successful and in initiating the SCOPES and the Map2MPSoCs series of workshops. He started to work on high-level synthesis in 1975 (in the context of the MIMOLA project) and focused on the synthesis of very long instruction word (VLIW) machines. Later, he added compilation for embedded processors (with emphasis on retargetability) to his scope. His projects also include synthesis of self-test programs for processors. Work on multimedia-based training led to the design of the levi multimedia units (see http://ls12-www.cs.tu-dortmund.de/teaching/download/index.html). He is a cluster leader for ArtistDesign, a European Network of Excellence on Embedded and Real-Time Systems. He is also leading projects on efficient compilation for embedded systems. The focus is on the exploitation of the memory architecture and timing predictability. He won the teaching award of his university in 2003.

Dr. Marwedel is an IEEE Fellow, a DATE Fellow, a senior member of ACM, and a member of Gesellschaft für Informatik (GI).

He is married and has two daughters and a son. His hobbies include model railways and photography.

E-mail: peter.marwedel@tu-dortmund.de

Web-site: http://ls12-www.cs.tu-dortmund.de/˜marwedel

P. Marwedel, *Embedded System Design*, Embedded Systems,
DOI 10.1007/978-94-007-0257-8, © Springer Science+Business Media B.V. 2011

List of Figures

Index